Build Your Own Medical Optometry Practice

Part 1: Medical Knowledge

SECOND EDITION

Jeffrey Sedgewick, O.D., M.D.

ISBN: 978-0-9968-1786-8 (sc)
ISBN: 978-0-9968-1787-5 (e)

Library of Congress Control Number: 2016921202

Lulu Publishing Services rev. date: 2/2/2016

On the cover:
Upper left: baseline photo of POAG suspect patient in 2010 with enlarged cup. Same patient for next 3 photos
Upper right: 2011 photo showing possible NFLD beginning inferior right quadrant and possible disk hemorrhage
Middle left: 2013 photo showing blood vessel changes lower temporal quadrant with large NFLD. Drops started
Middle right: continued blood vessel changes with obvious NFLD and disk hemorrhage
Lower left: hemi-retinal vein occlusion with hemorrhages and cotton wool spots.
Lower right: resolution of vein occlusion with collateral formation

CONTENTS

FOREWORD

I have one motive for writing this manual. To help *you* develop *your* Medical Optometry Practice as your ophthalmology consultant. The attachment of my name and credentials, however remote, to your charts is what stops most, if not all, ophthalmologists from offering this service. From my background as a practicing optometrist for 4 years and my sincere appreciation for the profession of optometry, I believe that large numbers of optometrists exclusively performing medical eye services is the next significant step forward for optometry. I also believe that the hurdles both in knowledge and in especially clinical experience, are what prevent most of you from starting your own Medical Optometry Practice. All of this can be overcome with the correct consultant. I am motivated by a desire to truly see you succeed as a deliverer of medical exams. I will avoid any of the political discussions between optometry and ophthalmology over scope of practice. A decision on your part to build your own Medical Optometry Practice is your first step forward. A solid bedrock of medical knowledge and how it applies clinically, is the second step and the reason for this manual. In it, I want to review the core research governing the medical treatment of eye diseases for the general eye doctor, either an optometrist or ophthalmologist. For brevities sake, I will try to have only one graph from each article display the main clinical take home point from that article. Again, I have chosen the articles that form the bedrock of medical treatment today.

If I had this consultant opportunity presented to me before I entered medical school, I would not have left the practice of optometry. While there is a lot more to developing your own Medical Optometry Practice than medical knowledge, everything starts there. If you and I enter into a formal consulting relationship, the other areas needed to develop your own Medical Optometry Practice will also be reviewed.

As a practicing ophthalmologist in Virginia, I occasionally see patients medically managed by optometrists from various practices. I see, in general, a poor level of clinical decision making while at the same time, observing subtle diseases far into the ora serrata. In short, the observation skills are there, it's just the clinical decision making skills that seem to be deficient in a significant number, but not in all, cases. I don't mean this as a criticism, just as an observation of the average optometrist's skill level in disease treatment. I believe that an honest discussion between an optometrist who has completed a rigorous VA residency in disease and the average optometrist will reveal pretty much the same thing. I believe that this can be remedied for the average optometrist with an ophthalmology consultant working behind, not in front of, you. Optometrists can also take advantage of the baby boomer's demand for the non-surgical eye care that most optometrists can legally perform. So, why don't more optometrists build an exclusively medical optometric practice? Surely, referrals from other optometrists are there waiting to be had, as long as the referring optometrist feels that the medical optometrist is competent and will refer back for refractive care. I think that the answer is a general lack of confidence in clinical decision making and the fact that few people in general, want to try something "out of the box". In any case, it occurred to me that this is a huge opportunity for a minority of optometrists to expand in a meaningful way into the medical treatment of ocular disease. My hope is that I might be a small part of developing a number of Medical Optometry Practices across the country.

DEDICATION AND ACKNOWLEDGEMENTS

This manual is dedicated to those optometrists who have a desire to build their own Medical Optometry Practice. Any task worth doing involves some uncertainty and hard work. A Medical Optometry Practice is no exception. Developing this program has been the center piece of my life for the past 2 to 3 years. Many thousands of hours and tens of thousands dollars have been spent putting this program together. I wish to thank my wife and children for their understanding as I have reduced my time spent with them in order to accomplish this self-inflicted goal. I also want to thank my two sons for "volunteering" to proof read this manual. When told that I might acknowledge their talents in the forward, they were very concerned that they might be blamed for any grammatical and spelling errors present in this manual. I assured them that I would assume all responsibility for any errors in grammar and facts that might be present. Any errors are totally my own. My goal is to see an entire network of optometrists operating their own Medical Optometry Practices across the entire United States, delivering truly competent and independent eye care. If I can contribute just a small piece towards the attainment of that goal for those optometrists, I am satisfied.

PERSONAL STORY FROM THE AUTHOR

I had a taste of what a Medical Optometry Practice would look like on my 2 month rotation through the Westside Veterans Hospital during my 4th year of Optometry school. The optometrists in charge of the eye service, the "chief" as that position was called, was known to be very difficult to work for and was probably partially the reason why only a dozen or so of the one hundred and fifty fourth year optometry students even raised their hand to attend the 2 month minimum rotation. Two months were necessary, the chief would like to say, so that he got at least 1 month of competent work out of us. The first month, according to him, we were not contributing, only taking his time training us. On this rotation, refractions became a nuisance to me, delaying and interfering with looking at the retina and optic nerve. I enjoyed seeing what disease the next patient might have, guessing what to do with the CRVO that just sat in my exam chair. I had another 18 to 20 patients to see that day and, while I could tell something was wrong with the bleeding I saw in the retina, I didn't know what the next step was. At least not initially. After I saw my fifth CRVO patient, I could pretty much tell a CRVO from PDR although I don't remember starting the CRVO patient on glaucoma drops prophylactically at that time. As I said, after a month, we became much more comfortable examining patients with eye diseases, although I won't comment on how competently.

One day during early in the rotation, I was late arriving at the VA hospital. Something that occasionally happened as I had to be there at 7:00 AM and traffic and parking were not easy. Keep in mind now that I was giving up my forth year rotation at the optometry clinic, consisting of seeing 2 patient a day, 2 to 3 days a week. This was voluntarily given up for the VA rotation, doing 8-9 hour days, being verbally abused at times, 5 days a week. And I raised my hand to volunteer for this? That morning, the O.D. in charge proceeded to ream me out in front of a patient for being late (about 15 minutes late). The veteran patient, most of who are no strangers to frank talk, believe me, volunteered to go to the O.D's boss and complain that no one had a right to talk to me that way. Picking my self-esteem off of the floor, I thanked him for his empathy and declined, saying that I was late and that it was my fault. Needless to say, I was not late again for the rest of the rotation and wound up taking the chief to lunch at the end of the rotation. (My apologies to the other optometry student on this rotation if I never paid you back for my half of the bill. I don't remember if I did or not).

I knew that there was a huge difference between the training I received in medical care at the optometry school and at the VA hospital. The optometry students that rotated back through the optometry school's clinics after completing the VA rotation were known by the optometry staff to be different as well. They were much better at disease than most if not all of the staff. I don't think too much has changed today.

Sitting in my apartment complex's huge Jacuzzi in San Jose, California late one February, I realized that I did not feel challenged enough working for other optometry offices, asking which was better, one or two. I knew that I would have to either own my own optometry practice and have the challenge of that be enough to satisfy me or go back to medical school and enter ophthalmology training in order to satisfy my urge for a challenge. (I didn't believe that an extra year at a VA hospital would truly prepare me for treating ocular disease. The University of Illinois' ophthalmology residency program did grand rounds through the Westside VA hospital and I saw the level of knowledge that the residents and ophthalmology staff had. I didn't believe that I could learn enough as an optometry resident to feel comfortable treating eye diseases at the level that I saw amongst the physicians at U of I). As I was not married yet and had few commitments,

I chose the medical school path. Leaving Silicon Valley, with all of the great weather, international people and excitement was not easy. I remember sitting in that Jacuzzi, after gaining acceptance to Georgetown's medical school and winning an air force scholarship to pay for it, trying to decide whether I should do this or not. I knew that it was a long tunnel I was entering with a trend of decreasing reimbursement for medical services, as one prominent cataract surgeon told me. But, I decided that I didn't want to be 70 years old and regret not having accomplished what I thought I was really capable of. Keep in mind, I was meeting people from all over the world, met one of the founders of Apple computer, Steve Wazniac, and was getting to do all of the fun things in and around Northern California. Giving that up was not something to be taken lightly let me tell you. In fact, I turned down Wayne State Medical School the year before when an invitation came to join the medical school just 3 days before the medical school year started. At that point in my life, I just hadn't had enough of the California lifestyle. That same year, I was wait listed at USC and might have accepted that invitation instead of turning down Wayne State since USC was still in California. Long story short, I got into Georgetown the next year and left California, promising to return as soon as possible. I haven't been back to live there yet.

I sometimes wonder if my round-about way of getting to where I am today has been time wasted. I don't think so. In fact, I believe that my 4 years of optometry school, 4 years of practicing optometry in various environments, 5 years of active duty seeing what leadership really is (and isn't), and building 2 of my own businesses from scratch, have all been necessary in order to put this program together. I don't see how anyone without those experiences can really put together a quality program for you. I hope that this program opens some doors for you and that you walk through those doors and build your own Medical Optometry Practice. If it had been presented to me, I would have jumped at this opportunity so many years ago. I have tried to put what I believe you need to practice as a true primary eye care doctor. An optometrist whose skills are good enough to treat their own family member's eye diseases with confidence. By saying that, I don't mean to slight those optometrists who are treating their own family member's glaucoma and I don't think that you have to complete this program to legally do so. However, I see patients that have been treated by optometrists for medical conditions all the time. I know that there are optometrists that can do a great job treating diseases. I also know that, for the bulk of you, a course like this could really increase your confidence level and may be what you need to build your own Medical Optometry Practice. It is up to each of you to judge the quality of my efforts and to let me know if I have succeeded and how I can improve this program. For those optometrists who do sign with me to be your consultant, part two of your training will be held in my office.

Jeffrey Sedgewick, O.D., M.D.
1985 O.D., Illinois College of Optometry
1993 M.D., Georgetown University
1999 Residency in Ophthalmology, Louisiana State University, Shreveport, LA

BUILD YOUR OWN MEDICAL OPTOMETRY PRACTICE,
PART 1, MEDICAL KNOWLEDGE

Mission:

My mission is to focus on helping *you* develop *your* Medical Optometry Practice. To have *you* attain the pride and economic freedom that comes from owning your own Medical Optometry Practice. To my knowledge, I am the only ophthalmology consultant willing to place *my* name on *your* charts. You own the practice, the charts, and you have all of the face time with the patients. I will review your charts via the internet and give pointers on what you need to be concerned with for each medical patient that you see. I will provide lectures (today), have you practice on a few procedures at my practice in Virginia, point out reference materials and perform a review of each medical visit you perform, guiding you so that your chances of successfully developing your Medical Optometry Practice are substantial and real. The ultimate success of your Medical Practice depends on you: the strength of your belief in yourself, your willingness to learn, and the energy that you contribute. Your Medical Optometry Practice should be centered on treating about 2 dozen common diseases and recognizing diseases that need to be referred out. There will be an emphasis on glaucoma since that will be the backbone of your practice (glaucoma diagnosis and treatment will probably be 75% of what you do medically.) You don't have to know everything about every ocular disease! You just need to recognize diseases that you do not feel comfortable treating and refer them out. Don't let any insecurity you might feel stop you from building your Medical Optometry Practice.

This manual and corresponding lecture is designed with this consultant relationship between you and me in mind. I have endeavored to include studies and topics that are useful in clinic and form the basic research that shapes our clinical decisions, setting aside interesting but less applicable information. A significantly increased level of comfort in you as you examine patients with diseases is the true test of whether I have succeeded here or not.

The "Why" before the "How":

I have come to appreciate that if you believe strongly enough in a goal, in other words the "why," the immense challenges that must be overcome in order to achieve that goal, or the "how to" will be surmountable. Without a strong enough "why" the many challenges of the "how to" will stop most people from accomplishing their goal.

Why and *why not* develop a Medical Optometry Practice?

To practice eye health, not sell glasses.
To improve your economic freedom. You do not really own the practice working for a chain.
To feel more like a professional, solving health issues, rather than selling glasses.

What are the downsides to developing a Medical Optometric Practice? In short, fear of the unknown as well as relinquishing a significant source of income, the sale of glasses. A Medical Optometry Practice needs to be perceived as a supplement, not a competitor, to the referring optometrists. Therefore, you cannot sell glasses in your practice. Selling glasses represents a significant revenue source for the average optometrist's practice, 62% of the gross revenue of the median Optometric Practice according to a 2008 Jobson study. So, giving this income source up can be frightening. You also have to be very careful about any financial relationship with the referring doctor as Stark Anti-kickback Laws significantly frown upon a financial "give and take" as it may increase the promotion of referrals. (Pete Stark is a nice but passionate guy, whom I had the pleasure of examining as an ophthalmologist at Andrews AFB one day). Inherent in the development of a Medical Optometry Practice is the non-financial *nurturing* of optometric referrals for medical conditions that the average Optometrist chooses not to follow.

You need to be seen by the referring optometrist as someone helping them take difficult medical patients off of their hands, all without violating the Stark Anti-kickback Laws. As such, the splitting of fees with the referring optometrist is not appropriate. Therefore, you would have to give up "co-managing" medical patients (LASIK is not included in the Stark Anti-Kickback Laws as long as insurance does not reimburse for the procedure). I do know of medical optometrists that travel to the referring optometrist's office to examine their medical patients. An optometrist that does this could easily land both doctors in trouble if not conducted strictly in accordance with federal and state anti-kickback laws. The risk is not worth the reward.

Most optometrists do not offer medical services. A 2012 Jobson publishing study found that only 17% of optometric office exams are medical. Why? I think the reasons include not only the discontinuation of glasses sales, but also a lack of knowledge, confidence, or experience in the treatment of ocular disease. That is what this course is for. The commitment to start a Medical Optometry Practice also involves the purchase of expensive equipment with only an in-the-future potential revenue source to fund those purchases. Reimbursement for medical services is not trending up either. Also, a lack of a desire to strike out in a new and challenging field, especially if you are making a comfortable income right now, can be unsettling. Whether you offer routine exams in your Medical Optometry Practice or not is up to you. The patients like the convenience but it may also be perceived by referring optometrists as a threat. **Serious consideration should be given to not offering routine eye exams or even refractions in your Medical Optometry Practice**. Referring optometrists tend to feel more comfortable trusting that you will not "steal" their contact lens and glasses patients if you don't. You decide which is best for you in your particular area.

Your goal in starting your own Optometry Medical Practice, then, is to have the 17% of total patient visits and 17% of revenue from medical eye care become essentially all of your income, as it is for most ophthalmologists. The aging demographics of the US population, the legal right to medically treat in all 50 states and the distribution of optometrists in all 50 states are all in the medical optometrists favor. And yet, the vast majority of optometrists have not traveled down this path. I recognize that most optometrists may not have the desire that it takes to develop a Medical Optometry Practice. But for those that do, I have developed this consulting business. You will own the practice, you will own the charts. I am the consultant helping *you* build *your* Medical Optometric Practice. This lecture is part one. Part two involves a trip to my private practice in Virginia to have you demonstrate gonioscopy and scleral depression among other techniques as well as to show you how a medical practice operates.

Critical Issues in Developing Your Medical Optometry Practice

1. **Develop your medical understanding of a disease to the point where you would feel comfortable treating your family for that disease. That is what this seminar is all about. My goal is to give you the medical background to form your own medical diagnosis and treatment opinions and not rely on other experts. Refer out diseases that you are not comfortable treating. You are no less of a doctor for referring! In fact, you are caring for your patients better for doing so.**
2. **Keep the patient's welfare as your highest priority at all times, even ahead of your own economic interests. Don't try to "sell" the patient on your expertise. Let your treatment of the patient, your explanation of the patient's medical condition, and your demeanor do all of the "talking." Your success is not about your office's surroundings or your location. Your success is about the patients and their perception of you as a skilled eye doctor. Keep your voice steady at all times, listen well, and never yell.**
3. **Become knowledgeable in malpractice law, labor law, how to lead your employees, and in investing your money for your future financial security. You are the only true advocate that your practice has and you need to be its spokesman. No one else will. The patients won't and your employees will eventually move on to another job. The practice feeds your employees as well as your family. Don't let it go bankrupt. The practice has a whole host of demands that, if not met, will terminate it.**
4. **Be perceived as an asset to all of the referring doctors, not as a competitor. This is very important. Hold local continuing medical education (CME) lectures in order to enhance your reputation as a Medical Optometrist.**
5. **Network for referrals by visiting referring doctors and explaining that you are there to *help them* with their medical patients, allowing them to concentrate on contact lens exams and glasses sales. Your biggest practice builders are 1) word of mouth, 2) your name on the insurance roster, and 3) your proximity to the patient. Probably in that order. Do not co-manage medical exams with other optometrists. Only accept referrals to your office. The average referring optometrist does not have all of the equipment and expertise necessary to examine medical patients. You do.**

The amount of knowledge to grasp in optometry and ophthalmology is immense. I attended an ophthalmology review course as a first year ophthalmology resident in Houston. It was 4 weeks long, 5 days a week for 8-9 hours each day of solid lectures. And it covered only the *basics* of ophthalmology. I think we had half a day of optics. The standardized textbooks published by the American Academy of Ophthalmology given to all first year ophthalmology residents are 13 dense volumes long. It only covers the basics of ophthalmology. Duke elder published an extensive review of ophthalmology in the 1950's and it was 14 volumes long then. No one can cover all of ophthalmology in a short course. My goal is to help you build a solid core of knowledge centered on the common diseases that a competent Medical Optometry Practice would need. Also, to recognize what you can and cannot treat and refer patients out that you are not comfortable treating. In my ophthalmology practice, I refer patients extensively to a number of specialists all of the time. Here, we will cover the basics of the eye exam, common disease and their treatments. We will go into detail on glaucoma detection and treatment as it will make up the bulk of your practice. We will use the Will's Eye Manual as our standard textbook. Additional reading in Neuro Ophthalmology (including an excellent section of visual fields) may be found in "The Neuro-Ophthalmology Board Review Manual" by Kline. I will also use as a reference the Preferred Practice Patterns (PPP) by the American Academy of Ophthalmology, various key articles, the Visual Field book by Heijl, and the Basic Clinical Science Course (BCSC) for various topics. I will also use my own patients as clinical examples. The Will's Eye Manual, the BCSC and the PPP are peer reviewed by many specialists in ophthalmology and are some of the most widely accepted sources for clinical treatment in general ophthalmology.

Billing: Medicare is the 800 pound gorilla in billing so we will concentrate on Medicare's rules, since insurance companies tend to follow Medicare's lead. Medicare does not hire doctors to review charts; they hire people who can count. They will count the number of items in each section as well as the number of days in a global period. When you agree to accept Medicare patients you agree to bill for procedures only when they are "medically indicted" or connected with the correct diagnosis from their list. You also agree to accept Medicare's payment (with the 20% from the secondary insurance) as payment in full and to get an advanced beneficiary notice (ABN) notification before a procedure is done from all patients if the procedure is even remotely not likely to be reimbursed by Medicare. Office visits are considered procedures. Each procedure code, including office visits have only a certain number of diagnoses that are acceptable for that particular procedure. If Medicare does not accept a diagnosis connected to a procedure, you will not get paid for that procedure and without an ABN, you cannot bill the patient for it either. There are no "screening" photos, visual fields, or other procedures in medical visits. In order to do a test, you have to have a diagnosis that qualifies for performing that test. The practice of charging for "screening" medical procedures, visual fields, or photos, is not allowed in medical visits. So, with that in mind, let's begin by reviewing the basics of an eye exam and then the most common diseases that you will see in your Medical Optometry Practice.

EXAM BASICS:

A. DISEASES YOU DO NOT WANT TO MISS.

Due to the special diseases and malpractice issues of pediatric ophthalmology, I do not examine patients 8 years old or younger. We will not cover those here.

Here are diseases you do not want to miss. Think of these diseases at every patient visit and make sure you have excluded them for each exam that you do:

4 A's: Amaurosis Fugax (monocular that loss of vision [LOV] that lasts minutes, not seconds. Will's eye manual page 283)

Anisocoria with neck pain (A Horner's with a dissecting carotid artery, death is imminent)

Angle closure glaucoma (covered later)

Arteritis or GCA (with a raised pale disk and profound vision loss. You are concerned about the other eye) (Will's eye manual page 274). Here is a photo of a temporal artery biopsy being performed (© American Academy of Ophthalmology or AAO 2014):

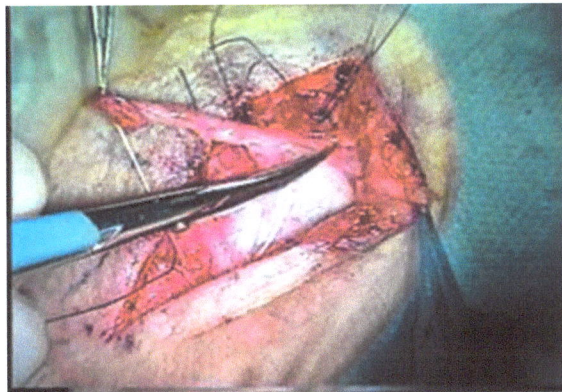

B scan when you cannot see the retina to R/O tumors. Here is a B-scan readout photo (© AAO 2014):

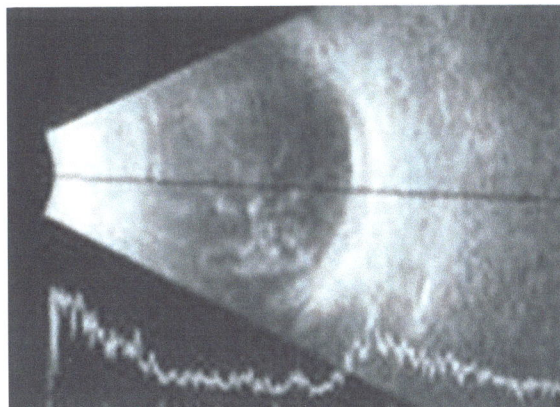

Cataract surgery/endophthalmitis. Here is a photo of a retained cataract that fell back into the vitreous when the posterior capsule was broken during cataract surgery. This can give rise to a vitritis that isn't an infectious endophthalmitis. This is a photo of a dropped nucleus and how a vitrectomy is performed (© AAO 2014):

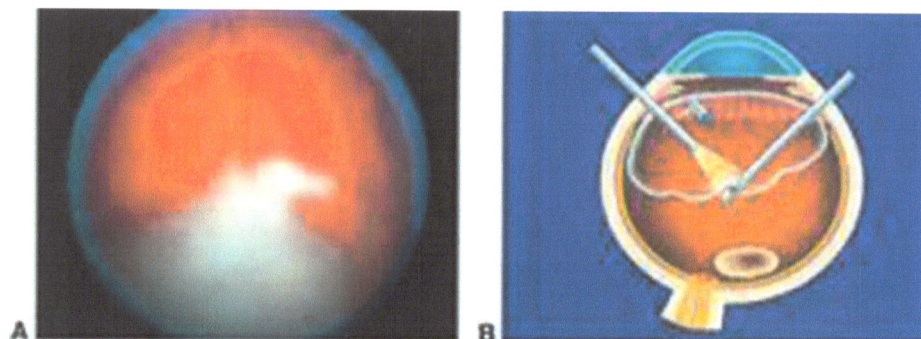

Here are 2 photos of an infectious endophthalmitis with a hypopyon (© AAO 2014):

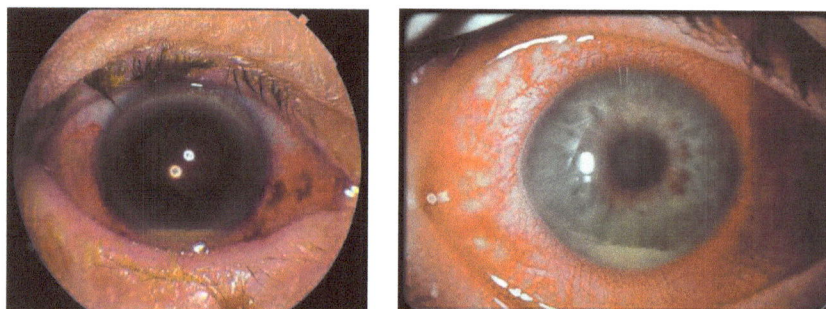

Detachment of the retina. Here is a photo of a retinal tear turning into an RRD (© AAO 2014):

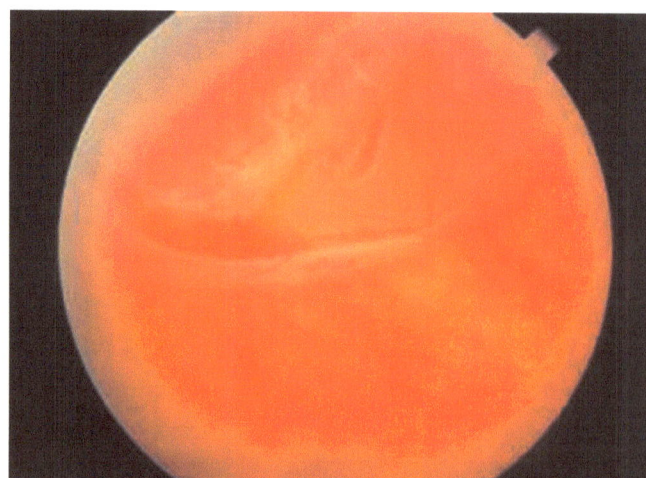

And 2 P's: Papilledema (raised injected disk). The symptom of papilledema is bilateral LOV that lasts for seconds. If you cannot see any optic nerve (ON) drusen, rule out buried drusen with a B- scan (Will's eye manual page 271) before you get a neurology referral. Here are two photos of ON drusen, the second showing the B-scan, CT scan appearance, and visual field (VF) defects from ON drusen (both © AAO 2014):

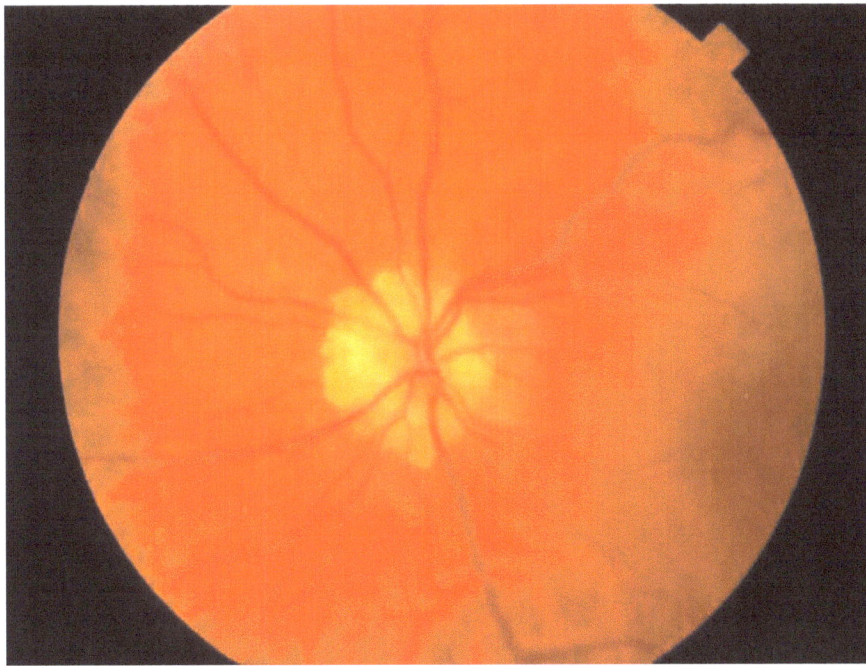

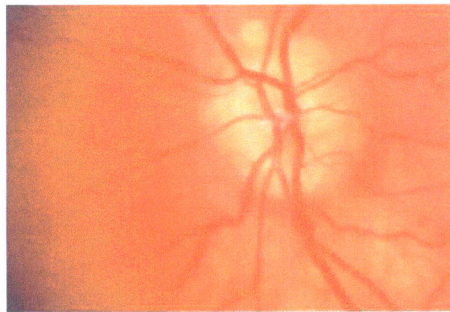

A

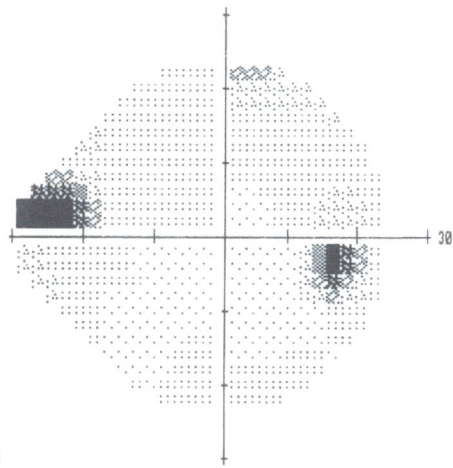

B

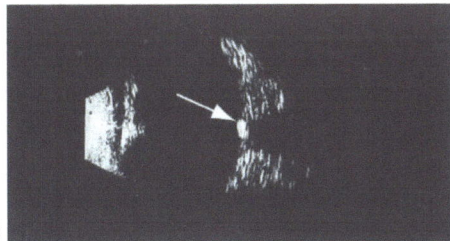

C

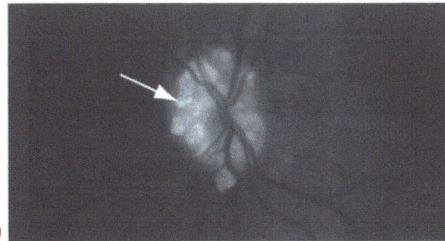

D

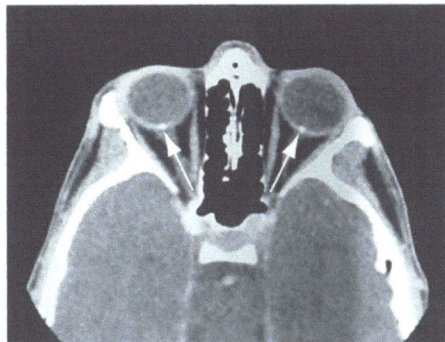

E

Pre-septal or worse, an Orbital cellulitis, with proptosis in diabetics; this could be mucormycosis. A fungal infection traveling from the sinuses into the orbit, which can kill a patient (Will's eye manual page 153). Here is a photo of a diabetic patient with a mucormycosis fungal infection extending from the palate into the orbit (© AAO 2014):

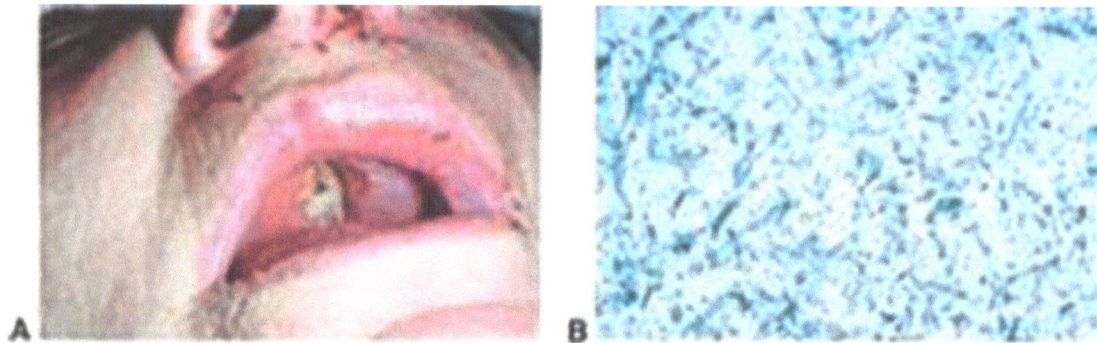

Causes of a swollen disk (Will's Eye pages 268 – 278)

After you have excluded optic atrophy, CRVO, malignant hypertensive retinopathy, and optic disk drusen.

parameter	optic neuritis	papilledema	non-arteritic AION	arteritic AION (GCA)
age	18-45	any	40-60	≥ 55
laterality	unilateral or significant asymmetry	bilateral	unilateral	unilateral
disk	swollen (1/3) normal (2/3)	injected →pale	injected→pale, with flame hemorrhage	pale from onset, with flame hemorrhage
BCVA	20/200 →CF	normal	20/20 →20/80	CF or worse
symptoms	pain with eye movement	TVOs, HAs, N/V	none altitudinal VF loss	scalp tenderness, wt. loss jaw claudication
APD	yes	no	yes	yes
color vision	bad	normal	abnormal	no vision
associated disease	ON #1 initial sign of multiple sclerosis	brain tumors, pseudo-tumor	HTN/ DM/ Cialis	none
tests you need to order	MRI w gad ?TFT's	MRI w gad and MR venography/LP	ESR/CRP/platelets should be NL	ESR/CRP/platelets, if high temporal artery biopsy

Ordering a CT/MRI (Wills eye 416 to 419):

When you want to image the orbit, order thin cuts, axial and coronal. When you want chiasmal and optic tract views, order a brain study with attention to the area you are concerned about. **Using contrast increases the risk of the imaging study**. An anaphylactic reaction is a rare but potentially fatal situation. Don't order contrast unless the radiology department requests it. CT contrast contains iodine and a reaction to iodine in the past means: don't use contrast, refer the patient out. Contrast for MRI uses gadolinium, a non-iodine material. Both contrast materials are excreted by the kidney and reduced kidney function is a problem when you want to use contrast. All diabetics and patients ≥ 60 yo need a blood creatinine test within 2 weeks of the imaging study. Diabetics using oral agents need to stop them after the test. Confirm the policy of the radiology center and tell the patient what it is. Refer out patient with abnormal creatinine results if the radiologist insists on contrast. MRI uses strong magnetic forces. All metallic devices can be a contraindication for an MRI study. This includes pacemakers, brain aneurysm clips, breast and penile implants, and intraocular FBs. Ask the patient if they have any of these devices and refer out those that do.

Optic atrophy:

Optic atrophy is a pale nerve rim tissue, not a pale cup. The lamina cribosa is always pale. Optic atrophy is accompanied by painless LOV, pale nerve rim tissue, an APD if unilateral (usually the case), and HVF losses. Usually, we think of optic atrophy in a patient where we don't know what is causing the reduction in BCVA and HVF losses. We then go back and look at the rim tissue in the photos again, looking for pale rim tissue.

> W/U includes:
> - serum heavy metal
> - B1, B12, folate
> - CBC (anemia)
> - HVF 24-2
> - ETOH abuse?
> - look for shunt vessels (orbital MRI, thin cuts axial and coronal, looking for meningioma)
> - review hereditary causes (Wills pg. 279)

B. BEST CORRECTED VISUAL ACUITY (BCVA):

Medically, best corrected visual acuity (BCVA) is measured to see if it is 20/20 or not. If the BCVA is not 20/20, pinhole (PH) the eye. If you do not obtain 20/20 with the pinhole, you *must* explain why it is not 20/20. Corneal distortion caused by LASIK or keratoconnus are two of the main reasons why some eyes cannot be refracted to 20/20 but PH to 20/20. The manual keratometer is an inexpensive device to determine corneal regularity or the degree of irregularity. I find that the number one reason for unexplained, mildly reduced BCVA *monocularly* is an undiagnosed monofixation syndrome. I use the Worth 4 dot test to detect a monofixation syndrome. I cannot tell you how many times I am the *first* to diagnose a young childhood onset monofixation syndrome in an adult patient. Near BCVA should not be any different than distance BCVA and are generally not measured in a medical exam.

C. PUPILS:

The pupil size is written as dim light pupil size -> bright light pupil size with the OD written on top of the OS. You measure it as follows: with the room lights on low, have the patient look into the distance with your transilluminator on low and below the patient's gaze, about 3 feet away (closer if the iris is dark and you have a hard time seeing the pupil from the iris). You note the absolute size of each pupil and whether there is a difference between the two pupils. You can use the pupil sizes printed on near cards. You then turn the transilluminator on high, come close to both pupils, shining the light on both pupils at the same time, just below the patient's gaze, about 12 inches away and note the same two items.

An APD is the measure of one eye not reacting to light as well as the other eye (rarely both may be abnormal, but you are detecting if one is worse than the other). You measure for an afferent pupillary defect (**APD**) in the following manner: after you have measured the pupil's sizes under dim and light conditions, let the iris recover for 2-4 seconds. Turn the transilluminator on its brightest setting and hold it in front of the patient's left pupil about 6 inches away for 3 seconds. Count out loud "1001, 1002, 1003." Quickly, over the bridge of the nose and *in a straight line*, move the transilluminator to the patient's right pupil, noting the *initial* reaction of the right pupil. It should stop dilating and constrict. A stuttering hesitation and then a redilation of the pupil is a possible mild positive APD. A blatant APD is one that does not even hesitate to constrict and just keeps on dilating when the light first shines on it. You repeat for the left pupil by counting while you are over the right pupil "1001, 1002, 1003" and then move over to the left pupil, again looking for any non-constricting movement of the left pupil. APDs are estimated as 1-4 plus. (+,++,+++,++++).[1]

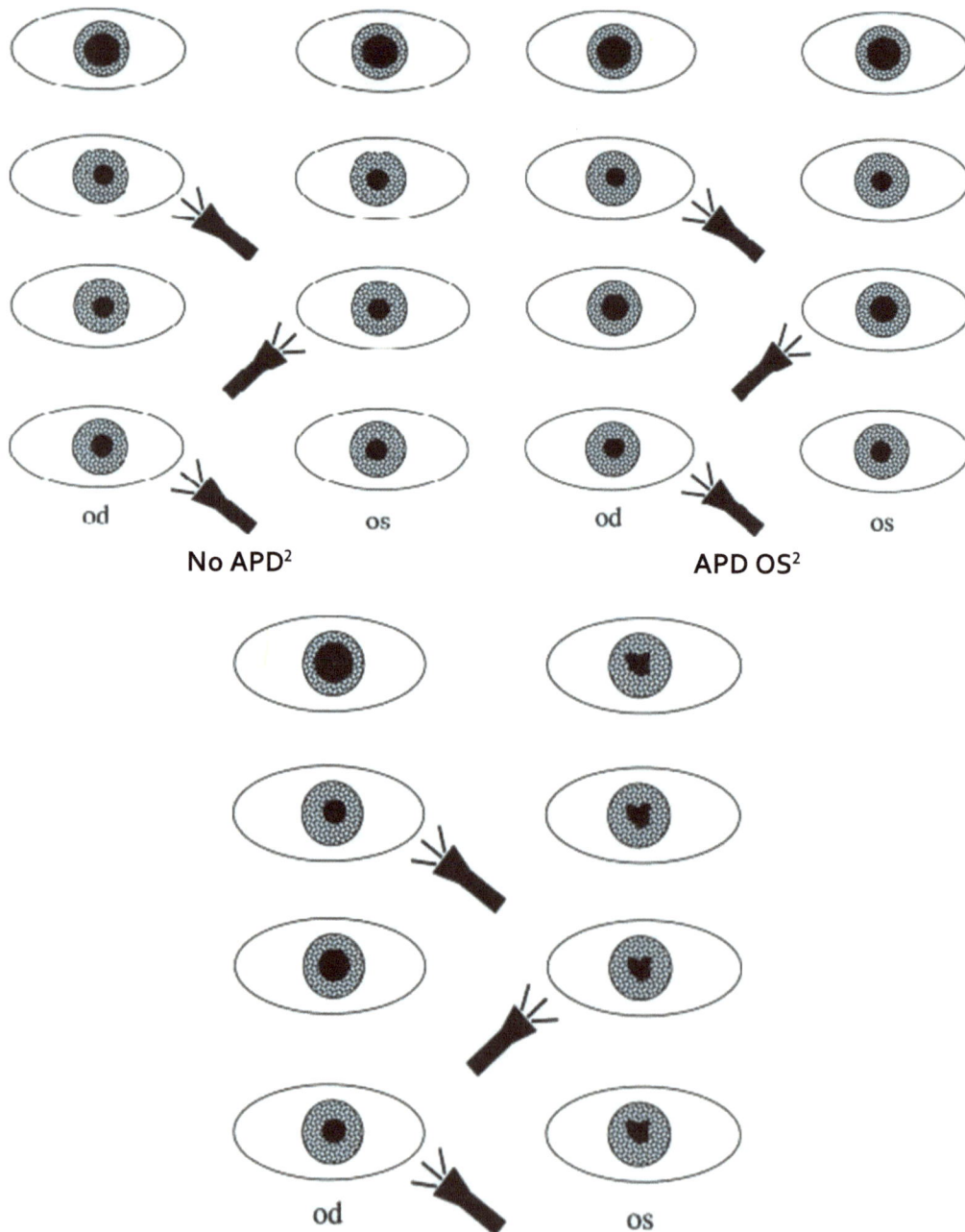

No APD[2]

APD OS[2]

APD test with fixed surgical pupil OS, showing a positive OS APD response by looking at the OD reaction to light shown on the OS[2]

The depth of an APD can be measured with a neutral density bar by trying to equalize the reaction to light between the two eyes. Since you cannot make the "bad eye" react more in order to equalize the reaction of the two eyes, you dim the light in the good eye to the level of the bad eye with the neutral density (ND) bar. The measurements are on a logarithmic scale with the lightest filter needed to equalize the pupil reaction between the two eyes noted as the depth of the APD. All APDs need to be worked up for the etiology and referred out.

Anisocoria: the most common cause is physiological (born with it) anisocoria.[3] The difference in pupil sizes stays the same in both dim and bright light. For example, if the difference between the pupils in dim light is 2 mm, it is also 2mm in bright light. If the difference is not the same in dim and bright light, then you are likely dealing with non-physiologic anisocoria.

The sympathetic nerve supplies the dilator muscle of the iris and Mueller's muscle in the upper and lower lid so that when you interrupt it (as in a Horner's syndrome), you get a constricted pupil that cannot dilate (making the anisocoria greater in dim light) as well as having a droopy upper lid on the miotic side and a slightly raised lower lid (reverse ptosis, which is easy to miss unless you look for it) on the side of the interruption. A new Horner's pupil needs a work up and a referral out. Causes of anisocoria *worse in dim light* include: [4]

- Horner's Syndrome
- use of pilocarpine drops
- iritis with posterior synechia (seen on slit lamp)
- Argyll-Robertson pupil (*always* bilateral with light near disassociation or LND, with CNS syphilis as the main concern. Get a serum RPR, FTA-abs)
- Adie's tonic pupil (vermiform iris movement and has LND). Get a serum RPR, FTA-abs if bilateral to r/o Argyll-Robertson pupil

A new onset Horner's syndrome with neck pain is a carotid artery dissection until proven otherwise and is an emergency referral.

Here is a photo of a right Horner's syndrome with a mild but significant ptosis OD (© AAO 2014):

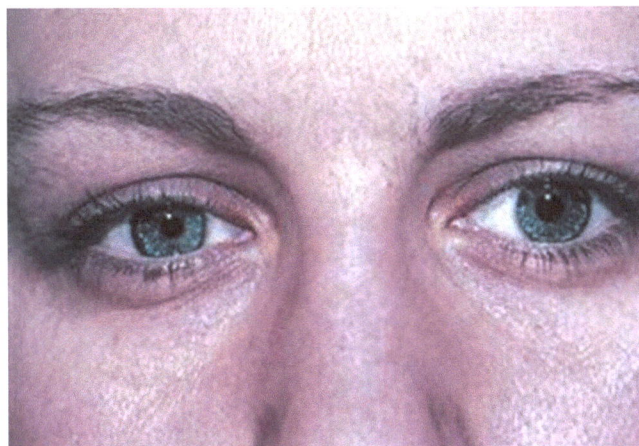

The parasympathetic autonomic fibers travel with the third CN and innervate the iris constrictor muscle, the ciliary body muscle for accommodation and the voluntary muscle fibers innervating all of the EOMs except the superior oblique and lateral rectus (SO4, LR6), as well as the levator muscle in the upper eyelid. The parasympathetic nerve travels on the outside of CN 3 and synapses in the ciliary ganglion, which is important in the understanding of Adie's Tonic Pupil. When you interrupt the parasympathetic nerve, you sometimes have an interruption of the other CN 3 functions as well. A complete third nerve palsy therefor has a dilated pupil that cannot constrict (with anisocoria greater in light), binocular diplopia, and a ptosis on the same side as the dilated pupil. Other causes of anisocoria worse in light *without a CN 3 palsy* include: [4]

- Posterior synechia
- Adie's tonic pupil (LND and vermiform or worm like movement)
- torn sphincter (seen at the slit lamp as a loss of a round pupil)
- the use of dilating drops

Pressure from a tumor or aneurysm affects the outside of the third nerve first, resulting in a pupil *involved* third nerve palsy (an internal or inside of the eye ophthalmoplegia since the pupil is one the inside of the eye), a much more dangerous condition than an ischemic third nerve palsy. Ischemic diseases (diabetes and hypertension) affect the inner part of the nerve first due to a relatively maintained oxygen supply on the outside of the nerve and a reduced oxygen delivery to the inner part of the nerve. This results in a pupil sparing third nerve palsy (an external or outside the eye ophthalmoplegia). A pupil sparing third nerve palsy can develop into a pupil involving third nerve palsy over a short period of time. This is why pupil sparing third nerve palsies need to be followed every day for 7 days and then every 4-6 weeks for 3 months, looking for pupil involvement. Ischemic palsies typically resolve spontaneously within 3 months. Pupil involvement mandates an intensive work up.

Light/near disassociation (LND): A normal response to light between the two pupils means there is no reason to test the near response.[5] If one or both of the pupils do not respond to light, however, they may respond to near accommodation via the near synkinetic response. For a patient with a non-reactive pupil, have them look into the distance and then at your finger 12 inches in front of them. If the pupil constricts more to near than to light, then you have light/near disassociation (LND). Adie's Tonic Pupil is a common cause of this and in Adie's, the pupil edge will exhibit a *slow, vermiform* movement when light from a slit lamp is directed from off to on over the pupil. The ciliary muscle fibers send adherent nerve fibers to the pupil muscle resulting in preserved miosis with accommodation. Typically, the Adie's patient requires no further W/U.[3] Here is a photo of a patient with a left Adie's tonic pupil. The first photo shows the dilated and non-reactive left pupil. The second photo shows a significant constriction OS to a near stimulus. The third photo shows the pupil size in dim illumination with the left pupil smaller than the right pupil (© AAO 2014):

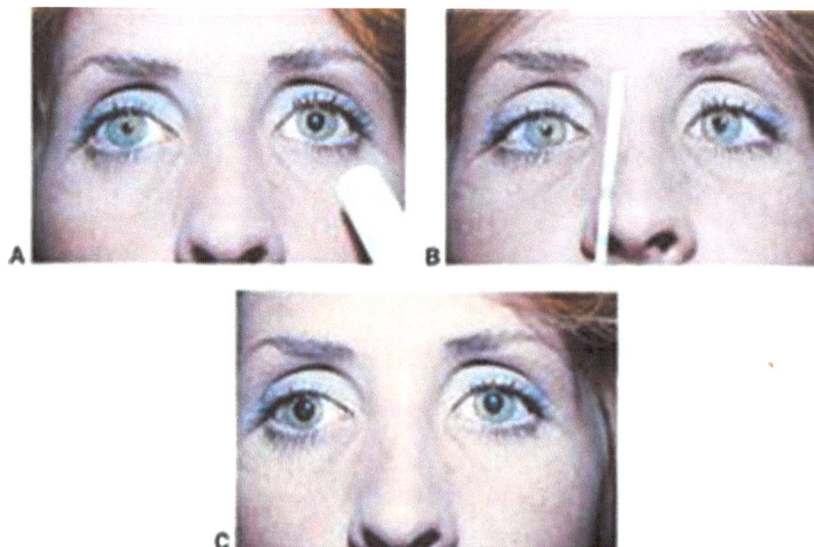

Here is a photo of a patient with a right Adie's tonic pupil. The first photo is a baseline showing mild anisocoria. The next two photos show the right pupil not reacting to light with a normal reaction to light OS. The forth photo shows the initial reaction to near OU which is a larger response than to light in the OD. The fifth photo shows continued pupil constriction OD after a little time has passed since a near object was presented, hence the term tonic. (© AAO 2014)

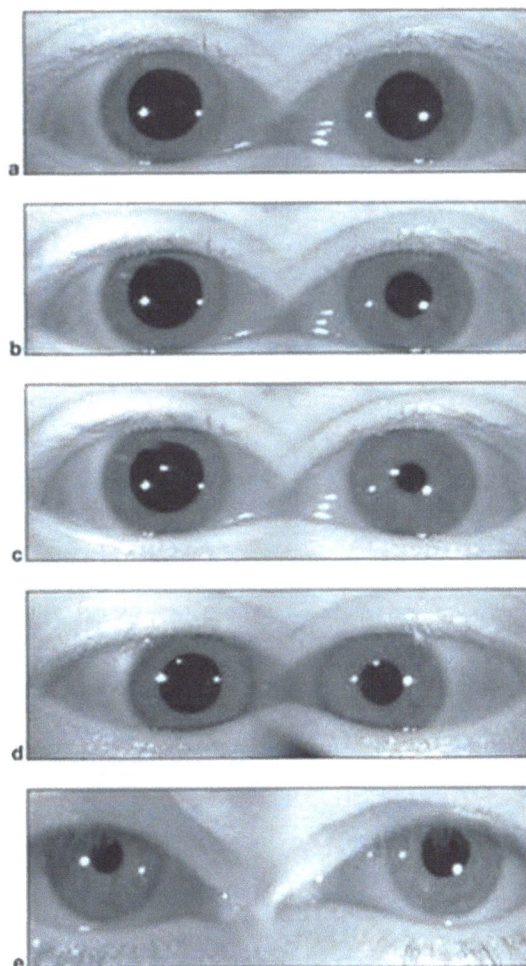

Other causes of LND are aside from an Adie's tonic pupil include:[6]

- severe optic neuropathy or retinopathy seen on exam. #1 cause outside of Adie's.
- Argyll-Robertson (syphilitic) pupil: miotic irregular pupil with normal VA and LND. Causes include syphilis with interstitial keratitis, diabetes mellitus, MS, alcoholism. Get an RPR/ FTA-ABS and referral to PCP to R/O tertiary syphilis. You do not treat syphilis, the PCP will.
- Parinaud's (Dorsal Midbrain) Syndrome: Rare. Has LND with convergence retraction nystagmus and skew deviation. Etiology includes tumors so this patient needs a referral to Neuro-ophthalmology for an MRI of the brain.

You are looking for:

- *ptosis*. There are 5 diseases that must be ruled out initially in every ptosis patient.[188] These are:

  - Horner's syndrome (anisocoria with the miotic pupil on the ptosis side)
  - 3rd cranial nerve palsy (binocular diplopia and restriction of EOM motility in one eye)
  - Myasthenia Gravis (ptosis worse in PM or with effort. Get serum acetylcholine receptor antibodies)
  - an orbital mass causing a ptosis or a mass causing a proptosis and making the normal eye look ptotic
  - CPEO (bilateral restriction of EOMs, no diplopia, normal pupils, and family history of CPEO)

Lift both of the eyebrows to reduce the effect of brow lifting hiding a contralateral eyelid ptosis. The right eyelid is ptotic in this photo. Lifting the left eyebrow allows the right eyelid go to its true position. The doctor raising the left eyebrow eliminates the eyebrow lift to the right eyelid, revealing the true ptosis in the right eyelid (both © 2014 AAO)

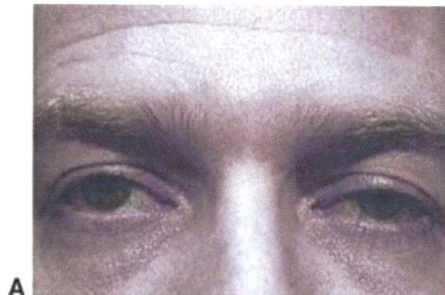

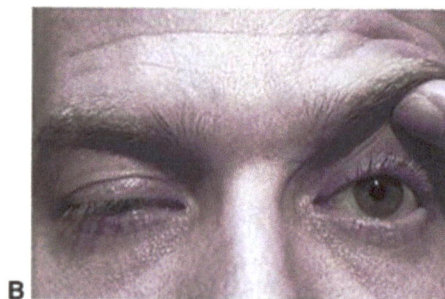

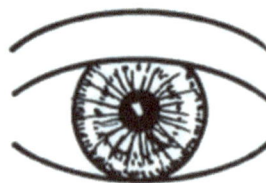

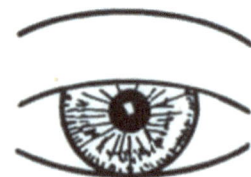

Palpebral Fissure Height	9.5	7.5
Margin–Reflex Distance	+4	+2
Upper Eyelid Crease	8	11
Levator Function	15	14

-proptosis: look at the position of the lower lid margin in relation to the limbus. It should be the same OD/OS with no inferior scleral show. Do a Hertel measurement if you are unsure. The upper limit of a Hertel is

22 mm in whites and 24 mm in blacks with no more than 2 mm difference between the two eyes. Review the Will's eye work up for proptosis, page 153.

-growths on the lids and on the exposed conjunctiva (tumors are covered later), the position of the eyelids (ectropion, entropion), lacrimal gland bulges, asymmetry of the facial features (Bell's palsy).

-meibomianitis/blepharitis. Press on the lid margins with a 6 inch Q-tip to express the meibomian glands. Tell the patient to perform daily WC/LS QD <u>for life</u> if they are plugged up. Meibomianitis is an amazingly common disorder. You can also have the patient use OTC Cliradex QD for 6 days to kill any Demodex mites that might be there.

You can test the other non-EOM cranial nerves to check any abnormality found on the external exam. (Or at any other time during the exam. I frequently go back and check CN 5 and 7 if I find something abnormal during the medical history review or upon entertaining a different diagnosis later in the exam). If I find one CN palsy, looking for other CN palsies is critical to determine if the patient has a cavernous sinus or orbital syndrome, which has an entirely different differential diagnosis list than a single CN palsy.

-testing of CN 2 through 7:

> CN 2 is measured with the BCVA, pupil exam and VFCF exam

> CN 3 is measured on the motility exam, pupil exam (parasympathetic nerve) and looking for ptosis (via the levator muscle)

> CN 4 is measured on the motility exam with vertical diplopia in the primary gaze as the primary sign of this palsy.

> CN 5 is measured by checking the sensitivity of the facial skin of the forehead (V1), upper cheek (V2) and the jawline (V3) area to finger touch. The corneal sensitivity can also be checked for decreased sensitivity from HSV infection but I rarely do so routinely. Here is a picture of the 5th cranial nerve's dermatomes and herpes zoster clinically demonstrating the V1 dermatome (all 3 © AAO 2014):

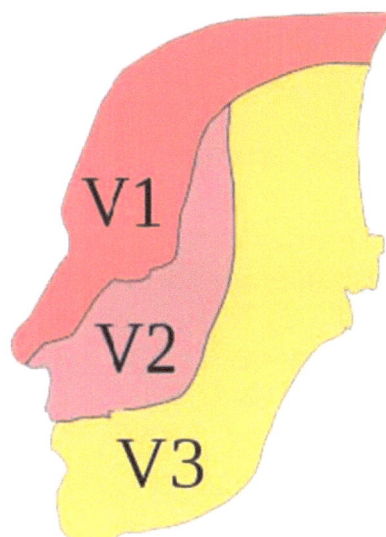

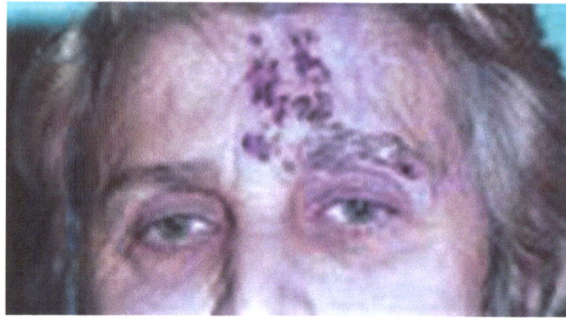

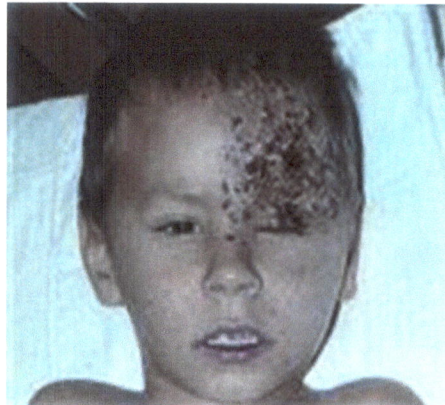

CN 6 is measured on the motility exam looking for horizontal diplopia worse in lateral gaze. Here is a patient with a left sixth nerve palsy showing the deficit in primary gaze position (© AAO 2014):

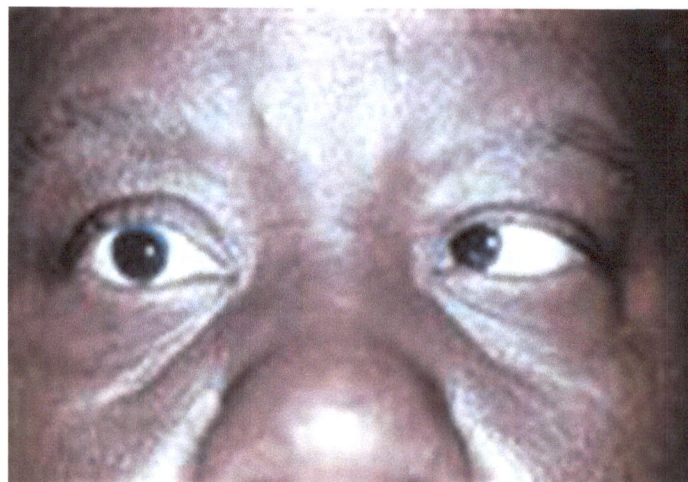

CN 7 is measured by checking the muscle function of the face. Have the patient raise their eyebrows and lips as well as have them blow out of their closed lips, checking for air escaping at the edge of the lips. (Bell's palsy). Here is a photo of a left Bell's palsy. Notice the reverse ptosis OS in photo A and the complete inability to close the orbicularis oculi muscles when asked to do so in photo B, raising the risk of exposure keratitis (© AAO 2014):

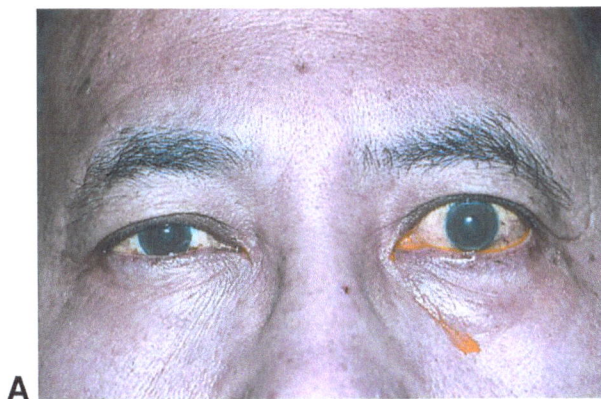 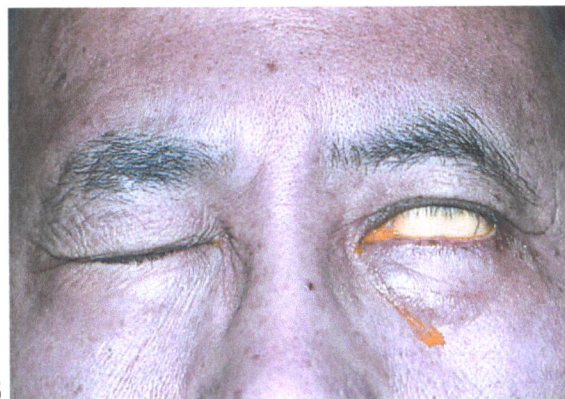

Blepharospasm is an involuntary closure of both eyelids that disappears during sleep and looks like this (© AAO 2014). No work is typically needed:[186]

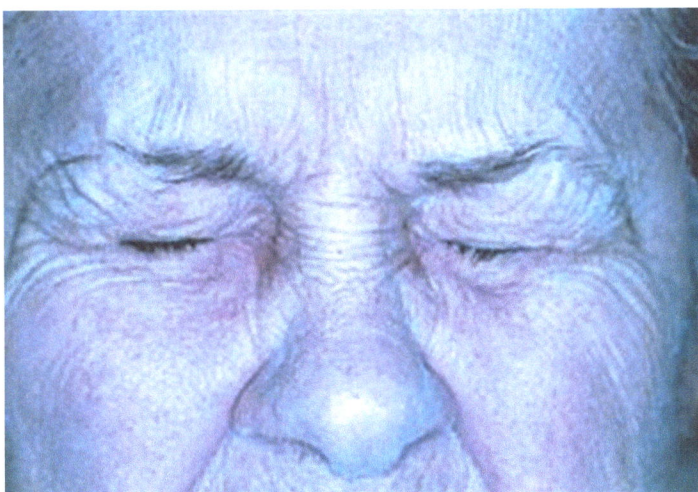

Hemi facial spasm is similar to blepharospasm but involves the entire side of the face and does not go away during sleep (© AAO 2014). This condition requires an MRI of the cerebellopontine angle to rule out a mass: [186]

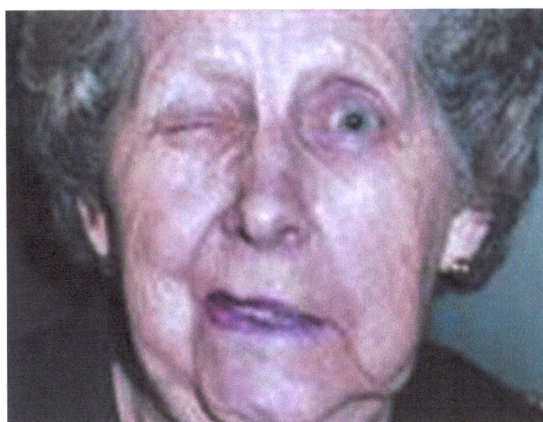

There are 6 extraocular muscles. Each muscle is the primary mover of each eye in one of the 6 main positions of gaze. These are the 6 cardinal positions of gaze. Each eye should be checked in the 6 cardinal positions of gaze with a transilluminator pointed at the cornea OU. Also, check the eye in primary, near gaze and with head tilt (if the patient has a vertical deviation where the oblique muscles need to be measured). Imagine this diagram superimposed on the face above, so right and left are as the patient would orient it, not the observer (© AAO 2014).

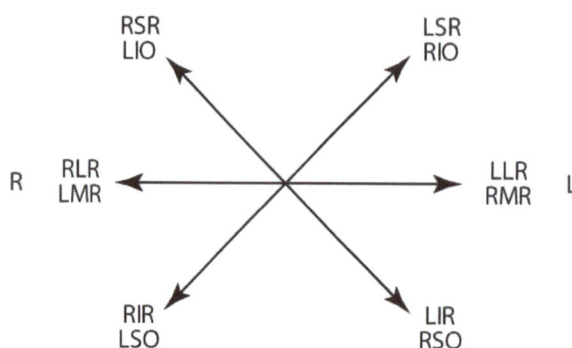

This forms a single large "H" in front of the patient. It is noted on the exam form as if the patient is looking at you. (The patient's right eye is on the left side of the exam form. This is opposite to the recording of the VF test which is recorded with the patient's right eye response on the right side of the form, as if you are recording your own VF responses). Results are recorded as 1 to 4 plus for overacting positions or 1 to 4 minus for underacting positions. Normal responses are noted as "0". If a patient has a defect in a position of gaze, you measure the deviation in primary gaze, head turn to the left and right, head tilt to the right and left, and at near. Note if the magnitude is the same in all *distance* positions. If the deviation is the same in all distance positions of gaze, it is called a commitant deviation. If the deviation is not the same in all distance positions of gaze, then it is a non-commitant deviations. Non-committant deviations are usually due to a paresis of an EOM or a restrictive pathology. You can measure the deviation by moving the patient's head to simulate the different positions of gaze. Here is a diagram of a **right** globe with the lateral wall removed showing the superior oblique insertion on the posterior, upper lateral quadrant of the globe: (© AAO 2014).

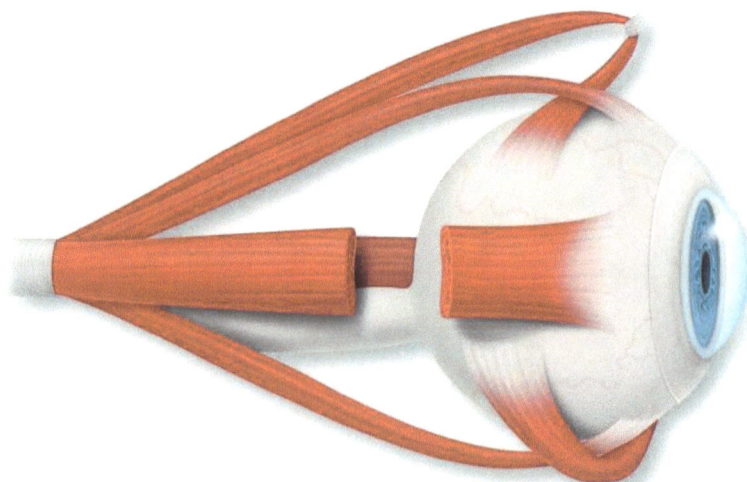

Young children adapt to binocular diplopia with suppression very quickly and therefore it is rare for children to present with diplopia from a disease that has been there for a while. Adults do not adapt well to new onset EOM palsies or restrictive diseases and tend to have diplopia for long periods of time after the onset of the disease.

The following data and graph is from the Olmstead study of adult (defined as >19 YO) onset diplopia from a non-referral, general practice setting (n=753 new onset adult strabismus patients from a population of 175,000 representing almost all of the people in a county examined by one health care system, the Mayo Clinic, from 1985 through 2004).[7] *86% of the patients had diplopia on initial presentation.* Interestingly, there wasn't any data from optometrists available for this study due to the medical records not having any optometric exams in it! The O.D.s probably would have made referrals to M.D.s and those patients would have been picked up that way. As you can see, the incidence of adult onset diplopia increases dramatically with age:

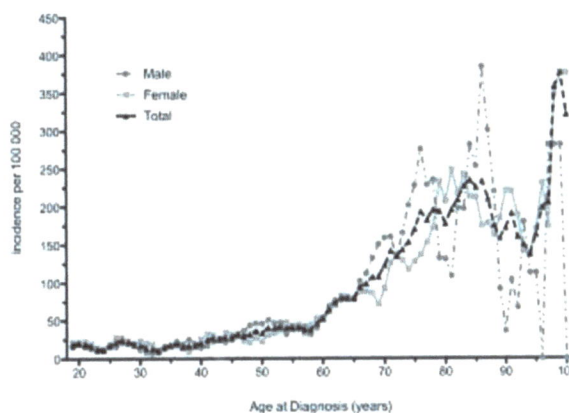

The study included the following definitions of the various causes of diplopia:

Table 2. Diagnostic Criteria for New-Onset Adult Strabismus Cases

#1 Cause is Paralytic strabismus: 333/753=44% of total, ***R/O DM and HTN.*** (Will's eye pg. 250,253, 255.)

1. Sixth nerve palsy: Lateral rectus under action and esotropia on affected lateral gaze at least 6 PD or more than in primary gaze or esotropia with worsening diplopia on corresponding lateral gaze. 10.4/23.9 = 44% of paralytic cases. (These numbers are extrapolated per 100,000 population, hence the decimals in the numbers above).
2. Fourth nerve palsy: Superior oblique under action and hypertropia of at least 2 PD or more in contralateral horizontal gaze and ipsilateral head tilt or hypertropia with worsening diplopia on contralateral horizontal gaze and ipsilateral head tilt. 6.3/23.9 = 26% of paralytic cases.
3. Third nerve palsy: Under action of any third nerve innervated muscles with or without complete or partial ptosis of affected eye and exotropia, hypotropia, or both of affected eyes. 3.5/23.9 = 15% of paralytic cases.
4. Internuclear ophthalmoplegia: Isolated adduction impairment of affected eye, with worse diplopia on contralateral gaze and associated abducting nystagmus of the contralateral eye. INO and MG account for 15% of paralytic cases in this study (3.6/23.9). You will notice that skew and other supranuclear causes are not mentioned specifically, probably due to their rarity. The tests to differential skew from 4th CN palsies weren't performed, preventing classification of some of these

cases as skew. So, we won't go into detail about these other causes of diplopia which amounted to 2.1% of all cases in this study. If the palsy is not clear, refer it out!

5. Myasthenia gravis: Myasthenia gravis diagnosis with subsequent or concurrent under action of the extraocular muscles and diplopia not resulting from another cause.

#2 Cause is Convergence insufficiency: 118/753=16% of total

Diplopia while reading with an exophoria/tropia and absence of diplopia at distance or exophoria/tropia ≥10 PD at near vision with orthophoria or small (<10 PD) phoria at distance.

#3 Cause is Small-angle hypertropia: 99/753= 13% of total.

This is defined as a commitant hypertropia <10 PD or prism prescription >3 PD on initial examination and symptoms of diplopia with no evidence of oblique muscle dysfunction. This group represents a breakdown in the compensation of a pre-existing phoria but could be a skew deviation as well.

#4 Cause is Divergence insufficiency: 80/753=11% of total diplopia.

This is defined as diplopia at distance only with an esophoria/tropia greater in distance than near and absence of double vision at near.

Restrictive strabismus: 47/753=6% of total.

Recent or pre-existing diagnosis of Graves' disease, trauma, or ocular or facial surgery with evidence of extraocular muscle restriction on ductions.

Sensory strabismus: 27/753=4% of total.

Pre-existing unilateral ocular condition affecting visual acuity with subsequent strabismus not resulting from another cause.

Another way of displaying the data from the same study is as follows: [7]

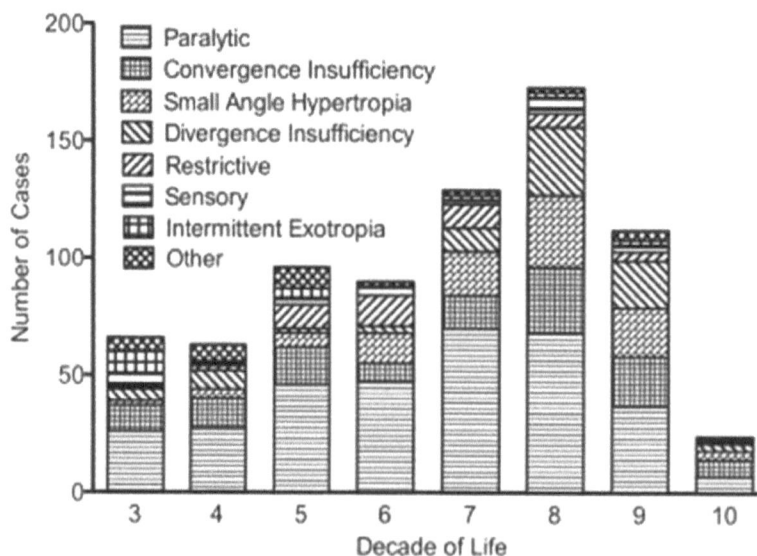

Forty-four percent of all adult cases in this study were non-commitant due to a paralysis of CN 3, 4, or 6, usually due to diabetes. Another 6% of cases were due to a "restrictive factor". So, about half of the cases are paresis or restrictive in etiology. These are the ones that I call "medical" in nature since it will involve a medical work up. These usually also present with diplopia at both near and far in most positions of gaze. The other half were commitant causes (same deviation in all 6 positions of gaze *at distance*) like convergence insufficiency (diplopia only at near), divergence insufficiency (diplopia only at distance), or pre-existing phorias that broke down with age. Skew can present as committant and does require a work-up however. One third of the total cases presented as esotropia, one third as exotropia, and one third as hypertropia. There were no cases of torsional diplopia.

(Remember, heterophorias are commitant, with the same deviation in all positions of gaze in the distance and do not typically have diplopia as a symptom, unless they are decompensated and then it will usually present as a commitant intermittent diplopia). New ocular palsy deviations have diplopia as the presenting symptom with the deviation different in different positions of gaze (non-committant). Think of a left CN 6 palsy where the OS cannot look left at all but the left eye can move to the right normally. You will get markedly different deviations in right and left gazes. Patients with a history of strabismus surgery as a child will not usually have diplopia as a symptom even though they may have residual strabismus (strabismus surgery in an adult is an exception but they will have a history of the surgery). Patients who have had strabismus surgery can have odd deviations such as DVD.

When presented with a potentially new palsy, look for other CN deficiencies, the pupil response, and lid position. This is crucial since if you have more than 1 CN palsy, you probably have pathology somewhere behind the eyeball involving the brain or another area of the CNS.

Every CN 4 palsy I have seen has a vertical deviation as the primary component. CN 3 and 6 have a horizontal deviation as the primary component. Do the 3 step test to confirm CN 4 from an incomplete superior CN 3. (Again, we have also checked the pupils and the lids). The eye with the CN 4 palsy will be hypertropic. For example, let's say a right CN 4 palsy walks in your door. Typically there will be a definite date when the diplopia was first noticed by the patient. The *right* eye will be hyper. This is the first step of the 3 step test. The second test is to measure the deviation looking laterally. To do this you turn the patient's head to the right to measure the left gaze deviation and turn the patient's head to the left to measure the right gaze deviation. In our example, the deviation will be worse with a left gaze or when the patient turns their head to the *right*. The third step is to measure deviation with head tilt. Tell the patient to place their right ear on their shoulder and then the left ear on their shoulder as best they can. In our example, the vertical deviation will be worse with *right* head tilt. You will notice that with a right CN 4 palsy, a hyper vertical deviation is in the right eye in primary gaze, worse in right turning of the head (which is left gaze) and worse with right head tilt. *Right/right/right* in a right CN 4 palsy. Just don't confuse head turning with gaze position and it is easy to tell if it is a superior oblique palsy and which eye has the palsy. A superior third nerve palsy will have vertical diplopia and will not have this right/right/right pattern.

Review the work up in Will's eye for a CN 4 palsy. (Will's eye pg. 252) Here is a photo of a right forth nerve palsy showing the up and in position of the affected eye as the third cranial nerve is unopposed (© AAO 2014):

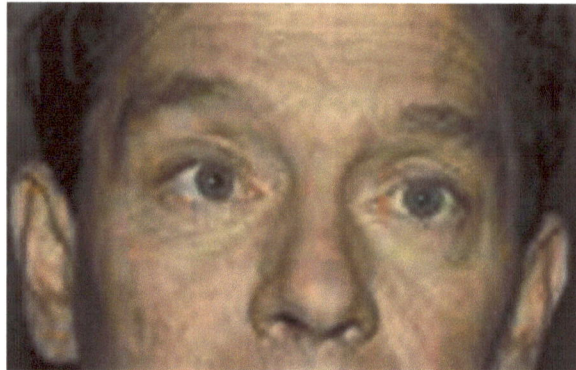

Review why an incomplete (pupil sparing) CN 3 palsy should make you nervous. (Will's eye pg. 250). Here is a photo of a right third nerve palsy showing the down and out position of the affected eye as well as a ptosis. The aneurysm causing the palsy is shown as well (© AAO 2014):

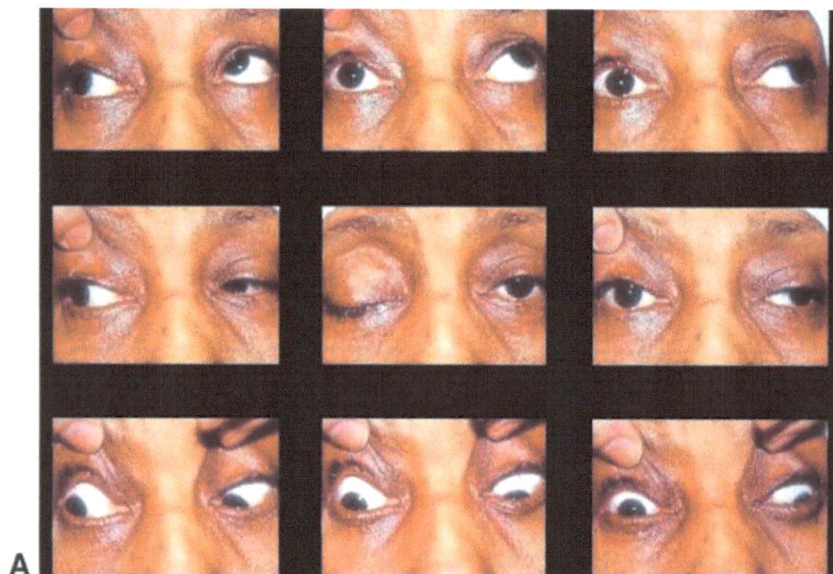

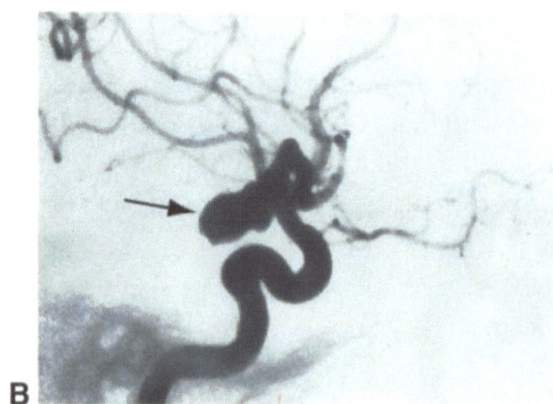

Review the work up for a CN 6 palsy. (Will's eye pg. 254)

Supranuclear gaze pathway disorders are complex and get referred out. For instance, a skew deviation is a vertical deviation caused by a supranuclear disorder.[8] How do you test for that? Frankly, I don't know. If the motility is unusual, refer it out. In fact, to be on the safe side, refer out all acute onset palsies once you think you have identified what it is.

F. VISUAL FIELDS (VF):

An entire day could be devoted to a discussion of visual fields.

Here is a diagram[9] showing the distribution of the ganglion cell axons in the OD retina. Notice the horizontal raphe (HR) and the papillo-macular (PM) bundle. Shown are defects in the arcuate fibers as they course around the PM bundle. The human optic nerve has 1.2 million[10] ganglion cell axons in it. The PM bundle has probably 90% of these fibers. The crowding of the arcuate fibers caused by the PM bundle at the optic nerve head is probably why the arcuate fibers are so sensitive to glaucoma damage.

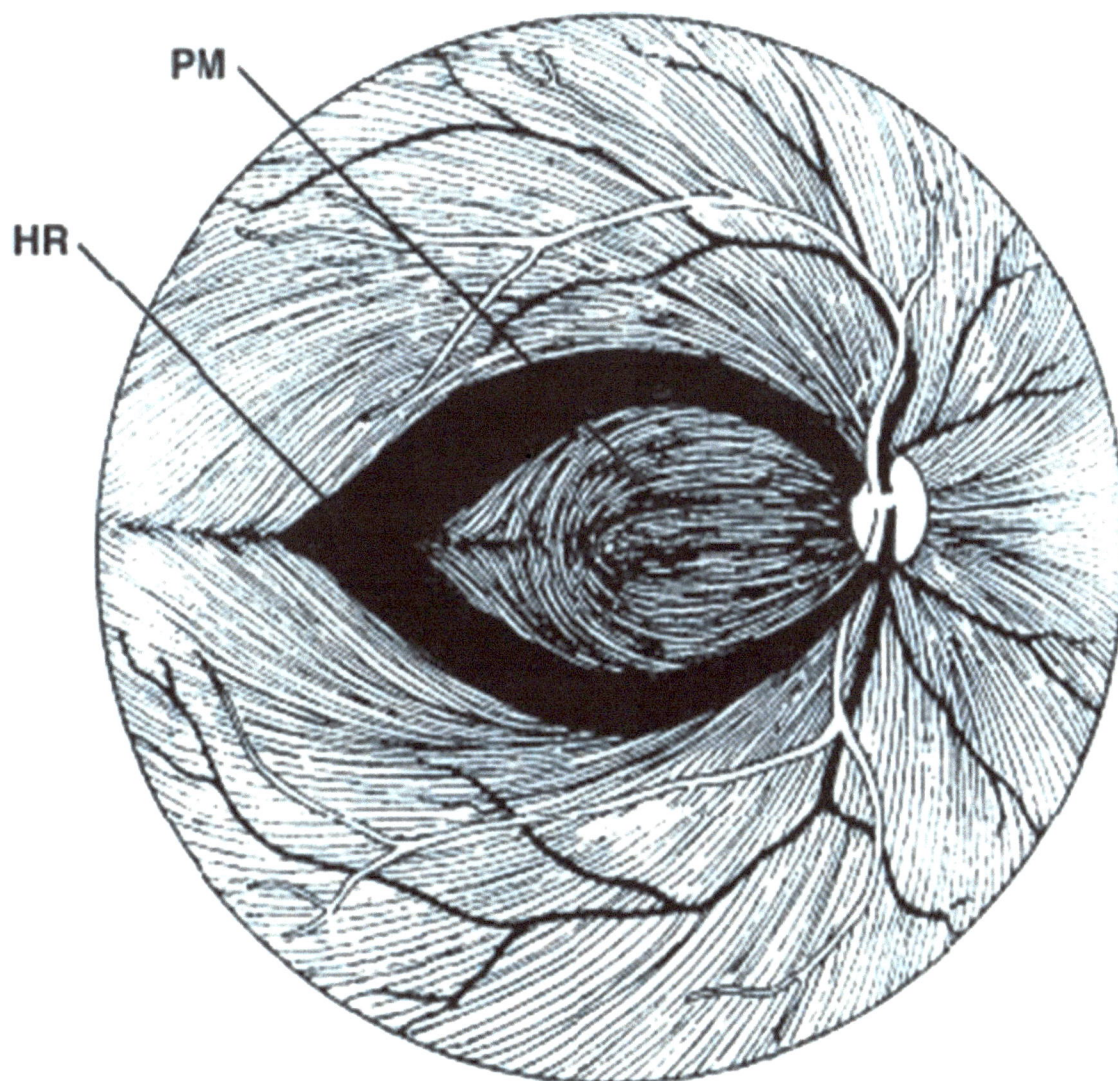

Here is another diagram showing a typical glaucoma damage site in the left eye and the corresponding visual field defect (© AAO 2014):

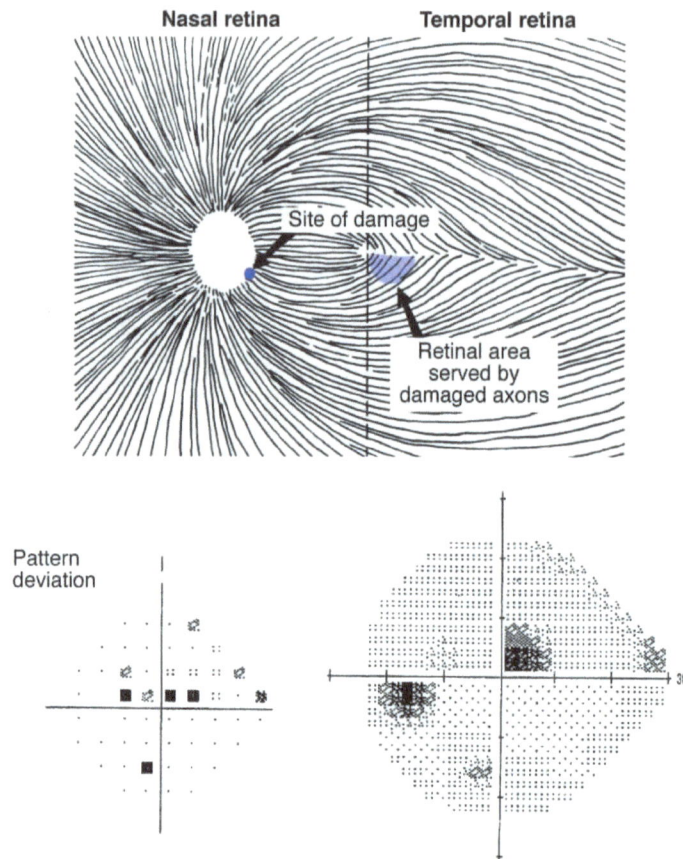

Nasal retina   Temporal retina

Site of damage

Retinal area
served by
damaged axons

Pattern
deviation

VF defects:

To do the visual fields to counting fingers (VFCF) test, have the patient close their left eye with their hand or an occluder. In all of our testing, we test the right eye first in order to make it easier to remember test results before we write them down on the exam form or EMR. Have them look at your left eye. Hold up both of your hands halfway between your eye and their right eye, both just above the horizontal midline and just on either side of the vertical midline. Ask them how many fingers they can see. Do it with one hand at a time if the patient is confused about adding both of your hands together. Do the same for the two inferior quadrants of the right eye. Repeat for the left eye having them look at your right eye. You do not come in from the side along the midline horizontal line since this will test both the superior and inferior VF at the same time and you will miss quadratic defects. Only the DMV is interested in testing VF this way, not you on your medical exam.

Any deficits or questionable results on the confrontational test should be tested formally with a Humphrey 24-2 automated threshold static perimeter with white on white points. We record the screening visual fields to counting fingers (VFCF) result on our exam form with the patient's right eye results on the right hand grid, just as if you were recording your own VF results. Recording the results can be confusing on the exam form, but imagine having the patient's formal VF results laid out in front of you with the right eye on the right and the left eye on the left. This is the opposite orientation of the patient when they are looking at you during the VFCT test. You have to then turn the deficit around in your mind to get it in the correct quadrant on the exam form. Think of losses as being nasal and temporal to help get it correct on the exam form.

Since the optic fibers from the same area of each eye are next to each other from the chiasm and further back, lesions affecting the chiasm and back will have bilateral deficits that are on the same side of the visual field for both eyes. For instance, a right cortical lesion will affect the right temporal optic nerve fibers and the left nasal optic nerve fibers giving a left VF defect in each eye. The closer to the visual cortex the lesion is, the more the VF defects are similar between the eyes. This is called congruity. Therefore, the more posteriorly the lesion is towards the occipital cortex, the more congruous the VF defects are between the two eyes. Exam findings that support an optic nerve disease as opposed to an optic tract disease include:[11,12]

- VF defects that affect only one eye
- VF defects with reduced BCVA
- VF defects with an APD. Significant retinal diseases can also cause an APD but you should be able to see this on DFE. A red saturation test using a mydriacyl bottle top will be desaturated in ON diseases and normal in retinal diseases.

Here is a diagram showing various VF defects at different levels of damage to the optic nerves and optic tracts (© AAO 2014):

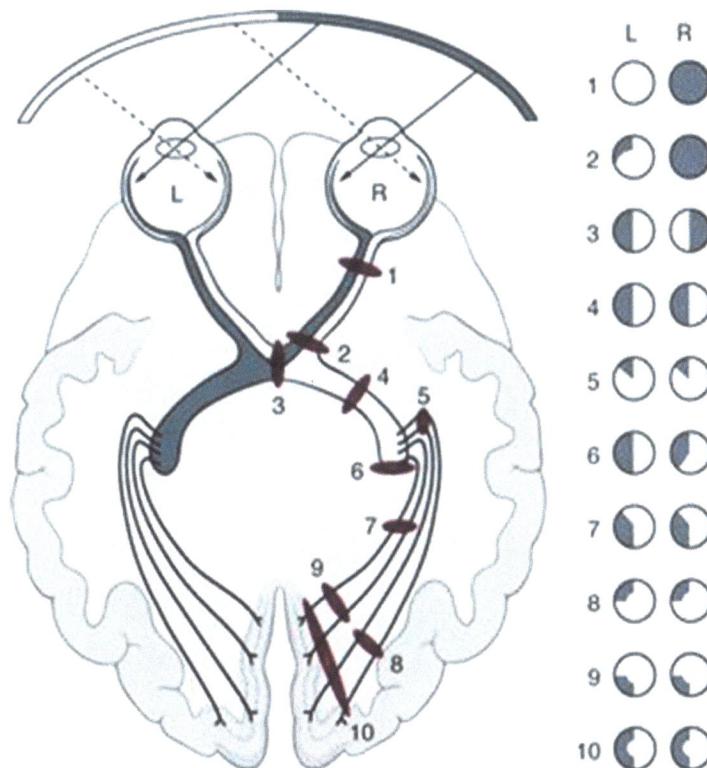

We need to condense it down to the basics. Due to the organization of the optic nerve fibers as they course from the eye to the brain, **optic *nerve* VF defects**, those involving the optic nerve in front of the chiasm, **respect the horizontal meridian** but not the vertical meridian.[11] **Optic *tract* VF defects**, those at and behind the chiasm, **respect the vertical meridian** but not the horizontal meridian.[12] Chiasmal lesions give rise to bitemperal defects. The vertical meridian goes through the fixation point at the *fovea*, not the optic nerve head. We detect VF defects during the exam with the presentation of our fingers in the 4 quadrants of the patient's visual field. This is done *monocularly* and is the visual field to counting fingers test (VFCF). Again, these quadrants are centered on the fovea, not the optic nerve.

Glaucoma VF defects will be what most of your practice will be dealing with. They occur due to damage of the optic nerve at the level of the optic disk itself. [13,14] The main VF defects in glaucoma occur in the arcuate ganglion cell fiber bundles that start in the retina lateral to the fovea, run over the fovea, around the PM bundle, and enter the optic nerve in the superior and inferior temporal areas. This area has the densest number of ganglion cells and helps explain why this area is so sensitive to damage from pressure at the optic disk and why they tend to be affected first in glaucoma. The area of the visual field corresponding to the arcuate fibers is called the Bjerrum area[15] and injury to these arcuate fibers give rise to the classic defects seen in glaucoma including the nasal step, arcuate losses (complete/altitudinal and partial/paracentral) and blind spot enlargement or Seidel defect. [16] The optic nerve superior fibers do not cross over to the inferior retina and vice versa giving rise to the rule that an injury at the level of the ON respects (or does not cross) the horizontal line. Glaucoma VF defects tend to start in the medial quadrants first (from damage to the temporal retinal area) and a deficit there, in the absence of retinal pathology, strongly suggests glaucoma.

Here are two examples of an optic *nerve* defect, respecting the horizontal line, the first one from glaucoma, the second one from AION. Both are optic *nerve* disorders and should look similar:

Glaucoma:

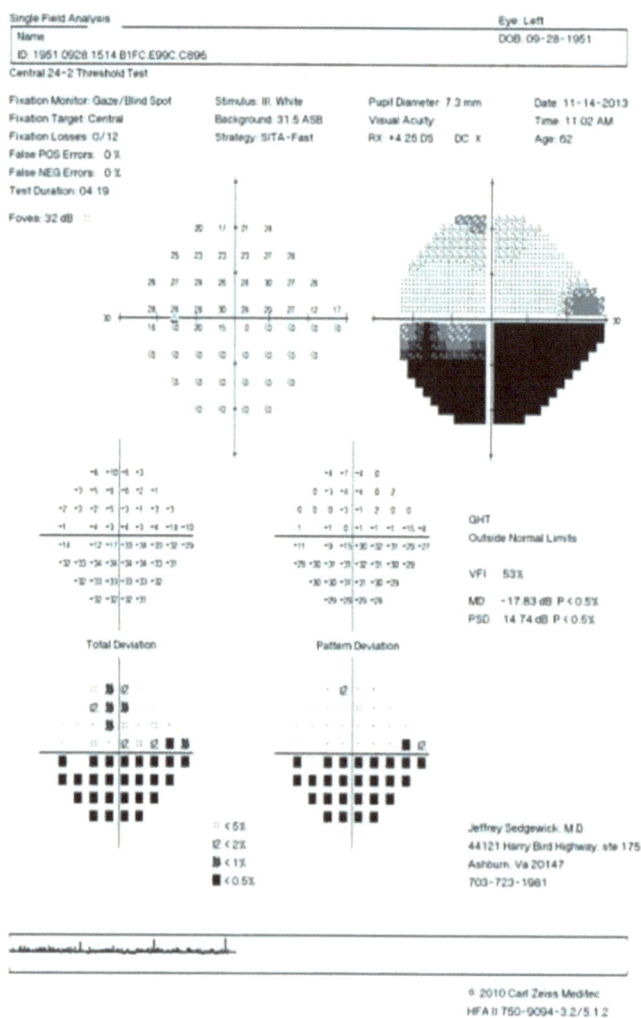

AION, which looks just like moderate glaucoma loss, which is consistent with an optic *nerve* injury that respects the horizontal line:

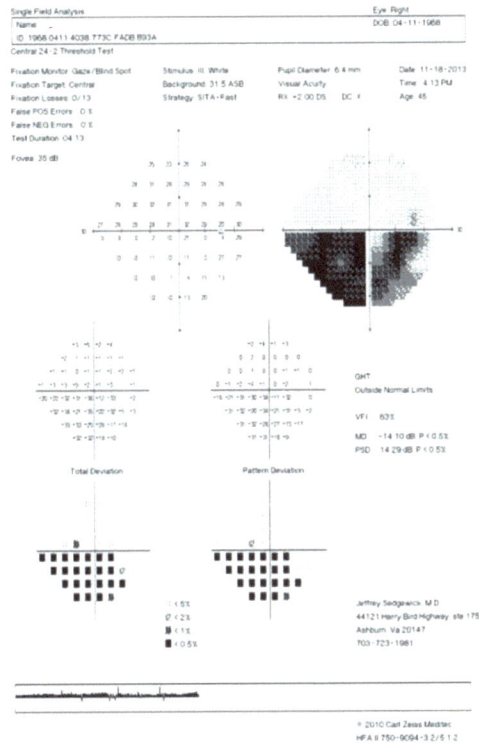

2 more optic nerve loss patterns from POAG:

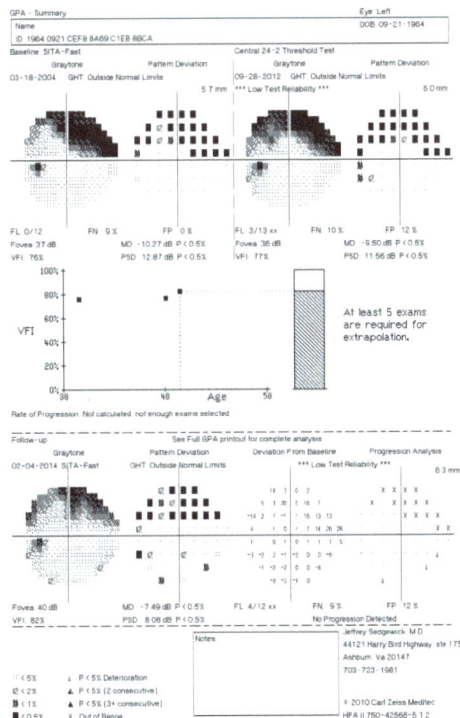

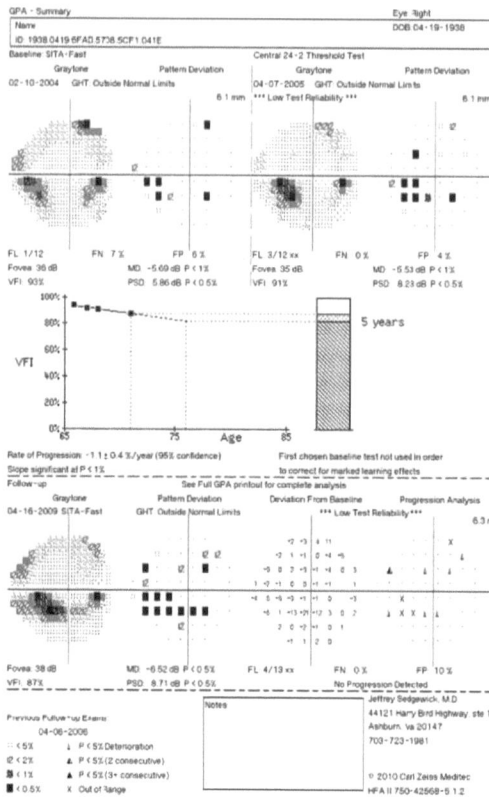

Here is a VF from an optic *chiasm/posterior optic nerve* defect that respects the vertical line. This VF was from a pituitary tumor patient that I diagnosed from this bilateral VF loss that respected the vertical line OS:

OS   **OD**

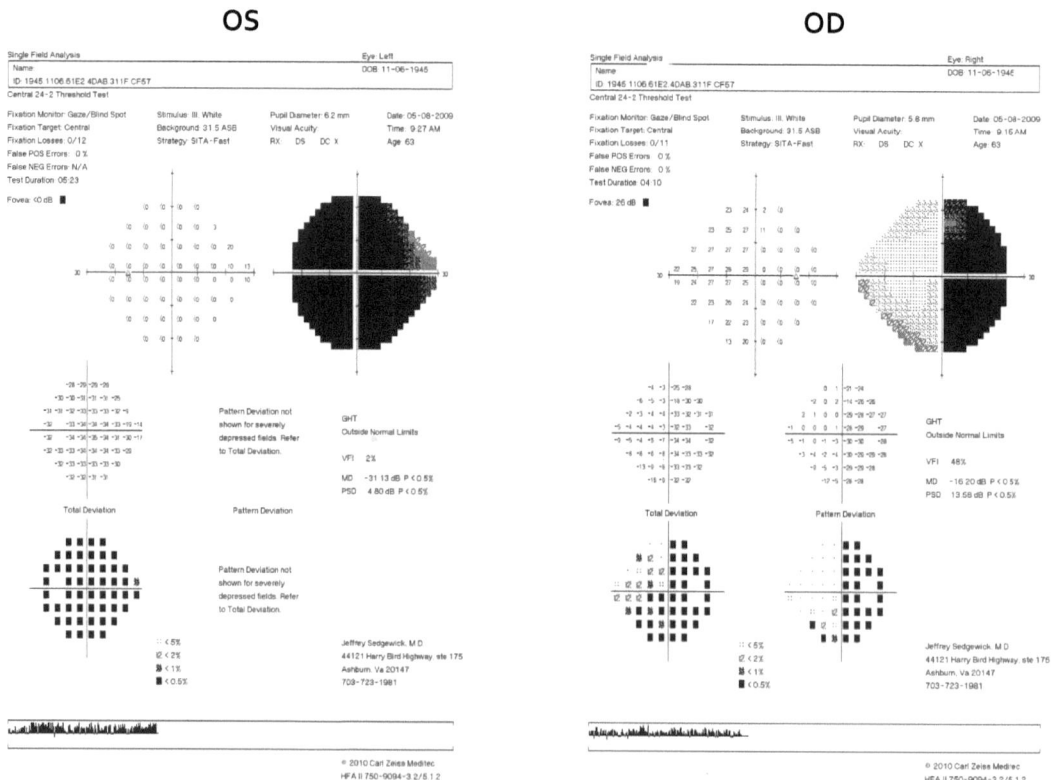

Here is another optic chiasm loss in a patient with an unknown pituitary adenoma that I diagnosed on MRI. I ordered the MRI based on this VF test:

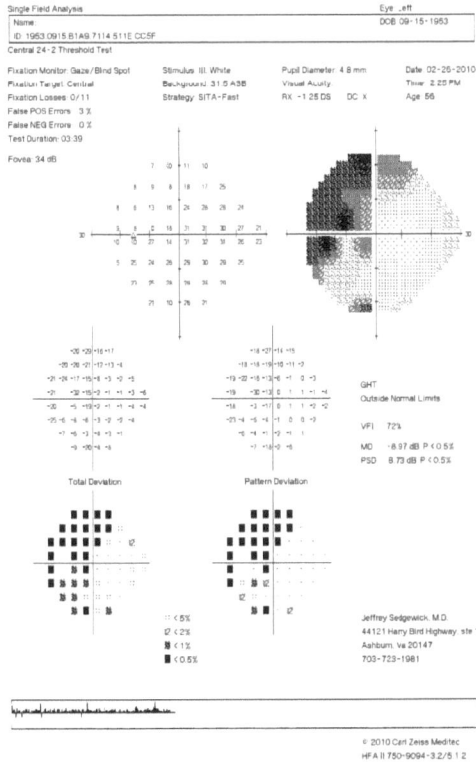

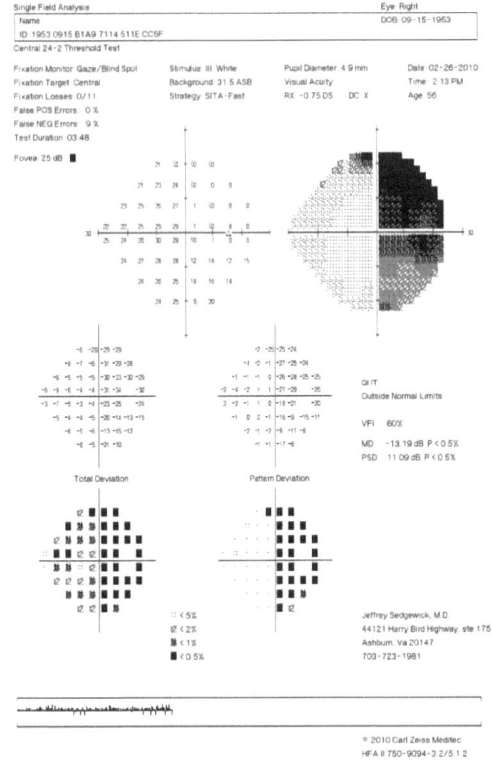

Here is a patient with optic *tract* losses that improved with time:

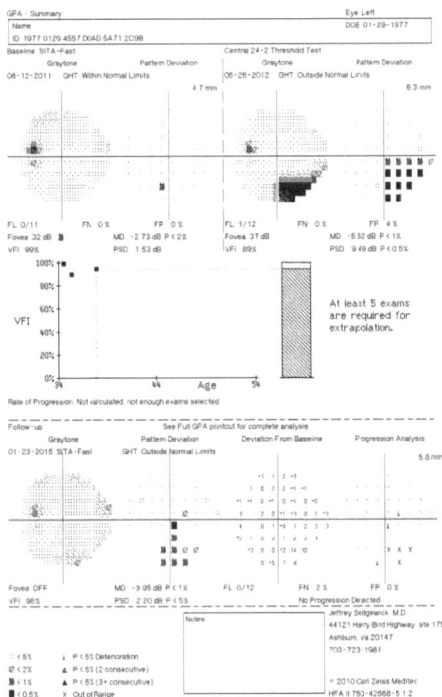

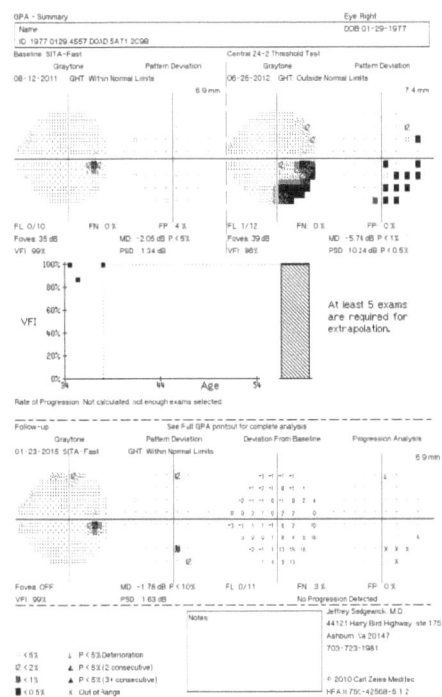

Here is another optic *tract* defect from a patient who suffered a right CNS CVA. Notice the congruity of the defects on this first, reliable HVF 24-2. The right sided VF defects are probably rim artifacts:

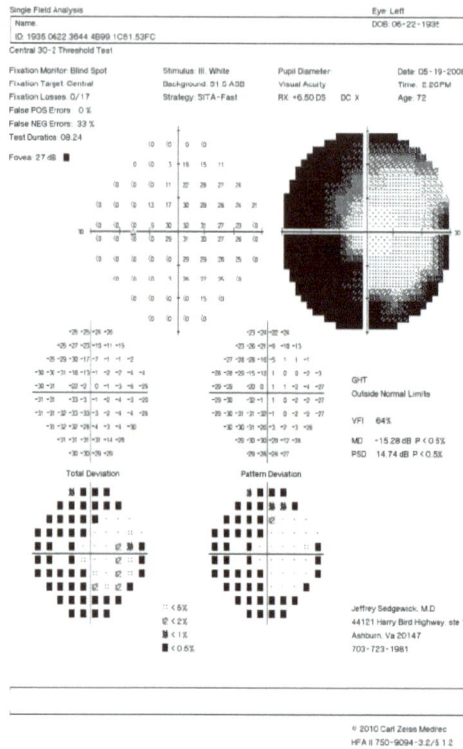

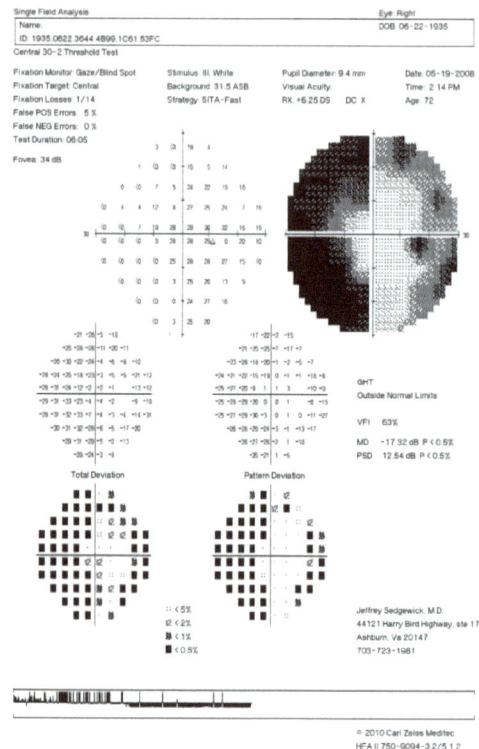

Here is another somewhat incongruous optic tract loss pattern:

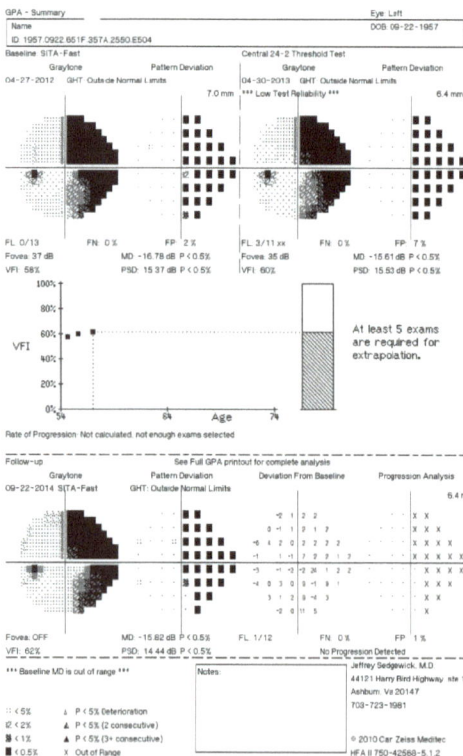

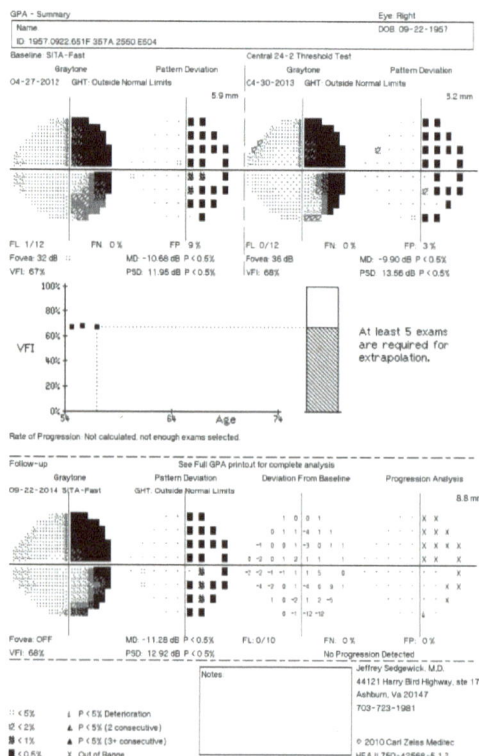

Here is an optic tract defect HVF from a patient with atrial fibrillation that developed a CVA. Notice the rim artifact in the first HVF OS that went away on the second test:

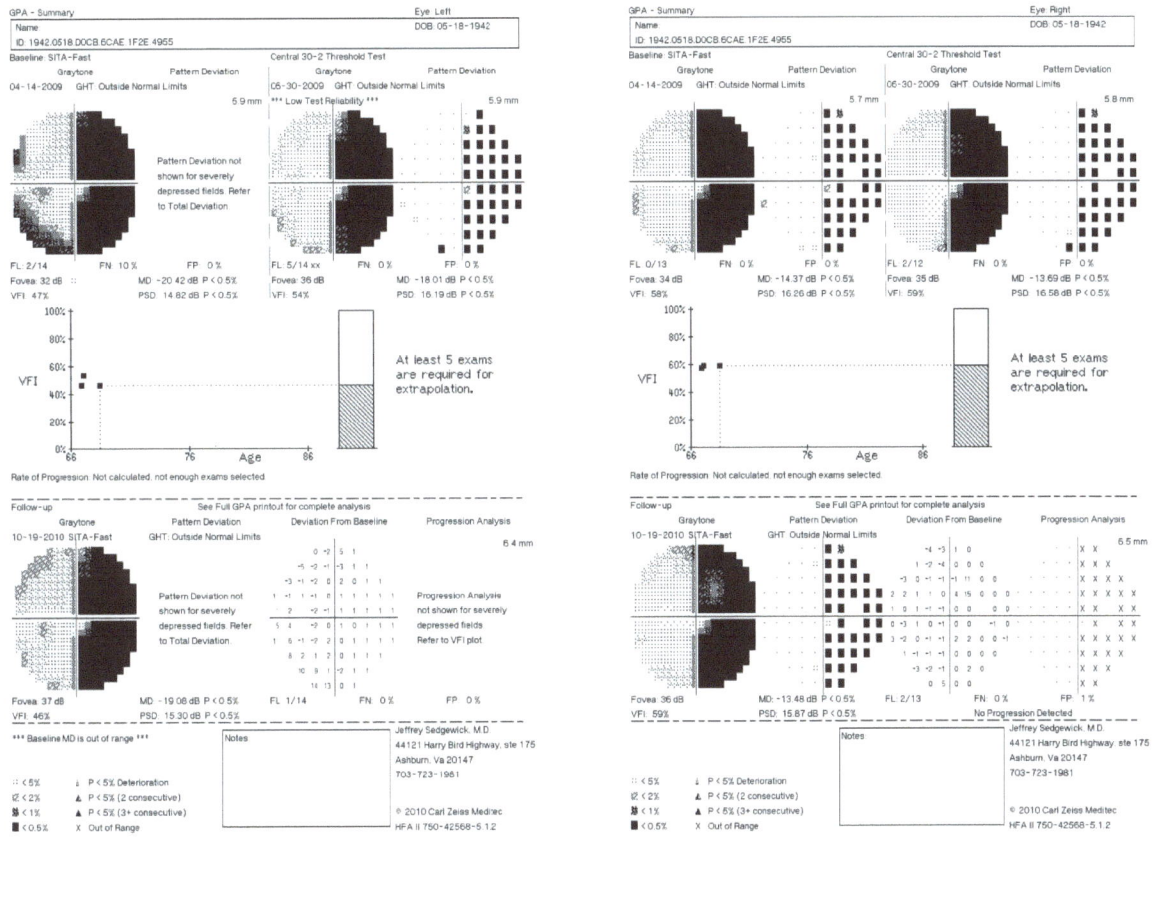

Pseudo VF defects, usually outside of the Bjerrum area, include[17]

- rim artifacts. This is the #1 artifact that you will have to take into account on your VF interpretations.
- droopy eyelid artifacts
- refractive errors
- patient inattention
- defects from retinal diseases

Artificial or *false* VF defects can through you off. **Rim artifacts**, the test **learning curve**, droopy lids, refractive errors (rarer), **patient inattention**, and retinal pathology make up most of the artifacts you will see. Below is a rim artifact from a POAG suspect patient. The latest exam is at the bottom of the printout. You will notice that the first two tests are normal. The patient's 6th test, on an unreliable exam, shows a dense rim artifact that decreases the VFI graph (explained later). A **rim artifact** can throw the VFI slope off, making it look like the VFs are getting worse when they are not.

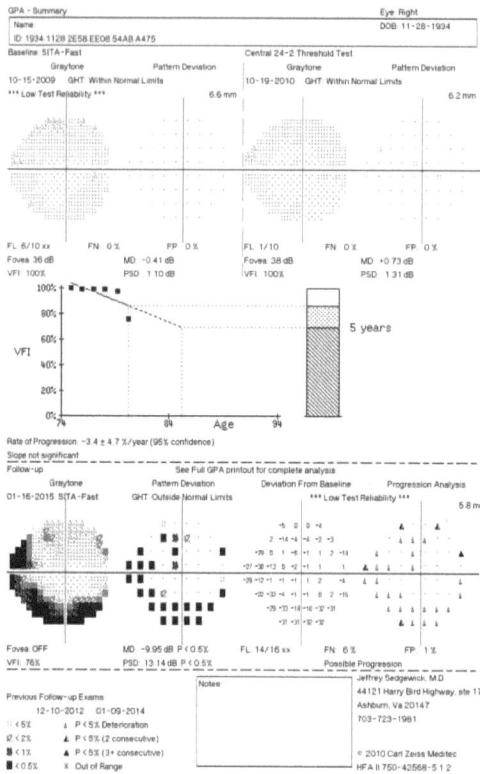

Here is another rim artifact on the second HVF 24-2, falsely decreasing the VFI:

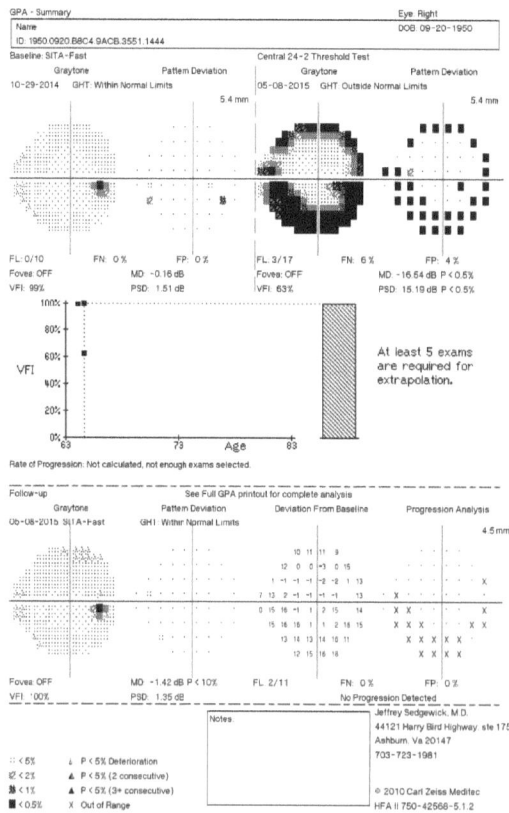

Three more examples of rim artifacts falsely depressing the VFI in glaucoma and glaucoma suspect patients. The first patient has a rim artifact falsely depressing the third test OS (left) and the first test OD (right):

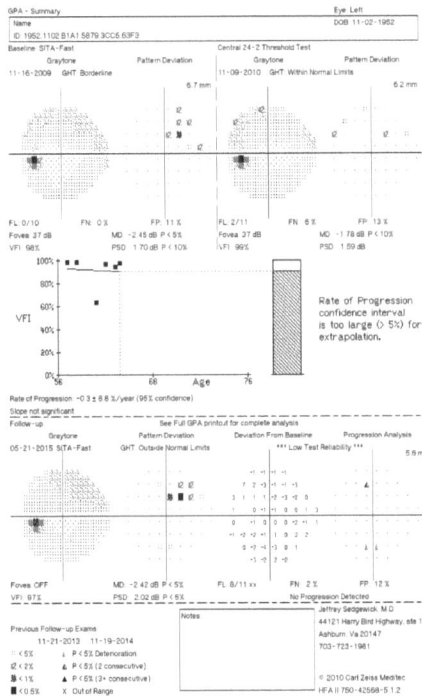

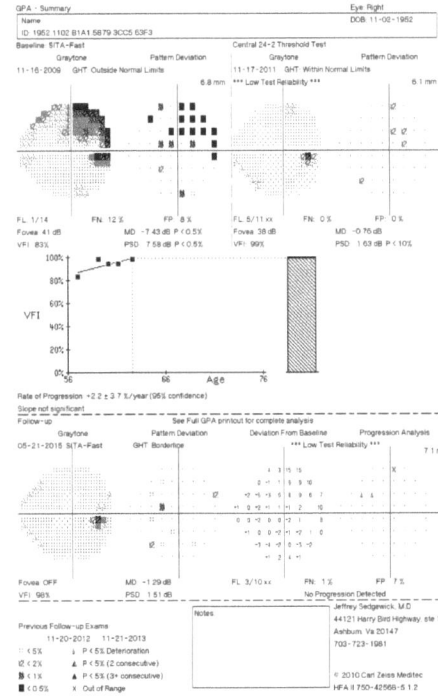

Here are two different patients both expressing rim artifacts that falsely depress the VFI. Both of these are a POAG suspect patients. The right hand VFI depression is subtle and has a significant depression (p < 0.1 %) that may not be due to the rim artifact in the last HVF 24-2, making the clinical picture even more confusing:

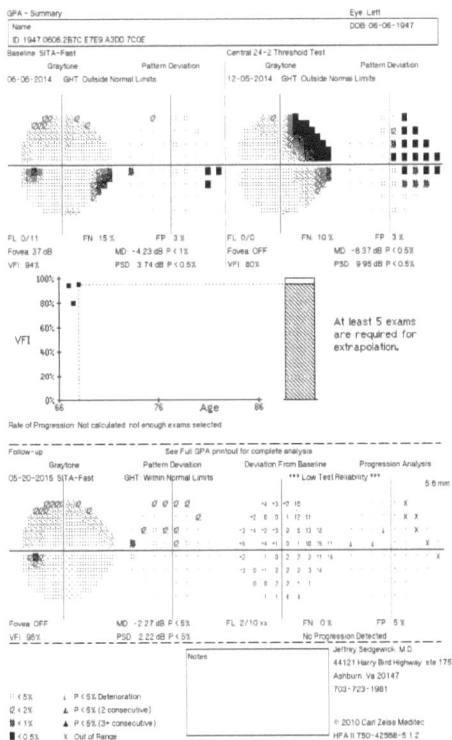

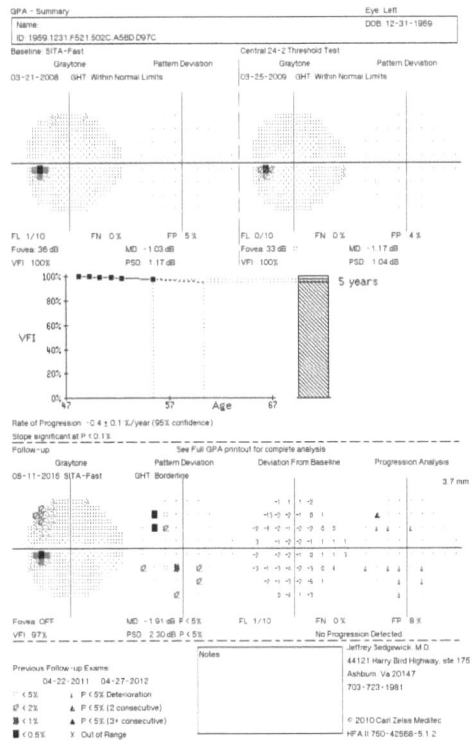

There is a significant learning curve in all of the tests we do. Variability in HVF/HRT/IOP test results are common, to be expected and will make your job more difficult. A learning curve is evident in the *majority* of glaucoma patients, with a **mean increase of 2.81 dB between the first and second HVF test (P < .001) but no significant increase between the second and fifth tests.**[190] **86% (604/703) of the abnormalities on the initial and deemed reliable HVF tests in the Ocular Hypertensive Treatment Study (OHTS) were not confirmed abnormal upon retesting.**[18] Here is an example of a large learning curve in a POAG suspect patient:

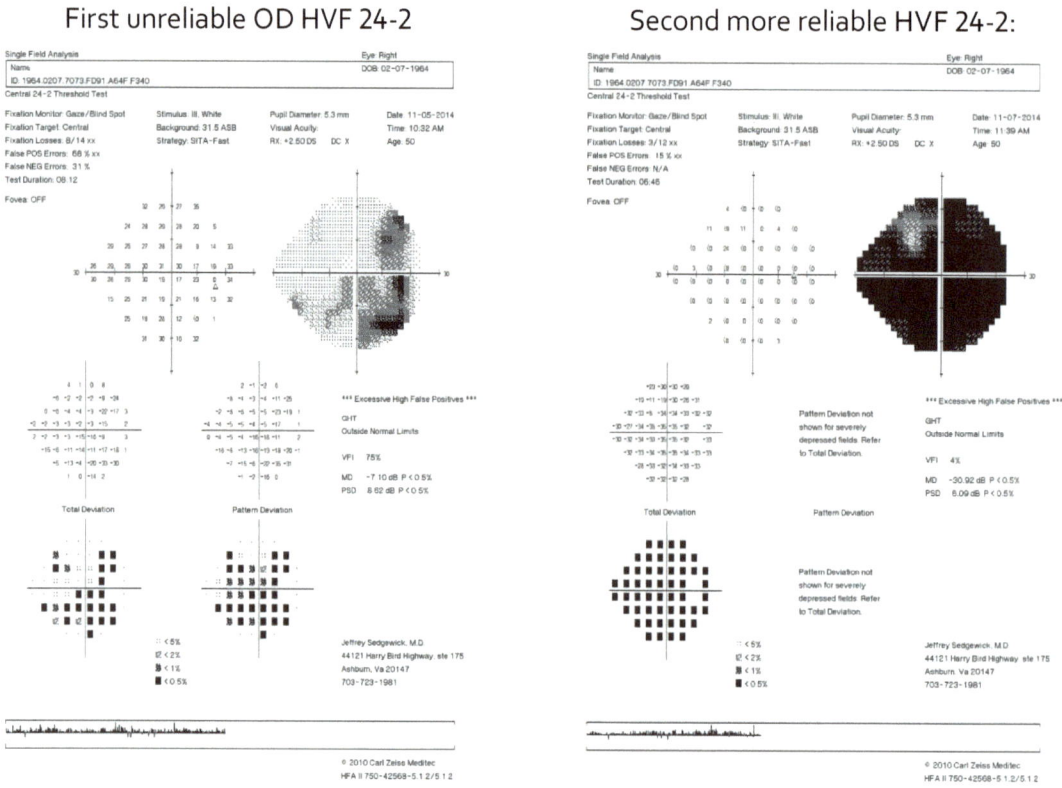

First unreliable OD HVF 24-2

Second more reliable HVF 24-2:

Third reliable HVF 24-2 OD:

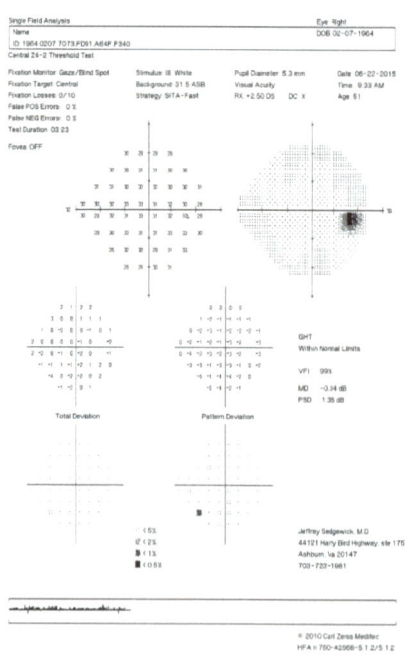

First HVF 24-2 OS w high FP, "washed out":

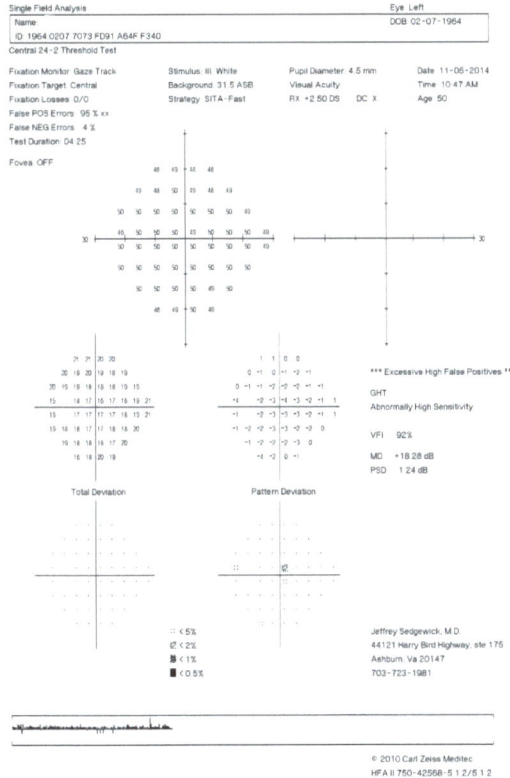

Second HVF 24-2 OS w high FP:

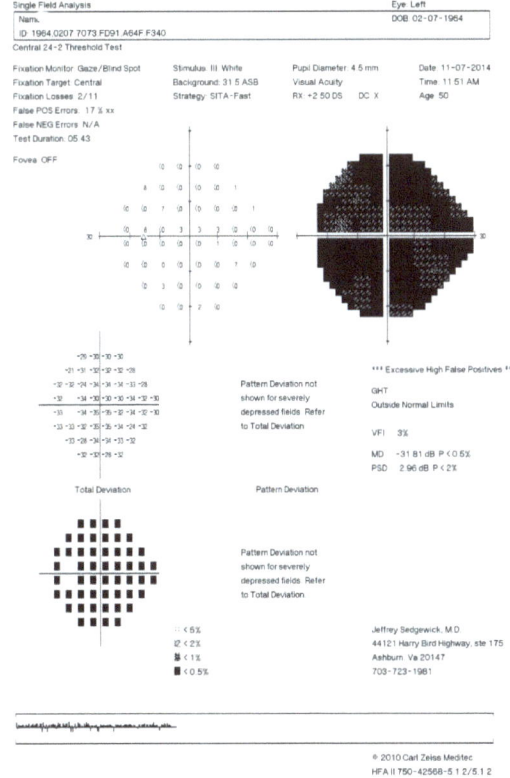

Third reliable HVF 24-2 OS:

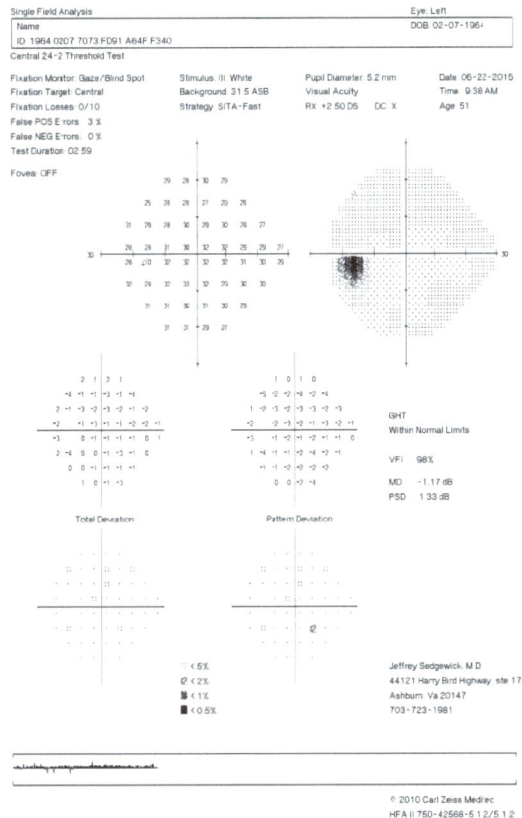

Here is another falsely decreasing VFI slope from glaucoma in a patient that was diagnosed and treated with brachytherapy OS for a choroidal melanoma between the first two VF and the last HVF. Her also being a POAG suspect patient further confuses the clinical picture:

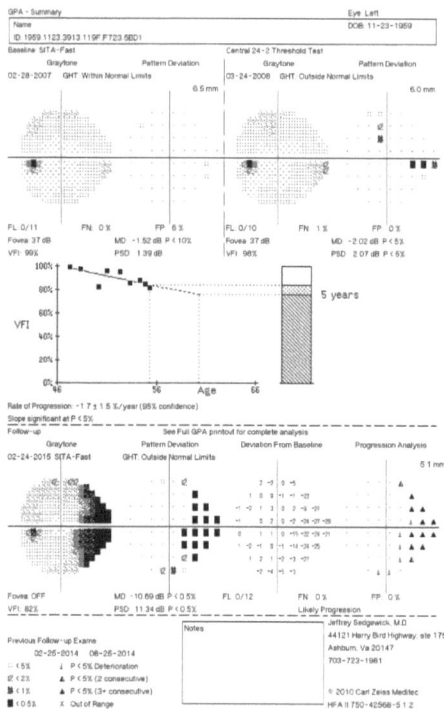

Here is another patient with an RRD treated with a scleral buckle (SB), coincidentally giving a true nasal step, but not from POAG in a patient also followed as a POAG suspect. This is their second HVF 24-2:

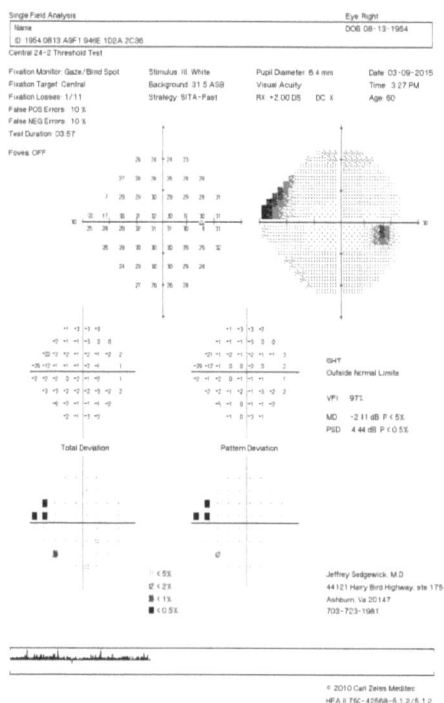

Another POAG suspect patient with a SB giving another true defect that looks like glaucoma but isn't:

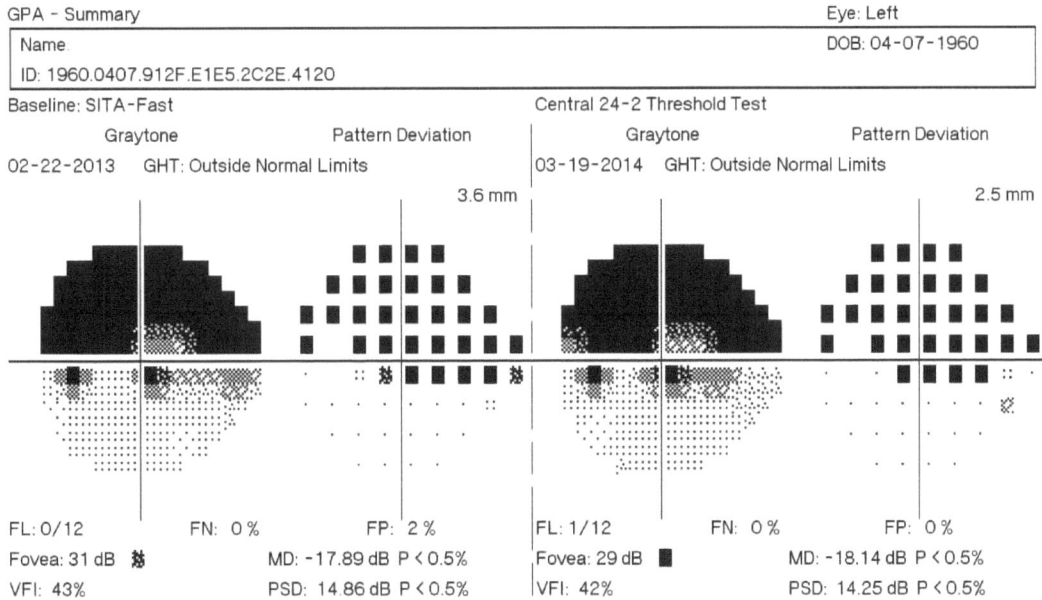

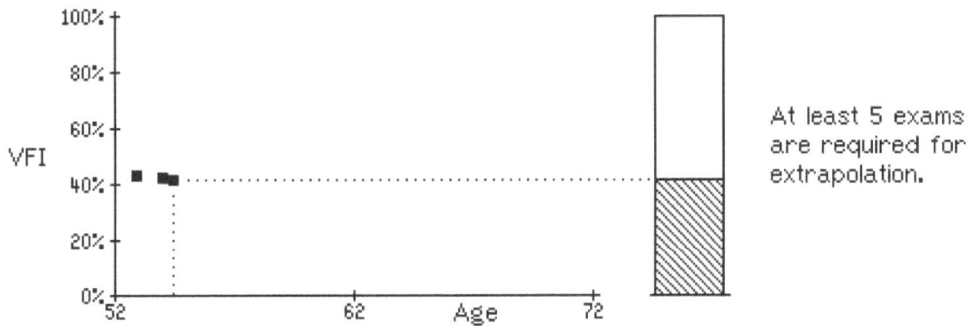

GPA - Summary

Eye: Left

Name.
DOB: 04-07-1960
ID: 1960.0407.912F.E1E5.2C2E.4120

Baseline: SITA-Fast — Central 24-2 Threshold Test

Graytone	Pattern Deviation	Graytone	Pattern Deviation
02-22-2013 GHT: Outside Normal Limits		03-19-2014 GHT: Outside Normal Limits	

3.6 mm

2.5 mm

FL: 0/12 · FN: 0% · FP: 2%
Fovea: 31 dB · MD: -17.89 dB P < 0.5%
VFI: 43% · PSD: 14.86 dB P < 0.5%

FL: 1/12 · FN: 0% · FP: 0%
Fovea: 29 dB · MD: -18.14 dB P < 0.5%
VFI: 42% · PSD: 14.25 dB P < 0.5%

At least 5 exams are required for extrapolation.

VFI chart — Age 52 to 72

Rate of Progression: Not calculated, not enough exams selected.

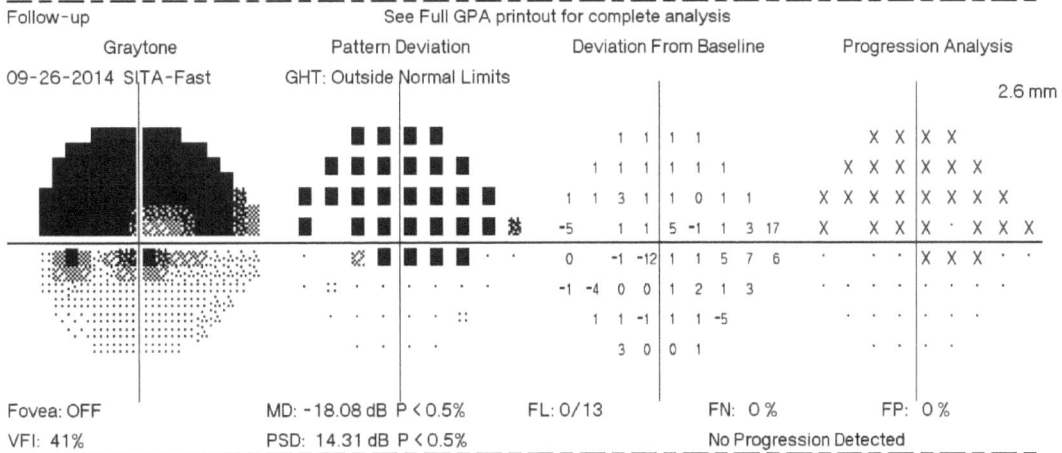

Follow-up — See Full GPA printout for complete analysis

Graytone	Pattern Deviation	Deviation From Baseline	Progression Analysis
09-26-2014 SITA-Fast	GHT: Outside Normal Limits		

2.6 mm

Fovea: OFF · MD: -18.08 dB P < 0.5% · FL: 0/13 · FN: 0% · FP: 0%
VFI: 41% · PSD: 14.31 dB P < 0.5% · No Progression Detected

*** Baseline MD is out of range ***

Notes:

Jeffrey Sedgewick, M.D.
44121 Harry Bird Highway, ste 175
Ashburn, Va 20147
703-723-1981

© 2010 Carl Zeiss Meditec
HFA II 750-42568-5.1.2

:: < 5%   P < 5% Deterioration
< 2%   P < 5% (2 consecutive)
< 1%   P < 5% (3+ consecutive)
■ < 0.5%   X Out of Range

Considerable test variability from one HVF test to another is a common occurrence in glaucomatous loss [16] and can mislead you into start glaucoma drops too soon. Again, I want to make the following points once more:

A *learning curve* is also evident in the majority of glaucoma patients, with a mean increase of 2.81 dB between the first and second HVF test (P < .001) but no significant increase was seen between the second and fifth tests. [190]

86% (604/703) of the abnormalities on the initial and deemed *reliable* HVF tests in the Ocular Hypertensive Treatment Study (OHTS) were _not confirmed abnormal upon retesting_.[18]

I find this to be true in my clinic as well and it is a significant factor in all glaucoma patients. Variable reduction of sensitivity in the same area, but not necessarily at the same point, commonly precedes real glaucomatous loss and occurs more commonly in diseased areas than in normal areas.[19] Glaucomatous losses typically develop over a period of years and are variable in expression until they finally become reproducible defects on HVF testing.[20] A number of HVF tests are usually required in order to insure that a VF loss is really there.[21] Judging whether a glaucoma loss has progressed usually requires at least 3 to 4 tests for comparison. Fortunately, glaucoma usually progresses slowly enough so that it allows time for subsequent testing to occur.[22] Large sudden visual field changes are not typical of glaucoma.[22]

Learning curve errors can look a lot like glaucoma at first. Here is an initial HVF 24-2 on a POAG suspect patient. It looks like early glaucoma on a *reliable* first test:

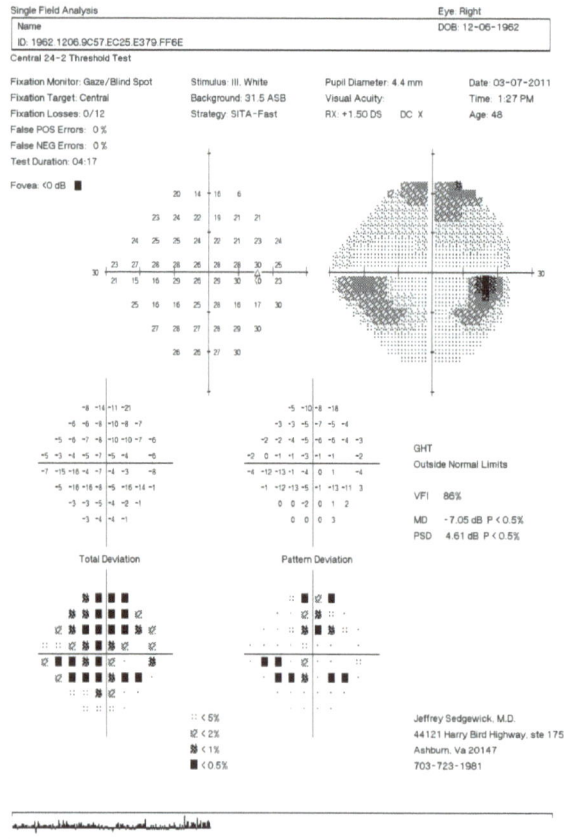

The second HVF 24-2 done 2 years later is normal (the patient no-showed their 6 month appointment), underlying the importance of A) not making a diagnosis too quickly and B) taking into account variability in the interpretation of HVFs:

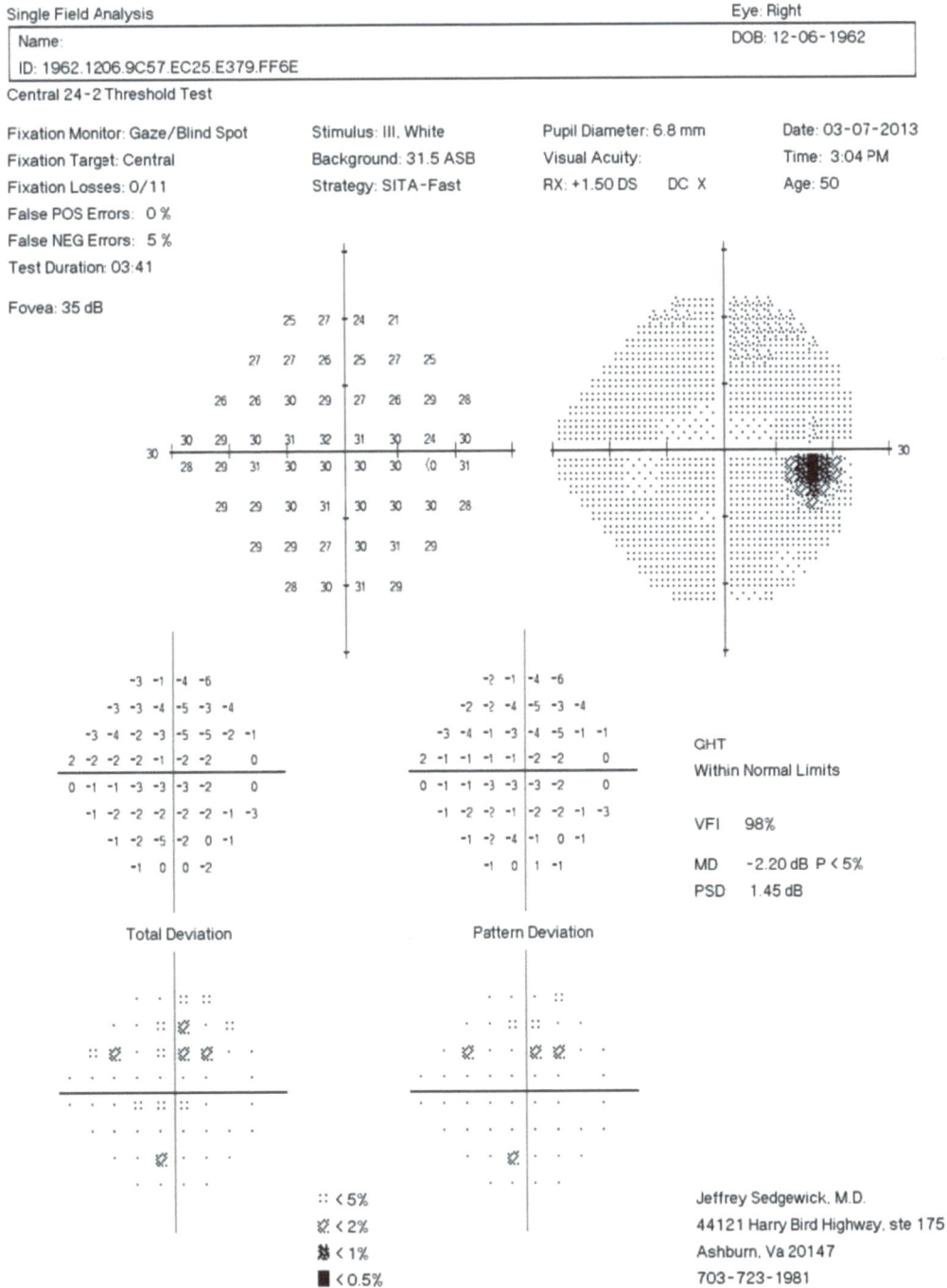

Single Field Analysis

Eye: Right

Name:

DOB: 12-06-1962

ID: 1962.1206.9C57.EC25.E379.FF6E

Central 24-2 Threshold Test

Fixation Monitor: Gaze/Blind Spot	Stimulus: III, White	Pupil Diameter: 6.8 mm	Date: 03-07-2013
Fixation Target: Central	Background: 31.5 ASB	Visual Acuity:	Time: 3:04 PM
Fixation Losses: 0/11	Strategy: SITA-Fast	RX: +1.50 DS DC X	Age: 50
False POS Errors: 0 %			
False NEG Errors: 5 %			
Test Duration: 03:41			

Fovea: 35 dB

```
            25  27   24  21
        27  27  26  25  27  25
    26  26  30  29  27  26  29  28
30  29  30  31  32  31  30  24  30
28  29  31  30  30  30  30  <0  31
    29  29  30  31  30  30  30  28
        29  29  27  30  31  29
            28  30   31  29
```

Total Deviation
```
        -3  -1  -4  -6
     -3  -3  -4  -5  -3  -4
  -3  -4  -2  -3  -5  -5  -2  -1
2  -2  -2  -2  -1  -2  -2      0
0  -1  -1  -3  -3  -3  -2      0
  -1  -2  -2  -2  -2  -2  -1  -3
     -1  -2  -5  -2   0  -1
        -1   0   0  -2
```

Pattern Deviation
```
        -2  -1  -4  -6
     -2  -2  -4  -5  -3  -4
  -3  -4  -1  -3  -4  -5  -1  -1
2  -1  -1  -1  -1  -2  -2      0
0  -1  -1  -3  -3  -3  -2      0
  -1  -2  -1  -1  -2  -2  -1  -3
     -1  -2  -4  -1   0  -1
        -1   0   1  -1
```

GHT
Within Normal Limits

VFI   98%

MD   -2.20 dB P < 5%
PSD   1.45 dB

:: < 5%
✗ < 2%
▨ < 1%
■ < 0.5%

Jeffrey Sedgewick, M.D.
44121 Harry Bird Highway, ste 175
Ashburn, Va 20147
703-723-1981

Here is another first test on a glaucoma suspect with a defect that looks like glaucoma on a reliable test. The second test, taken 2 years later is normal:

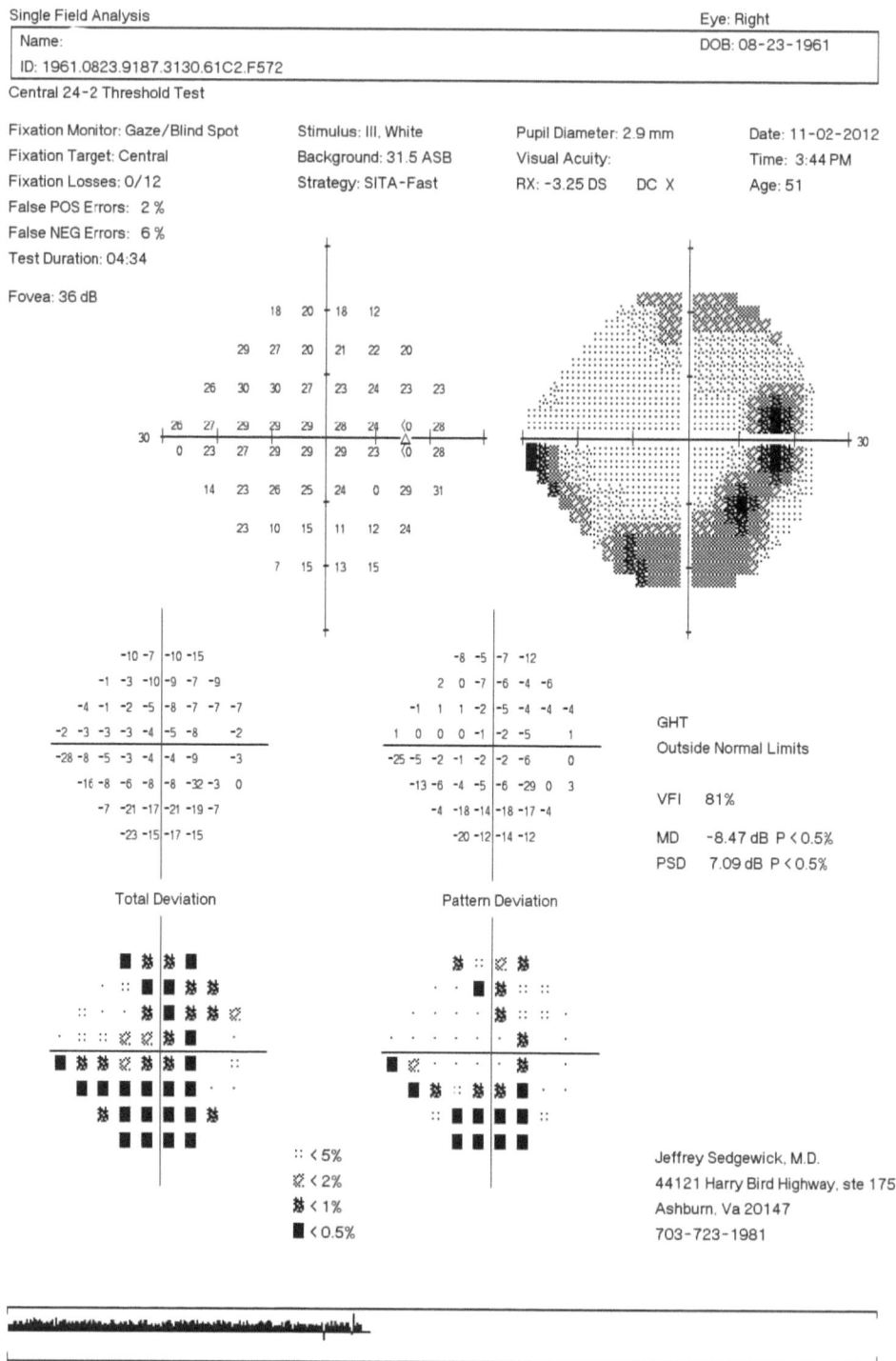

Single Field Analysis   Eye: Right

Name: DOB: 08-23-1961

ID: 1961.0823.9187.3130.61C2.F572

Central 24-2 Threshold Test

Fixation Monitor: Gaze/Blind Spot	Stimulus: III, White	Pupil Diameter: 2.9 mm	Date: 11-02-2012
Fixation Target: Central	Background: 31.5 ASB	Visual Acuity:	Time: 3:44 PM
Fixation Losses: 0/12	Strategy: SITA-Fast	RX: -3.25 DS DC X	Age: 51
False POS Errors: 2 %			
False NEG Errors: 6 %			
Test Duration: 04:34			

Fovea: 36 dB

```
              18   20   18   12

         29   27   20   21   22   20

      26   30   30   27   23   24   23   23

   26   27   29   29   28   24   <0   28
 30 +
   0    23   27   29   29   29   23   <0   28              30

      14   23   26   25   24   0    29   31

         23   10   15   11   12   24

              7    15   13   15
```

```
Total Deviation

     -10  -7  -10 -15
  -1  -3 -10 -9  -7  -9
-4 -1 -2 -5 -8 -7 -7 -7
-2 -3 -3 -3 -4 -5 -8   -2
-28 -8 -5 -3 -4 -4 -9    -3
-16 -8 -6 -8 -8 -32 -3  0
  -7 -21 -17 -21 -19 -7
     -23 -15 -17 -15
```

```
Pattern Deviation

          -8  -5  -7 -12
        2   0  -7 -6  -4  -6
     -1   1   1  -2 -5  -4  -4  -4
   1   0   0   0  -1 -2  -5     1
 -25 -5  -2  -1  -2 -2  -6     0
 -13 -6  -4  -5 -6 -29  0   3
    -4 -18 -14 -18 -17 -4
      -20 -12 -14 -12
```

GHT
Outside Normal Limits

VFI   81%

MD   -8.47 dB P < 0.5%
PSD   7.09 dB P < 0.5%

Total Deviation   Pattern Deviation

:: < 5%
⬚ < 2%
▩ < 1%
■ < 0.5%

Jeffrey Sedgewick, M.D.
44121 Harry Bird Highway, ste 175
Ashburn, Va 20147
703-723-1981

© 2010 Carl Zeiss Meditec
HFA II 750-9094-3.2/5.1.2

The second HVF 24-2 test also reliable showing an essentially normal result.

Single Field Analysis	Eye: Right
Name:	DOB: 08-23-1961
ID: 1961.0823.9187.3130.61C2.F572	

Central 24-2 Threshold Test

Fixation Monitor: Gaze/Blind Spot	Stimulus: III, White	Pupil Diameter: 3.9 mm	Date: 01-29-2014
Fixation Target: Central	Background: 31.5 ASB	Visual Acuity:	Time: 2:08 PM
Fixation Losses: 1/11	Strategy: SITA-Fast	RX: -3.25 DS DC X	Age: 52
False POS Errors: 7 %			
False NEG Errors: 7 %			
Test Duration: 03:17			

Fovea: 35 dB

```
            18  23  24  21
         26  27  25  26  24  27
      28  30  29  30  29  26  28  27
   28  30  31  31  31  31  27  30  30
30 27  29  31  31  31  32  30  (0  31      30
      26  19  30  30  29  30  30  29
         27  24  25  27  30  27
            16  23  25  28
```

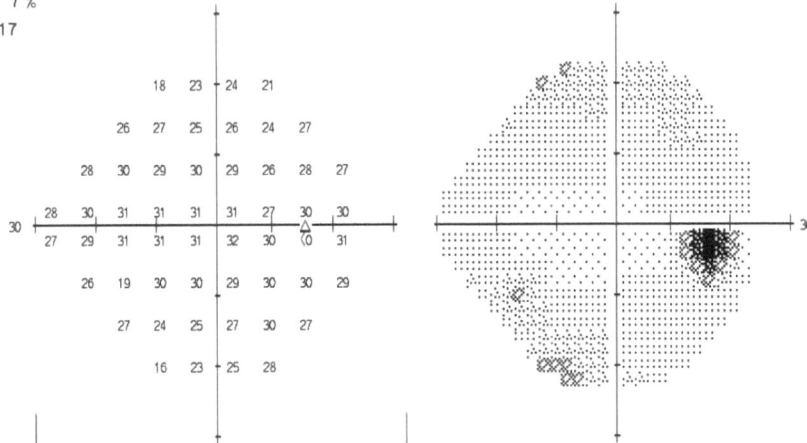

```
      -10 -5 -3  -6
    -4 -3 -5 -3 -6 -2
 -1 -1 -3 -2 -2 -5 -2 -2
 0 -1  0 -2 -2 -2 -5     0
-1 -1 -1 -2 -2 -1 -2     0
-4 -12 -2 -3 -3 -2 -1 -1
   -3 -7 -7 -5 -2 -4
     -13 -7 -5 -2
```

```
       -9 -4 -2 -5
     -3 -2 -4 -2 -5 -1
   0  0 -2 -1 -1 -4 -1 -1
 1  1  1  0 -1 -1 -4     1
 0  0  0 -1 -1  0 -1     1
  -3 -11 -1 -2 -2 -1  0  0
    -2 -6 -6 -3  0 -3
      -12 -6 -4 -1
```

GHT
Outside Normal Limits

VFI   95%

MD   -3.24 dB P < 1%
PSD   2.96 dB P < 2%

Total Deviation

Pattern Deviation

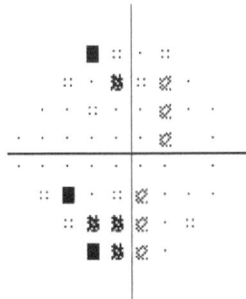

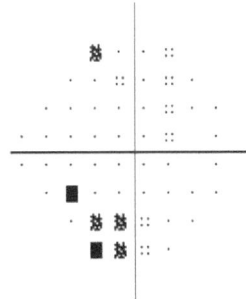

:: < 5%
▨ < 2%
▩ < 1%
■ < 0.5%

Jeffrey Sedgewick, M.D.
44121 Harry Bird Highway, ste 175
Ashburn, Va 20147
703-723-1981

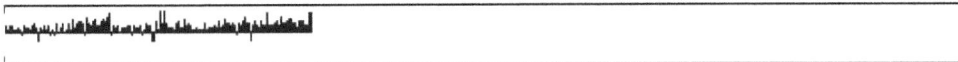

© 2010 Carl Zeiss Meditec
HFA II 750-9094-3.2/5.1.2

Another abnormal first test that gets better on the second test by a poor test taker:

Single Field Analysis   Eye: Left

| Name. | DOB: 06-03-1935 |
| ID: 1935.0603.AD61.7F11.8D34.9905 | |

Central 24-2 Threshold Test

Fixation Monitor: Gaze/Blind Spot   Stimulus: III, White   Pupil Diameter: 4.4 mm   Date: 07-15-2013
Fixation Target: Central   Background: 31.5 ASB   Visual Acuity: Time: 1:48 PM
Fixation Losses: 11/14 xx   Strategy: SITA-Fast   RX: -2.75 DS   DC X   Age: 78
False POS Errors: 9 %
False NEG Errors: 7 %
Test Duration: 05:20

Fovea: 30 dB

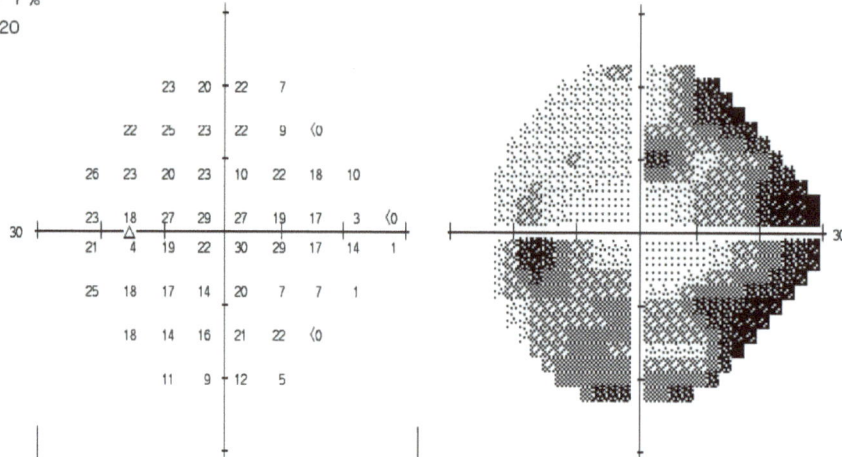

```
              23  20  22   7
          22  25  23  22   9  (0
      26  23  20  23  10  22  18  10
      23  18  27  29  27  19  17   3  (0
30 ----△------------------------------ 30
      21   4  19  22  30  29  17  14   1
      25  18  17  14  20   7   7   1
          18  14  16  21  22  (0
              11   9  12   5
```

```
Total Deviation                                    Pattern Deviation
        -2  -6 -4 -19                                      1  -3 -1 -16
    -5  -3  -5 -6 -19 -29                              -2   0 -2 -3 -16 -26
-2  -5  -9  -7 -20 -8 -11 -18                      1  -2  -6 -4 -17 -5  -8 -14
-5      -3  -2 -4 -12 -13 -26 -28                  -2       0  1 -1  -9 -10 -22 -25
-8     -12 -10 -2 -2 -14 -15 -25                   -5      -9 -6  1   1 -10 -12 -22
-4 -12 -14 -17 -11 -24 -23 -27                    -1  -9 -11 -14 -8 -21 -20 -24
   -11 -16 -14 -9  -8 -31                             -8 -13 -11 -6  -5 -28
       -18 -20 -17 -23                                  -15 -17 -14 -20
```

*** Low Test Reliability ***

GHT
Outside Normal Limits

VFI   69%

MD   -12.66 dB P < 0.5%
PSD   8.30 dB   P < 0.5%

Total Deviation   Pattern Deviation

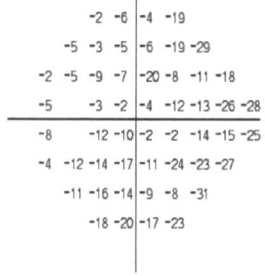   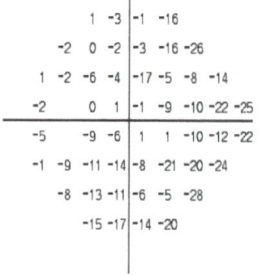

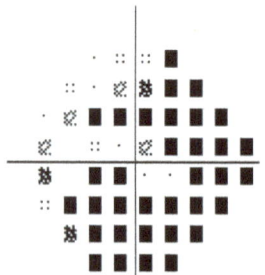   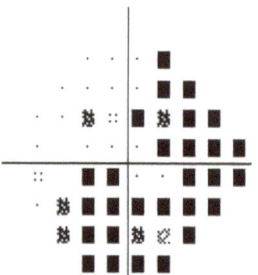

:: < 5%
※ < 2%
⛝ < 1%
■ < 0.5%

Jeffrey Sedgewick, M.D.
44121 Harry Bird Highway, ste 175
Ashburn, Va 20147
703-723-1981

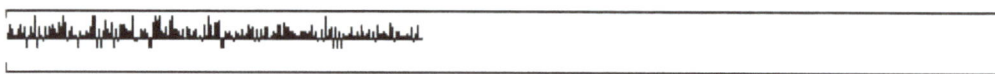

© 2010 Carl Zeiss Meditec
HFA II 750-9094-3.2/5.1.2

Name

DOB: 06-03-1935

ID: 1935.0603.AD61.7F11.8D34.9905

Central 24-2 Threshold Test

Fixation Monitor: Gaze/Blind Spot   Stimulus: III, White   Pupil Diameter: 5.0 mm   Date: 08-12-2014

Fixation Target: Central   Background: 31.5 ASB   Visual Acuity: Time: 2:41 PM

Fixation Losses: 9/12 xx   Strategy: SITA-Fast   RX: -2.75 DS   DC X   Age: 79

False POS Errors: 6 %

False NEG Errors: 14 %

Test Duration: 05:28

Fovea: OFF

```
              23  21 | 22  22
          26  25  24 | 27  27  25
      25  24  22  24 | 27  27  25  13
      23   8  25  27 | 29  29  25  20  19
  30 ─────────────△──┼──────────────────── 30
      27   8  24  25 | 28  29  26  21  16
      24  24  25  23 | 23  25  23  14
          24  24  26 | 22  23  19
              15  22 | 21  20
```

Total Deviation

```
           -2  -5 |-4  -4
       -1  -2  -4 |-2  -1  -2
   -3  -5  -7  -6 |-3  -3  -4 -14
   -6      -5  -4 |-2  -2  -5  -9  -7
   -2      -7  -7 |-4  -3  -4  -7 -10
   -5  -6  -6  -8 |-8  -6  -7 -14
       -6  -6  -5 |-8  -7 -10
          -14  -7 |-7  -8
```

Pattern Deviation

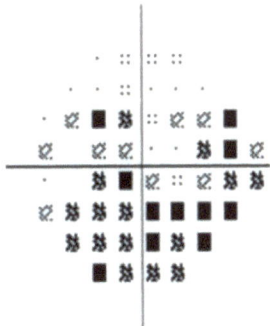

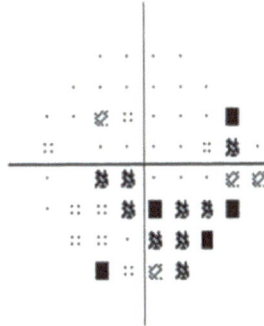

```
            0  -3 |-2  -2
        1   0  -2 | 0   1   3
   -1  -3  -5  -4 |-1  -1  -2 -12
   -4      -3  -2 | 0   0  -3  -7  -5
    0      -5  -5 |-2  -1  -2  -5  -8
   -3  -4  -4  -6 |-6  -4  -5 -12
       -4  -4  -3 |-6  -5  -8
          -12  -5 |-5  -6
```

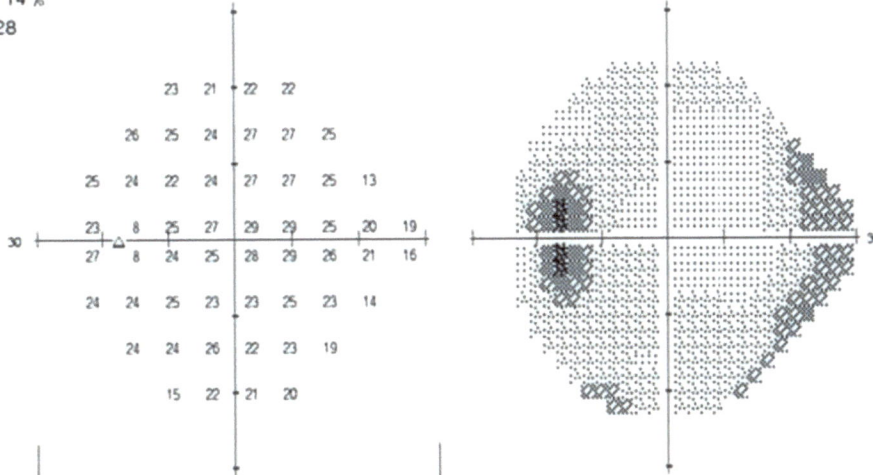

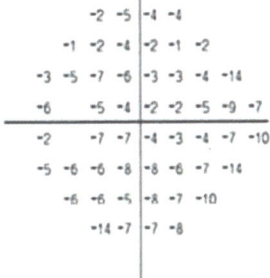

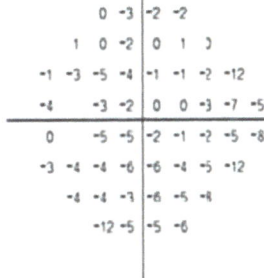

*** Low Test Reliability ***

GHT

Outside Normal Limits

VFI   89%

MD   -5.60 dB P < 1%

PSD   3.08 dB P < 2%

:: < 5%

⌀ < 2%

✗ < 1%

■ < 0.5%

Jeffrey Sedgewick, M.D.

44121 Harry Bird Highway, ste 175

Ashburn, Va 20147

703-723-1981

Another example of a patient with mild bilateral temporal losses on the first exam that got better on subsequent exams. I have not ordered an MRI of the pituitary area yet, but I might if the VFs get worse:

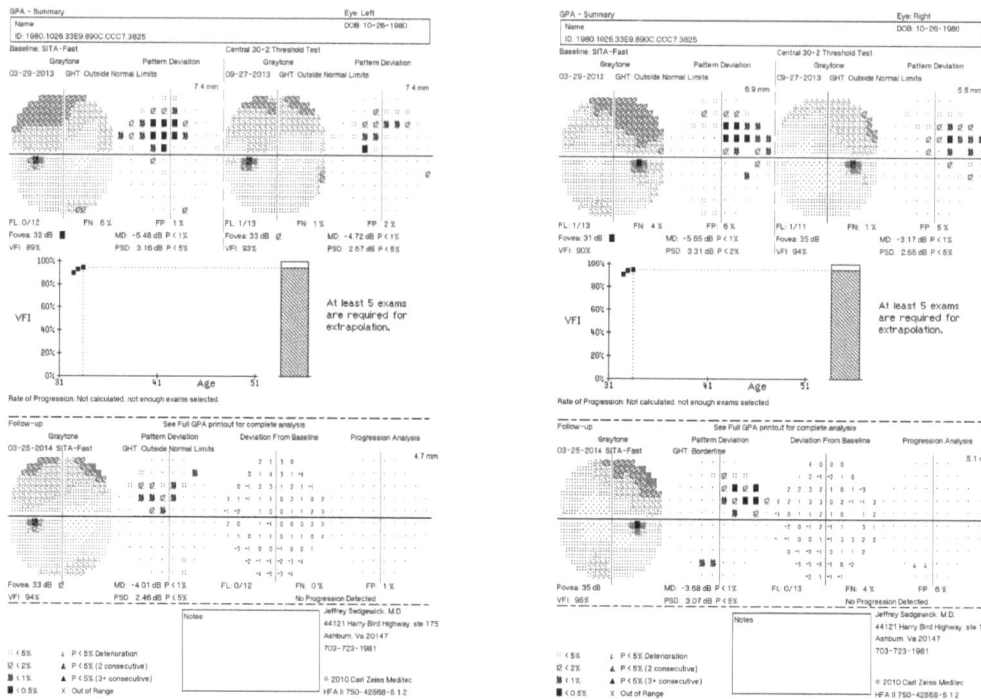

Here are more examples of learning curves seen in patients of mine. The first one is an extreme example of this with the OD shown first.

First unrelaible HVF OD with very high FP

Second unreliable HVF OD with high FL and FP

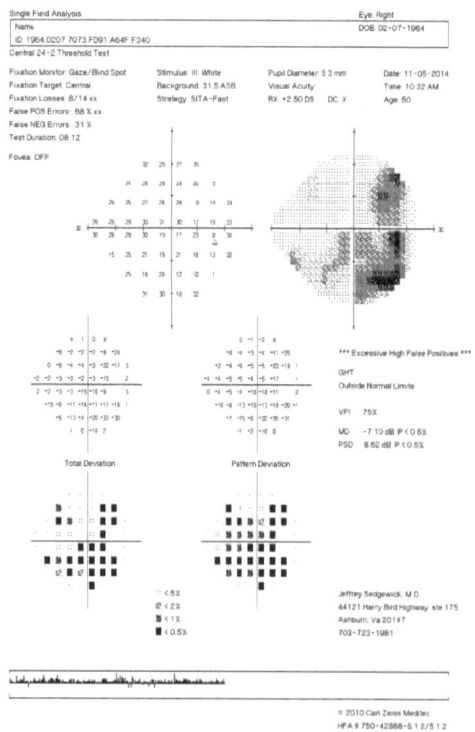

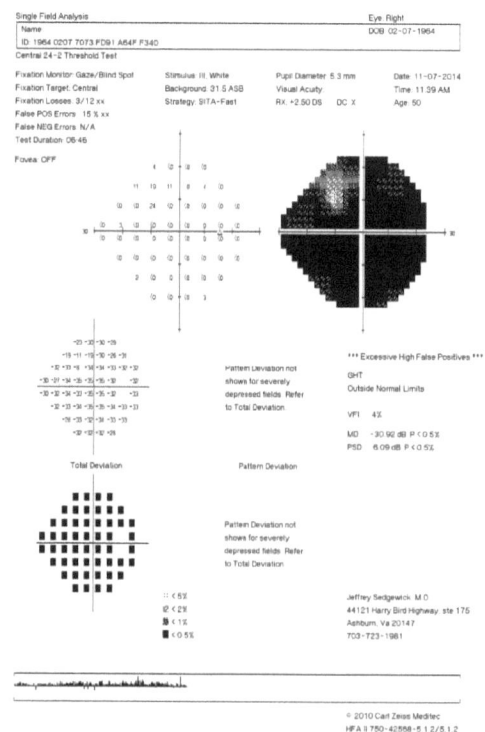

Third reliable HVF OD, normal!

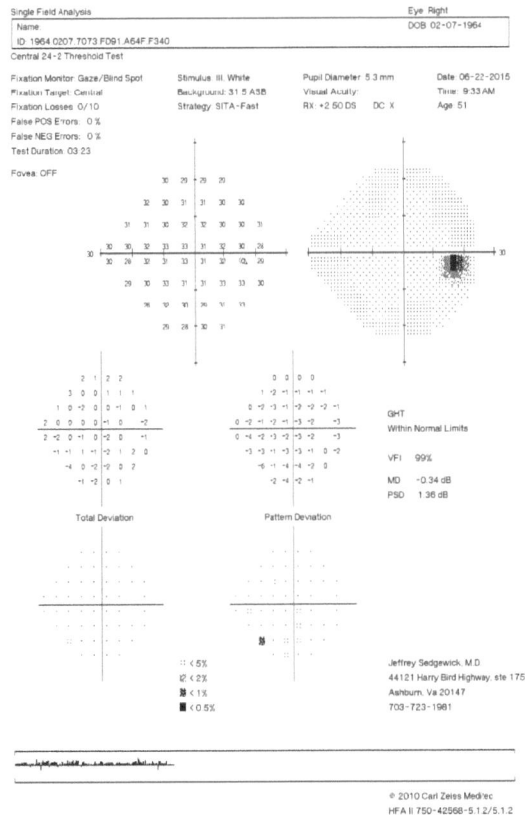

Same patient but in the OS.

First unreliable HVF OS with very high FPs

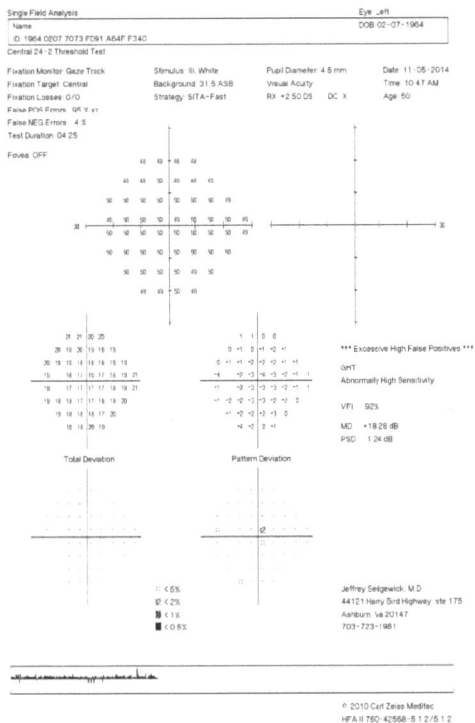

Second unreliable HVF OS

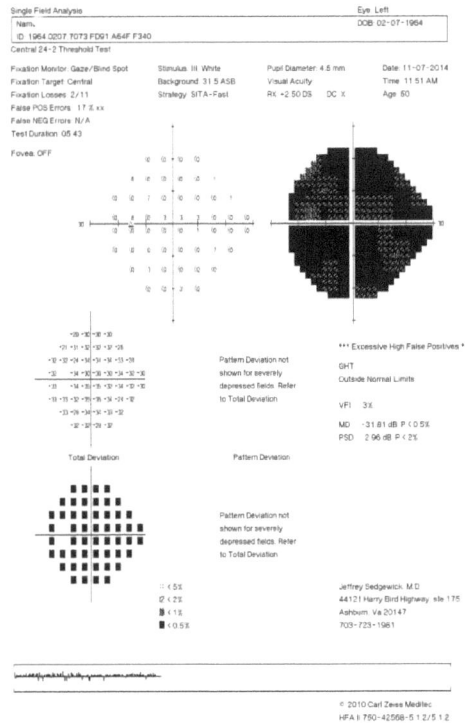

Third normal and reliable HVF OS!

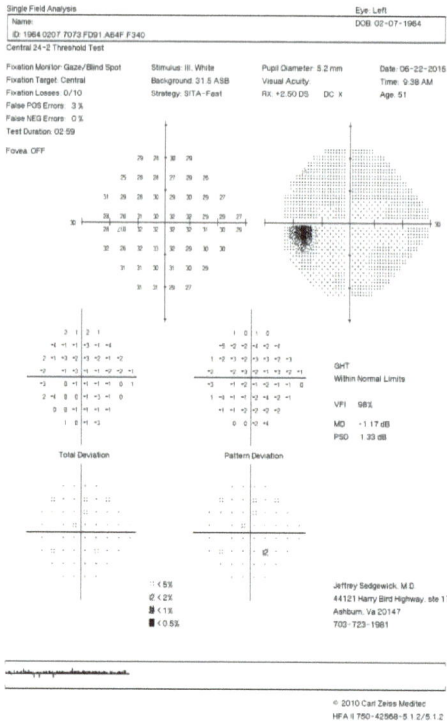

Another patient with mixed mechanism glaucoma, S/P LPI OU and SLT OU. Notice the improvement in their VFI OD over time in the VFI:

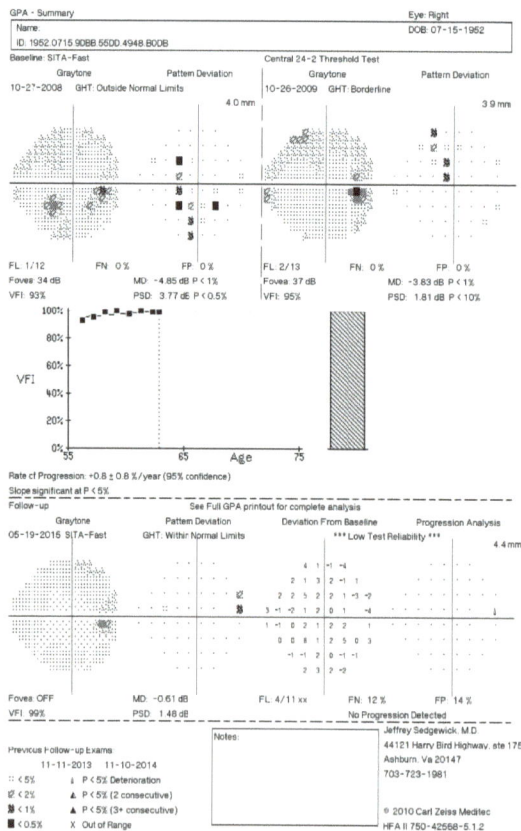

Another POAG patient of mine on Xalatan OU showing an improved VFI over time due to a disappearing rim artifact OD and and a rim artifact OS shown in the third HVF that wasn't there on subsequent exams:

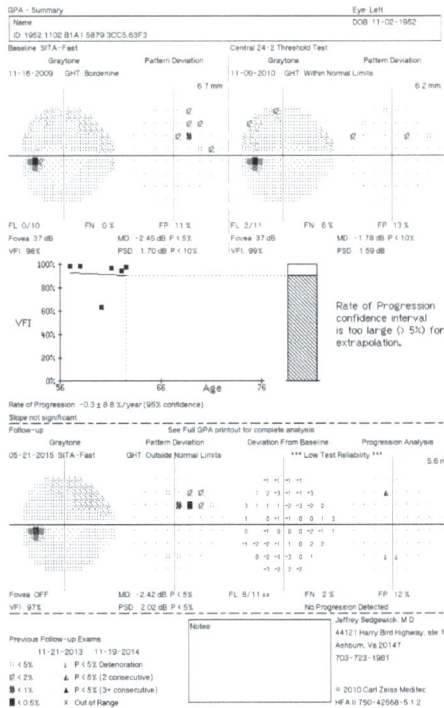

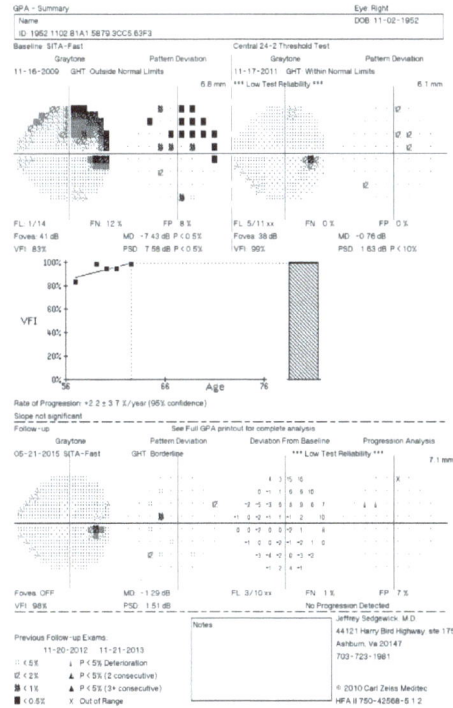

VFI readings can also look like they are going down but "not significantly". Here is an example of this in both eyes of a patient with POAG OU receiving adequet IOP lowering treatment:

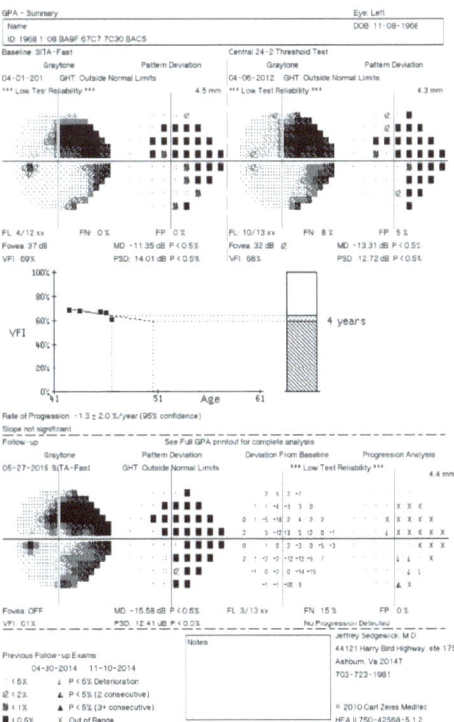

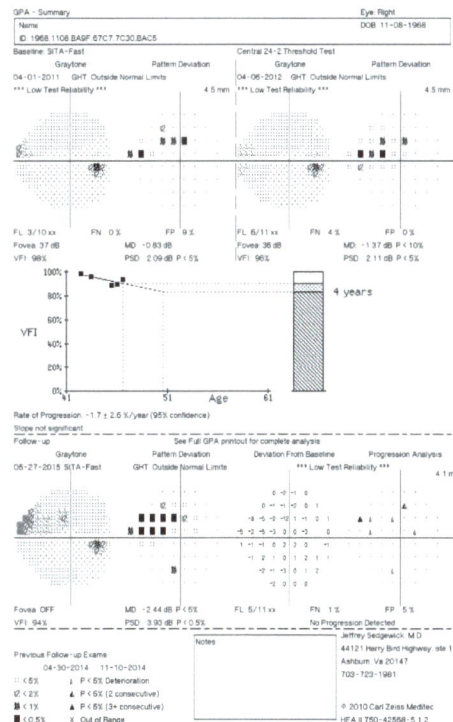

I hope that these multiple examples drives home the importance of taking into account the learning curve that patients have on the testing we do for them.

In the HVF analyzer:

HVF: The Humphrey VF (HVF) analyzer is mandatory for a Medical Optometric Practice. There is little agreement between the standard perimetry (HVF), SWAP, and frequency doubling machine[23] when it comes to VF defect detection so we use the standard VF machine, the Humphrey, using white on white testing.

On the 24-2 and the 30-2 tests, the light spots are 6 degrees apart starting 3 degrees above and 3 degrees below the horizontal meridian and 3 degrees on either side of the vertical meridian. The 10-2 test has light spots 2 degrees apart, splitting the horizontal and vertical meridians just like the 24-2 pattern. Below is a diagram showing the placement of the lights of a VHF 24-2 test (A) and a 30-2 (B). The light grey spots shown on the diagram below show the spots added to the 24-2 pattern to make the 30-2 pattern.[24] These additional spots don't add much to our testing but they do add time to the test, so we usually use the 24-2 test for all of our glaucoma VF tests. With the addition of VFI capability, all of the tests on a patient have to be the same type in order to analyze the results for progression. You definitely want to use the VFI capability for your own evaluation and to show to the patient what is happening with their VF results over time. The VFI graph is readily understandable by most patients. The "get it" quickly and appreciate the technology. Again, use the 24-2 for all of your glaucoma VF testing.

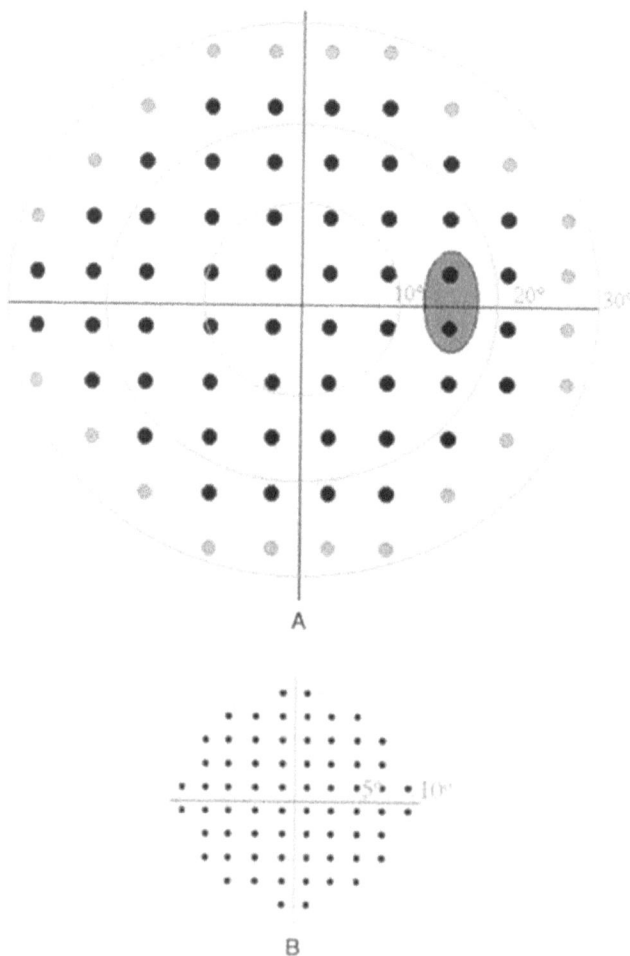

A

B

This is a diagram that shows the test spots of a right eye 24-2 VF actually on the right retina. Notice that the fovea is the center of the test, not the optic disk, and that the horizontal meridian does not bisect the disk but does bisect the fovea:[25]

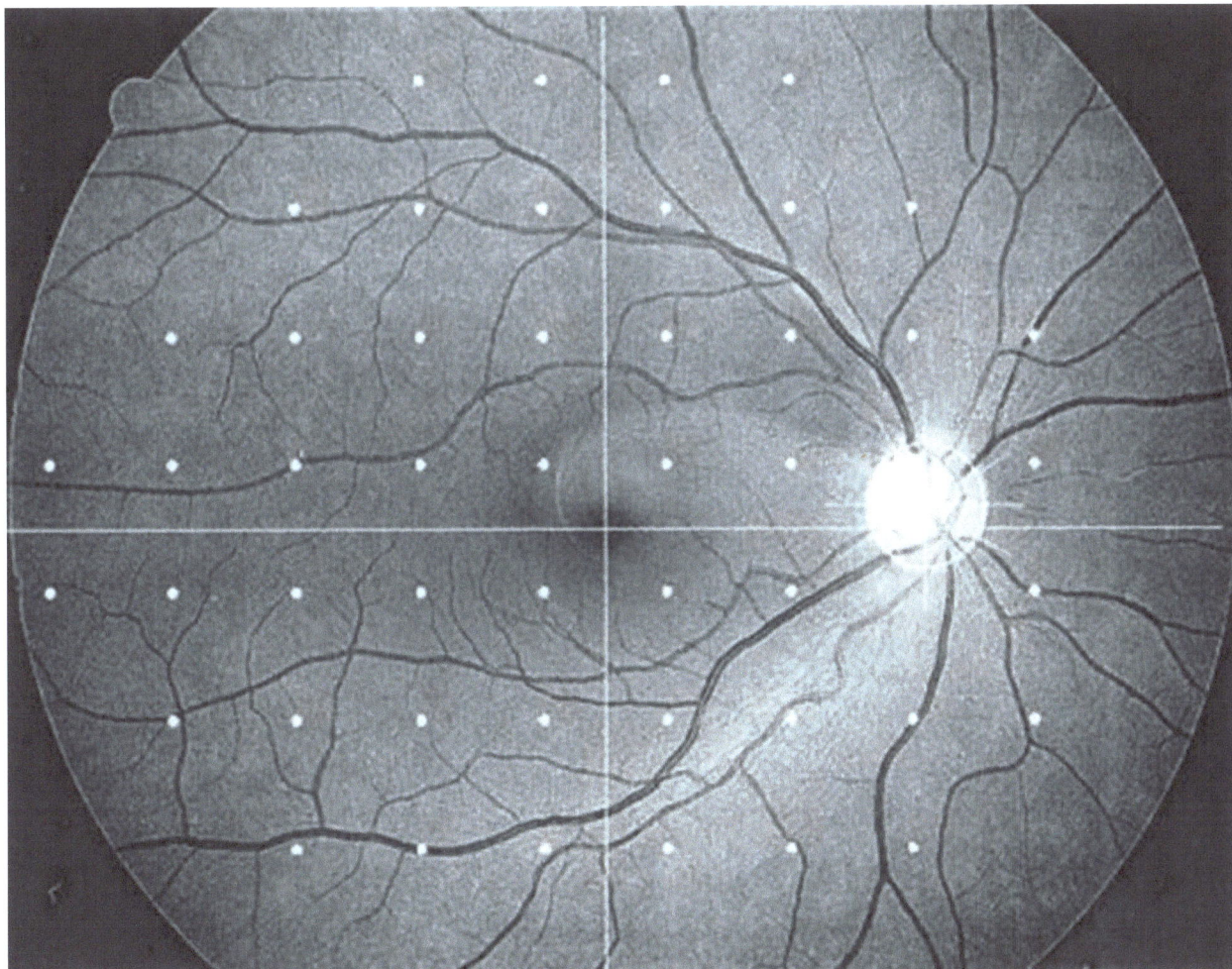

The testing distance for the HVF is about 30 cm which gives a +3.25 add for dilated patients and total presbyopes (use +3.00 for simple math).

The minimal light seen is measured in decibels of *retinal sensitivity*, not stimulus light sensitivity, so a higher decibel reading means a more sensitive area of the retina (a dimmer light is seen) while a lower decibel reading means a lower retinal sensitivity, or brighter light, is seen.[26] This is the opposite of what your intuition says. 0 db is the brightest light the machine can produce and, if it is not seen, corresponds to a blind area of the retina. 38 to 40 dB is the dimmest light a well-trained young adult can see so elderly patients should not have sensitivities at this level.[26]

The standard HVF spot is a white, size III stimulus, which is equivalent to a Goldmann III 4e spot size.[26] The size of the spot stays the same during a test, only the brightness changes in order to measure sensitivity.

The background illumination is 31.5 db, determined to be the minimum for photopic viewing, so the cones are tested during HVF tests.[19]

Threshold testing finds the dimmest light observable for each test location. Suprathreshold testing uses a very bright stimulus and is used to determine whether or not there is any vision at all at that location. Suprathreshold testing is typically used in advanced glaucoma patients and in screening for the DMV.

On the **24-2, 30-2, and newer VF 10-2** tests, decreased sensitivities are expressed on the total deviation (TD) and pattern deviation (PD) tests are *negative* db values. Decreased sensitivities on the older machines had the 10-2 test expressed as *positive* db numbers. **Careful:** older Humphrey machine 10-2 tests will express "defect depth" (not sensitivity defects as it does on the 24-2 test) on the printout where the total deviation printout is. Losses on the 10-2 "defect depth" plot show up as *positive* numbers. 24-2, 30-2, and the newer VFI 10-2 tests have TD and PD plots that show losses as *negative* numbers. The older 10-2 test has *positive* numbers on the depth plot as a bad result. *Negative* numbers on the 24-2, 30-2, and newer VFI 10-2 tests are a bad result.[27]

Glaucoma deficits start out as a relative loss of sensitivity and develop into an absolute loss of sensitivity to light. Therefore, threshold VFs are done, reserving suprathreshold testing for drivers licensing and advanced glaucoma patients. Use the **threshold 24-2** for all glaucoma testing with white on white spots in order to reduce the variability already present in HVF testing.[28] Again, VF defects in glaucoma are noted to have variability from test to test. Most of the variability occurs in diseased areas so variability from one test to another does not reduce your concern about glaucoma, it should increase it.[19] VF defects tend to occur slowly in time[19] so we have time to catch these changes with yearly or semi-yearly VF tests. VF changes should be re-checked before making a clinical decision.

Coach the patient during the VF exam. Tell them that they are not supposed to see about half of the lights and if they don't see a light, don't go looking for it. Reassure them that they aren't going to "fail" the test if they don't see any lights after an extended period of time. Just coach them to keep looking straight ahead and make sure their forehead is against the rest securely with the lens snug against the patient.

Really reduced HVF fields can be retested using the size V test spot on the 10-2 pattern looking for asymmetrical VF loss in the critical macular area. The last part of the VF you lose in glaucoma is the temporal area, leaving a "temporal island". The best test for these patients is looking for CF/HM in this temporal island area when you do the VFCF. *We want to pay close attention to VF losses close to the fovea as further loss here could snuff out BCVA. Recent research is focusing on using 10-2 VF testing in addition to 24-2 testing.*

How to read a Humphrey VF print out. (The 2012 edition of the Anders Heijl book *Effective Perimetry* has a new metric, the visual field index (**VFI**), which changes some of the priorities below):

1. Ensure the patient ID and date of exam is correct.
2. Look at the fixation losses (**FL**) and false positive (**FP**) numbers.
   FL: The machine uses the blind spot to check for FL. The first area that the machine locates during a 24-2 test is the blind spot. During the test, the machine randomly places a light into the blind spot that the patient isn't supposed to see. If they see the light, the machine knows that they are not looking straight ahead (or they have moved their head during the exam) and it counts this as a FL. The more FL there are, the less reliable the test is. FL above 20% is a concern for reduced reliability.[29] A patient that falls asleep after the beginning of the test results in a ***cloverleaf*** pattern (see below and the next page). Typically, the patient responds initially and then stops responding usually due to falling asleep during the rest of the exam. The machine initially tests the 4 areas where the patient responds which is why these areas have some responses. This test should be discarded and re-done.

Single Field Analysis   Eye: Left

| Name: | DOB: 11-13-1949 |

ID: 1949.1113.24D4.70FC.5A92.D164

Central 24-2 Threshold Test

Fixation Monitor: Gaze/Blind Spot   Stimulus: III, White   Pupil Diameter: 5.6 mm   Date: 11-08-2007
Fixation Target: Central   Background: 31.5 ASB   Visual Acuity: Time: 11:00 AM
Fixation Losses: 5/12 xx   Strategy: SITA-Fast   RX: +5.50 DS   DC X   Age: 57
False POS Errors: 10 %
False NEG Errors: N/A
Test Duration: 06:31

Fovea: 32 dB ::

```
                6   8  ⟨0   7
            12  1   5   0   0  ⟨0
        ⟨0  0   2   0   1  ⟨0  ⟨0  ⟨0
        0, ⟨0 |0  10   1   0   0   0   0
   30 ─────────────△──────────────────── 30
        13 ⟨0  0   0   2   8   3  18   4
        7   0   0   4  ⟨0  ⟨0   0   1
           ⟨0   0  13   7  ⟨0   3
                3   0   9   0
```

*** Low Test Reliability ***

GHT
Outside Normal Limits

VFI   10%

MD   -28.10 dB P < 0.5%
PSD   5.36 dB P < 0.5%

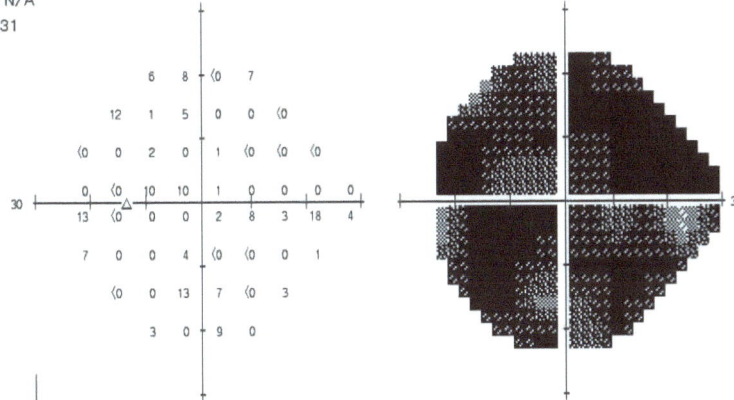

Total Deviation
```
        -21 -19 -29 -20
    -16 -28 -24 -30 -30 -31
-31 -30 -29 -31 -30 -33 -32 -31
-30     -22 -22 -32 -32 -31 -30 -27
    -17     -32 -33 -31 -25 -29 -12 -24
    -23 -31 -32 -28 -34 -34 -31 -28
        -32 -31 -18 -24 -33 -27
        -27 -30 -21 -29
```

Pattern Deviation

Pattern Deviation not
shown for severely
depressed fields. Refer
to Total Deviation.

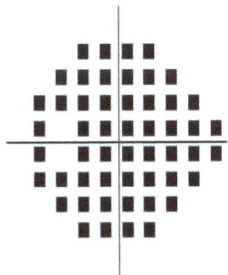

Pattern Deviation not
shown for severely
depressed fields. Refer
to Total Deviation.

:: < 5%
▨ < 2%
▩ < 1%
■ < 0.5%

Jeffrey Sedgewick, M.D.
44121 Harry Bird Highway, ste 175
Ashburn, Va 20147
703-723-1981

© 2010 Carl Zeiss Meditec
HFA II 750-9094-3.2/5.1.2

Another cloverleaf like pattern, where the patient is non-responsive initially and then performs better for the rest of the test. Kind of a reverse-cloverleaf like pattern.

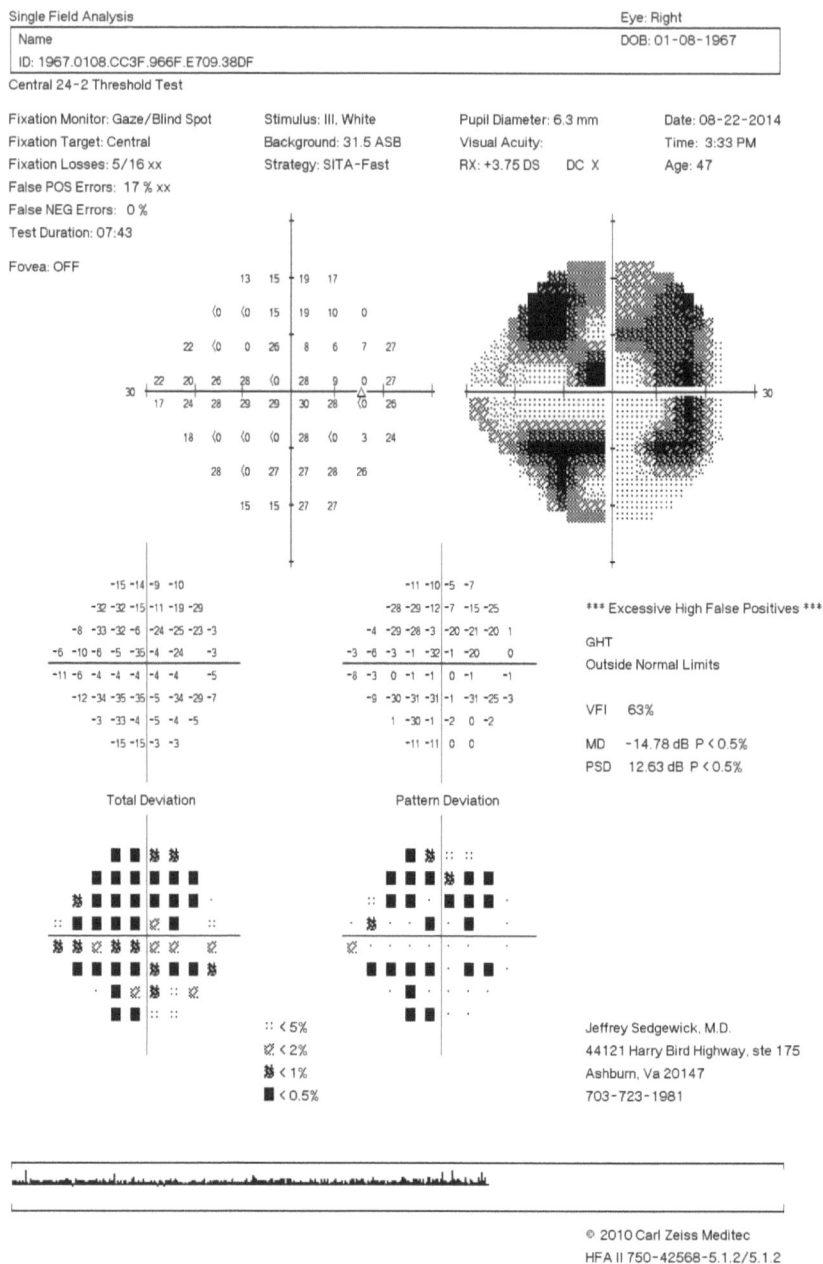

Single Field Analysis — Eye: Right

Central 24-2 Threshold Test

FP: FP occur when the patient is pushing the trigger when there isn't a test light shown, the so-called "*trigger happy*" patient. Usually this occurs when the patient is afraid of failing the test and wants to do well on it. They push the button by timing when they think a light should be presented. The result is a threshold that is way too sensitive and has high db numbers. This is also represented on the printout by:[25,34]

- a high number of FPs with white, washed out areas on the grey scale readout
- the Glaucoma Hemifield Test will say "abnormally high sensitivity". 38 - 40 dB is the dimmest light seen by a young, well trained adult,[26] so the elderly should not be close to this on their exam.

- the pattern deviation probability plot (PD) will be much worse than the total deviation plot. Here are three examples of this. The third one had both eyes give a "trigger happy" response:

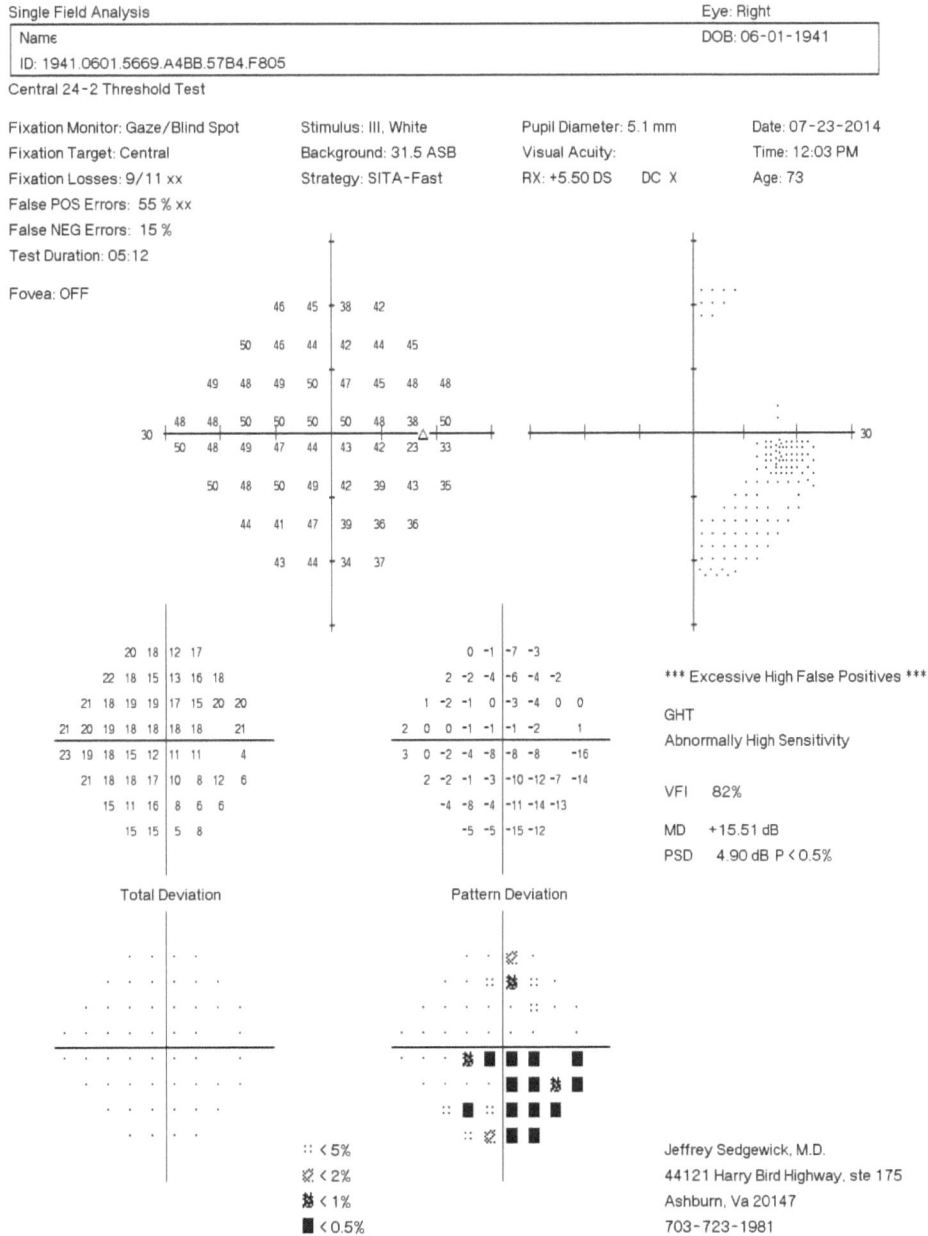

Single Field Analysis | Eye: Right

Name | DOB: 06-01-1941
ID: 1941.0601.5669.A4BB.57B4.F805

Central 24-2 Threshold Test

Fixation Monitor: Gaze/Blind Spot | Stimulus: III, White | Pupil Diameter: 5.1 mm | Date: 07-23-2014
Fixation Target: Central | Background: 31.5 ASB | Visual Acuity: | Time: 12:03 PM
Fixation Losses: 9/11 xx | Strategy: SITA-Fast | RX: +5.50 DS   DC X | Age: 73
False POS Errors: 55 % xx
False NEG Errors: 15 %
Test Duration: 05:12

Fovea: OFF

```
              46  45   38  42
          50  46  44  42  44  45
      49  48  49  50  47  45  48  48
  48  48  50  50  50  50  48  38  50
  50  48  49  47  44  43  42  23  33
      50  48  50  49  42  39  43  35
          44  41  47  39  36  36
              43  44  34  37
```

```
    20  18  12  17                          0  -1  -7  -3
  22  18  15  13  16  18                   2  -2  -4  -6  -4  -2
21  18  19  19  17  15  20  20          1  -2  -1   0  -3  -4   0   0
21  20  19  18  18  18  18      21      2   0   0  -1  -1  -1  -2       1
23  19  18  15  12  11  11       4      3   0  -2  -4  -8  -8  -8     -16
21  18  18  17  10   8  12   6          2  -2  -1  -3 -10 -12  -7  -14
  15  11  16   8   6   6                   -4  -8  -4 -11 -14 -13
      15  15   5   8                           -5  -5 -15 -12
```

Total Deviation | Pattern Deviation

*** Excessive High False Positives ***

GHT
Abnormally High Sensitivity

VFI   82%

MD   +15.51 dB
PSD   4.90 dB P < 0.5%

:: < 5%
※ < 2%
⊠ < 1%
■ < 0.5%

Jeffrey Sedgewick, M.D.
44121 Harry Bird Highway, ste 175
Ashburn, Va 20147
703-723-1981

© 2010 Carl Zeiss Meditec
HFA II 750-42568-5.1.2/5.1.2

Single Field Analysis   Eye: Right

Name: DOB: 05-13-1971
ID: 1971.0513.92C7.6FD6.6B70.DA3B

Central 24-2 Threshold Test

Fixation Monitor: Gaze/Blind Spot   Stimulus: III, White   Pupil Diameter: 5.1 mm   Date: 08-27-2012
Fixation Target: Central   Background: 31.5 ASB   Visual Acuity: Time: 9:03 AM
Fixation Losses: 6/11 xx   Strategy: SITA-Fast   RX: +1.50 DS   DC X   Age: 41
False POS Errors: 31 % xx
False NEG Errors: 16 %
Test Duration: 04:33

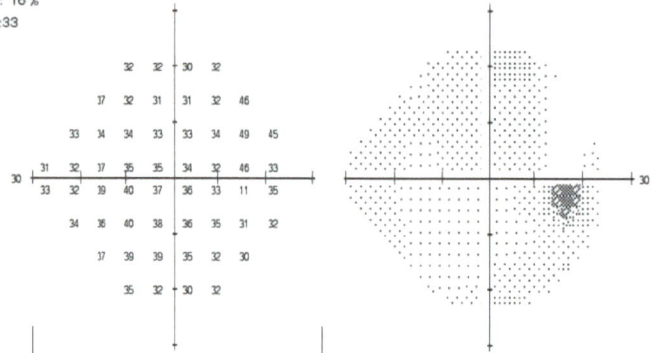

Fovea: 38 dB

```
              32  32    30  32
          37  32  31  31  32  46
      33  34  34  33  33  34  49  45
  31  32  37  35  35  34  32  46  33
  33  32  39  40  37  36  33  11  35
      34  36  40  38  36  35  31  32
          37  39  39  35  32  30
              35  32    30  32
```

```
      3  3  2  4                        -4  -3  -5  -3
   7  1  0  0  2  16                  0  -6  -7  -6  -5   9
3  2  2  1  1  2  18  15          -4  -5  -5  -6  -6  -4  11   8
3  1  5  2  1  0  -1     2        -4  -6  -2  -5  -6  -7  -8      -5
4  1  7  7  3  2  0     4         -3  -6  0  0  -3  -4  -6      -3
   3  4  7  5  3  2  -1  ·           -4  -3  0  -2  -4  -4  -8  -6
   6  7  7  2  1  -2                 -1  0  0  -4  -6  -9
   5  1  -1  1                          -2  -6  -8  -6
```

Total Deviation   Pattern Deviation

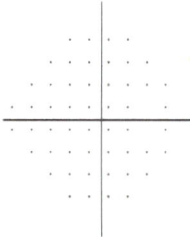

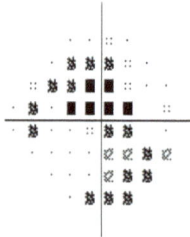

*** Excessive High False Positives ***

GHT

Abnormally High Sensitivity

VFI   97%

MD   +3.38 dB
PSD   3.85 dB P < 0.5%

```
::  < 5%
⁄2  < 2%
▨  < 1%
■  < 0.5%
```

Jeffrey Sedgewick, M.D.
44121 Harry Bird Highway, ste 175
Ashburn, Va 20147
703-723-1981

© 2010 Carl Zeiss Meditec
HFA II 750-9094-3.2/5.1.2

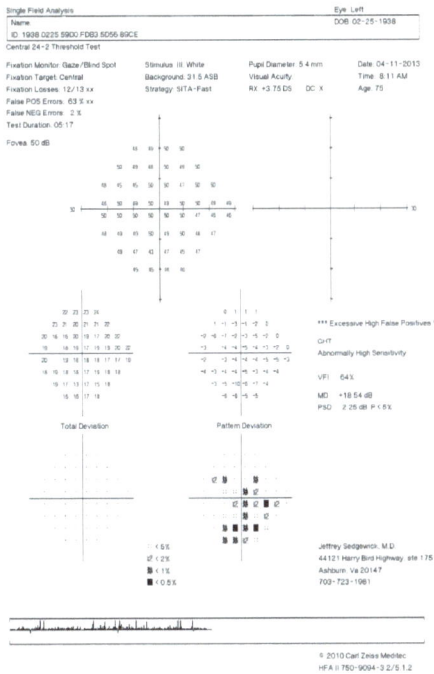

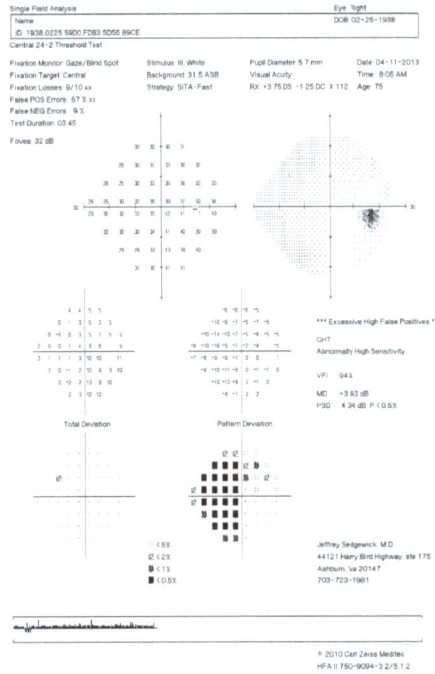

FPs above 15-20 % are a concern.[23,29] High numbers on the FL and FP indicate an unreliable test, which makes your task of determining if glaucoma is present or worsening more difficult.

FN: I don't see a HVF with high false negatives very often and it isn't described in the literature very often as well, but here is an example of a patient of mine that has high FN OU on 2 tests 5 months apart. Remember, a false negative is defined as a patient not pushing the button when a light is shown so deficits will be displayed that aren't really there. Their 2 OD HVFs are shown first:

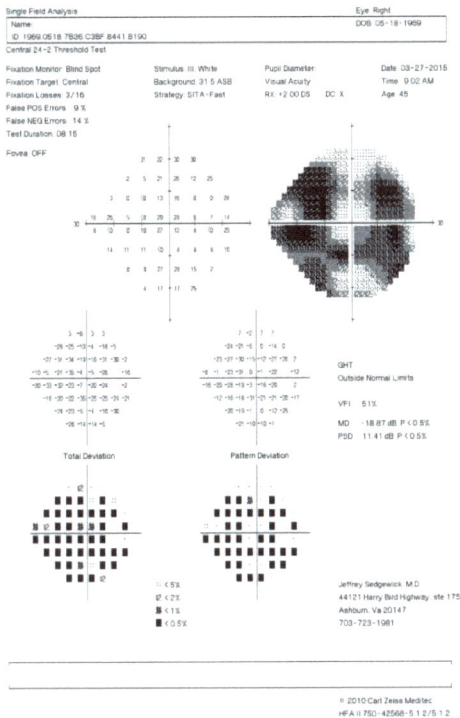

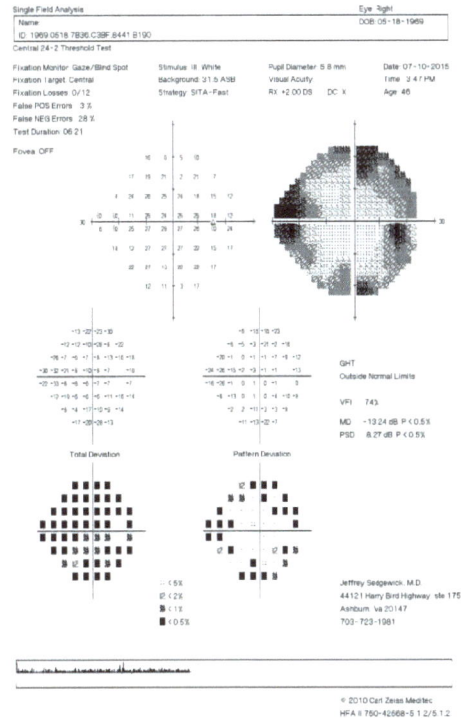

Single Field Analysis — Eye: Left
Name:
DOB: 05-18-1969
ID: 1969.0518.7B36.C3BF.8441.B190
Central 24-2 Threshold Test

Fixation Monitor: Blind Spot | Stimulus: III, White | Pupil Diameter: | Date: 03-27-2015
Fixation Target: Central | Background: 31.5 ASB | Visual Acuity: | Time: 9:14 AM
Fixation Losses: 1/1 xx | Strategy: SITA-Fast | RX: +2.00 DS DC X | Age: 46
False POS Errors: 65 % xx
False NEG Errors: 31 %
Test Duration: 04:41
Fovea: OFF

*** Excessive High False Positives ***
GHT Abnormally High Sensitivity
VFI 98%
MD +0.99 dB
PSD 2.99 dB P < 2%

Total Deviation   Pattern Deviation

:: < 5%
∅ < 2%
▨ < 1%
■ < 0.5%

Jeffrey Sedgewick, M.D.
44121 Harry Bird Highway, ste 175
Ashburn, Va 20147
703-723-1981

© 2010 Carl Zeiss Meditec
HFA II 750-42568-5.1.2/5.1.2

Single Field Analysis — Eye: Left
Name:
DOB: 05-18-1969
ID: 1969.0518.7B36.C3BF.8441.B190
Central 24-2 Threshold Test

Fixation Monitor: Blind Spot | Stimulus: III, White | Pupil Diameter: | Date: 07-10-2015
Fixation Target: Central | Background: 31.5 ASB | Visual Acuity: | Time: 3:56 PM
Fixation Losses: 1/13 | Strategy: SITA-Fast | RX: +2.00 DS DC X | Age: 46
False POS Errors: 0 %
False NEG Errors: 50 %
Test Duration: 05:30
Fovea: OFF

Pattern Deviation not shown for severely depressed fields. Refer to Total Deviation.
GHT Outside Normal Limits
VFI 20%
MD -25.79 dB P < 0.5%
PSD 6.89 dB P < 0.5%

Total Deviation   Pattern Deviation

Pattern Deviation not shown for severely depressed fields. Refer to Total Deviation.

:: < 5%
∅ < 2%
▨ < 1%
■ < 0.5%

Jeffrey Sedgewick, M.D.
44121 Harry Bird Highway, ste 175
Ashburn, Va 20147
703-723-1981

© 2010 Carl Zeiss Meditec
HFA II 750-42568-5.1.2/5.1.2

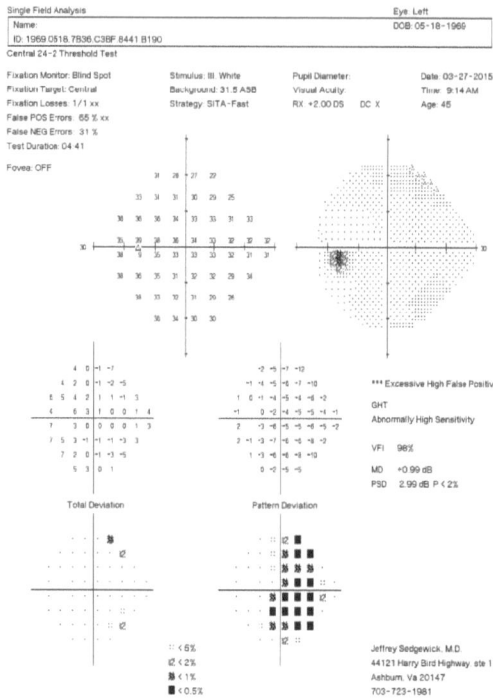

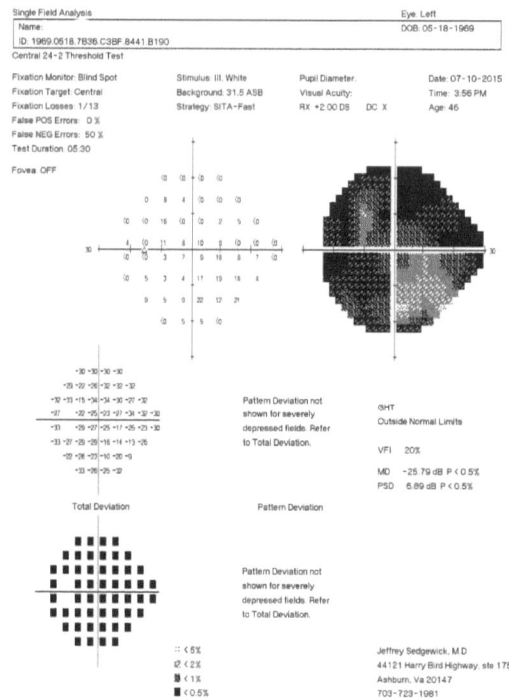

3. **PD: The most useful reading on the printout (pre-VFI) is the Pattern Deviation Probability Plot (PD)**[30] although the Glaucoma Hemifield Test (GHT) is also reported to be the most effective method (also pre- VFI) of VF analysis.[31,33] The PD highlights *age adjusted* localized losses after *global depressions*, such as the effects that cataracts or a small pupil might induce, have been filtered out of the total deviation plot.[27] Reduced sensitivities have a negative number. The great strength of the PD is that it highlights subtle defects, sometimes within the normal range, that might otherwise be missed as early glaucoma. Beginning defects can show up in the PD plot before the grey scale plot. The PD figure shows the statistical significance of the abnormal spot at each of the flagged points on the figure. A < 1% highlight means that there is less than a 1% chance that the abnormal spot is normal. I use a < 0.5% cut off in most of my analysis.

4. The **Glaucoma Hemifield Test (GHT)** compares deficits in corresponding areas above and below the horizontal midline, taking advantage of the difference between the upper and lower fields in optic nerve diseases.[33] It is read as within normal limits, borderline, or outside normal limits. It is probably the **second most useful result** and is widely used in clinic along with the PD plot. The graph below is from the advanced glaucoma intervention study (AGIS) and is an approximation of the areas in the GHT that are compared above and below the horizontal meridian.[181] The round black circle is the OD optic nerve. The two lines cross through the fovea. Notice that what is labeled as "nasal" is not referring to the nasal retina, rather it is referring to the nasal visual field corresponding to the temporal retinal area. Vice versa for the area labeled as "temporal". The numbers correspond to the minimal amount of depression in decibels that the HVF 24-2 will identify as defective.

AGIS Visual Field Test Scoring

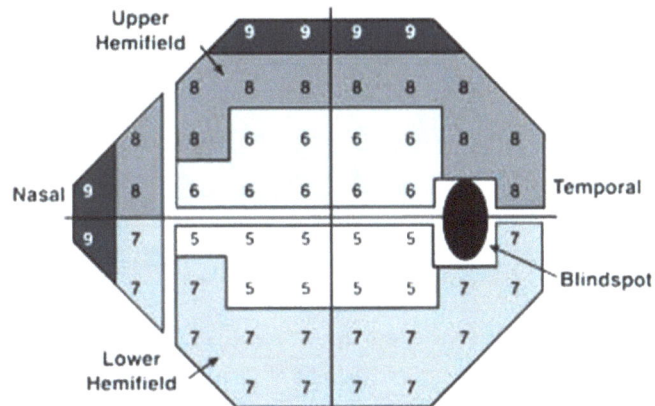

5. **Visual Field Index (VFI):** The VFI is a new metric looking at the rate of progression of visual field loss.[160] I expect the VFI, in one form or another to replace the PD and GHT as the primary method to judge the risk for glaucoma and of the effectiveness of your glaucoma treatment. The new Zeiss HVF machines uses the VFI. The VFI uses 2 baseline tests (we use 24-2 for all of our glaucoma tests) and compares them to at least 3 subsequent tests of the same type (5 reliable tests needed in total) to arrive at a diagnosis of "likely progression" or not giving a probability number of P < 5% or P < 1% that the slope is random and not significant. It plots the mean deviation from each test after correcting the results for *age*, *global depression* and giving *central VF* areas a greater weight due to the higher ganglion cell density there[161.] It graphs the results over the time period of eye exams and extends the trend out for 5 more years. The two baseline VFs, the most recent VF (with its FL, PD, and deviation from the baseline tests and numerical progression analysis) are displayed on one page as well as the VFI for each eye.[162] The reason for the development of this new metric is the discovery that rates of glaucoma progression vary significantly from patient to patient.[109,163] Some VFI slopes look significant but are not. Some VFI slopes don't look significant but are. Look at the probability number assigned to the slope to determine significance.

Anders Heijl, M.D. examined the untreated arm of the **early manifest glaucoma trial** (**EMGT**) patients.

> [The **EMGT** examined 44,243 patients for newly diagnosed OAG (n = 255, with 227 completing the median follow up of 67 months). They treated half of the patients with drops and/or ALT and withheld treatment for the other half.[106] I know, it sounds cruel by today's standards. The intent was to prove that lowering IOP with treatment would reduce the risk of VF loss, which was an unproven concept at that time. The results showed a 50% reduction in the risk of progression with treatment after a multivariate analysis was performed.][165]

In his review of the *untreated* arm of the EMGT, Heijl identified 118 patients with *untreated* open angle glaucoma (OAG) followed for at least 6 years.[109] 46 patients had high tension (uncorrected for CCT) OAG patients, 57 had NTG and 15 had pseudoexfoliation. Normal tension glaucoma patients had a yearly dB MD loss rate of -0.36 dB/year, ocular hypertension patients had a yearly loss of -1.31 dB/year and pseudoexfoliation patients had a yearly loss of -3.13 dB/year.[109] So, glaucoma progression rates of VF loss vary a lot from one type of untreated OAG to another.

Heijl also retrospectively reviewed an estimated 225,000 patient charts to find 583 patients with a diagnosis of OAG that had *received treatment* over a median follow up of 7.8 years.[166] He analyzed the rate of VF loss. A negative MD slope was found in 89% of the patients with a mean loss of -0.80 dB/year. 5.6% of the patients progressed at a rate of -2.5 dB/year or higher, with higher rates (on a third multivariate analysis) associated with increasing age, higher mean IOP and a more intensive treatment regimen. While 38% of the patients had pseudoexfoliation (PEX), the presence of PEX was not, by itself, associated with a faster rate of progression. (A more intensive treatment of OAG was statistically related to a faster rate of progression and the PEX patients may have required a more intensive treatment, confounding the results). This study shows that patients differ in their rate of glaucoma VF loss. This study also points out that glaucoma patients can have progressive optic nerve damage despite treatment, although the rate is much lower with treatment. This is the rational for more aggressive treatment of younger glaucoma patients.

In the **Advanced Glaucoma Intervention Study** (**AGIS**), a prospective, randomized, multicenter study, 161 eyes that were not controlled on topical drops for their OAG were randomized to 2 groups, the first group received the following surgeries in this order for IOP control; ALT- trabeculectomy-trabeculectomy, and the second received trabeculectomy-ALT-trabeculectomy. Germaine to our discussion here, the only significant difference (at < 0.1%) between the two groups after 8 years of follow-up on multivariate analysis was the slope of the rate of change, for reasons not identified.[163]

The bottom line is: progression rates for glaucoma loss can vary a lot from person to person. This lead to the development of the visual field index (VFI).

Here is the print out for the new VFI visual field program by Zeiss, showing the two baseline exams at the top, the current 24-2 at the bottom and the VFI graph in the middle. Again, the VFI will calculate for you the rate of progression and whether the slope is at a statistically significant level, noted as p < 5% or p < 1%. If the slope is not significant, it will state this also just below the VFI graph. This patient is being treated for POAG OU. The 5 year extension of this patient's graph loss is flat as well as stating that the slope is "not significant", indicating no projected progression. The VFI requires a minimum of 5 exams (11 exams have been done for this patient) in order to calculate if progression is present or not. Progression can be used for glaucoma and glaucoma suspect patients. The wider the 5 year band is, the worse the future loss is projected to be. I love the VFI and patients seem to understand the graph intuitively as well.

GPA - Summary | Eye: Right

Name: | DOB: 07-29-1935
ID: 1935.0729.1CC6.5A12.5FF9.F5D5

Baseline: SITA-Fast | Central 24-2 Threshold Test

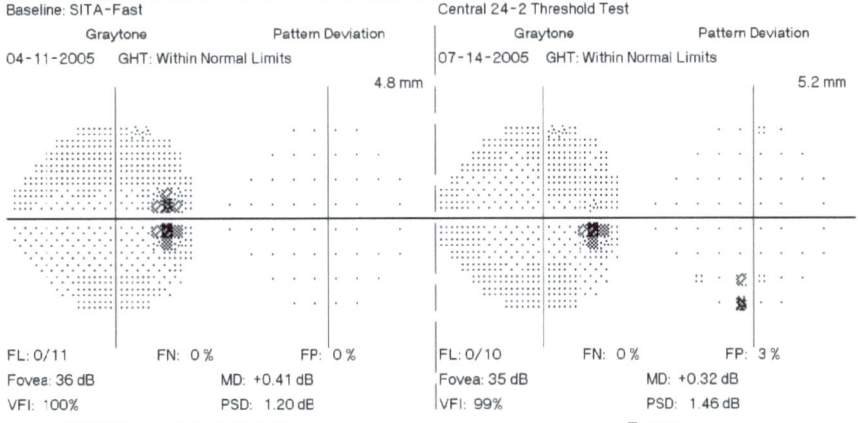

Graytone   Pattern Deviation   Graytone   Pattern Deviation

04-11-2005   GHT: Within Normal Limits   07-14-2005   GHT: Within Normal Limits

4.8 mm   5.2 mm

FL: 0/11   FN: 0 %   FP: 0 %   FL: 0/10   FN: 0 %   FP: 3 %
Fovea: 36 dB   MD: +0.41 dB   Fovea: 35 dB   MD: +0.32 dB
VFI: 100%   PSD: 1.20 dB   VFI: 99%   PSD: 1.46 dB

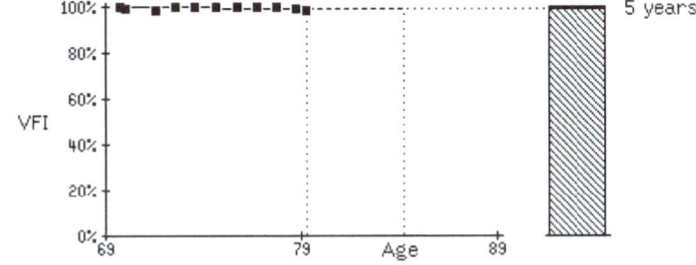

VFI — 5 years

Rate of Progression: +0.0 ± 0.1 %/year (95% confidence)

Slope not significant

Follow-up   See Full GPA printout for complete analysis

Graytone   Pattern Deviation   Deviation From Baseline   Progression Analysis

10-21-2014 SITA-Fast   GHT: Within Normal Limits   *** Low Test Reliability ***   4.6 mm

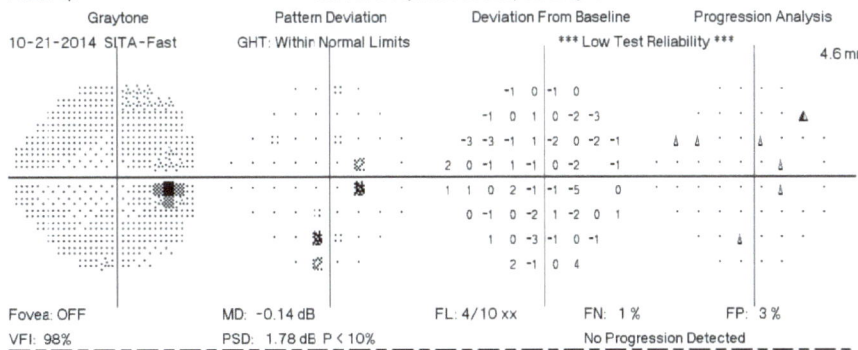

Fovea: OFF   MD: -0.14 dB   FL: 4/10 xx   FN: 1 %   FP: 3 %
VFI: 98%   PSD: 1.78 dB P < 10%   No Progression Detected

Previous Follow-up Exams:
     04-18-2013   04-18-2014

∷ < 5%   ⏷ P < 5% Deterioration
⌘ < 2%   ▲ P < 5% (2 consecutive)
⌘ < 1%   ▲ P < 5% (3+ consecutive)
■ < 0.5%   X Out of Range

Notes:

Jeffrey Sedgewick, M.D.
44121 Harry Bird Highway, ste 175
Ashburn, Va 20147
703-723-1981

© 2010 Carl Zeiss Meditec
HFA II 750-42568-5.1.2

Same patient but in the left eye. Notice the learning curve improvement over time:

GPA – Summary   Eye: Left

Name: DOB: 07-29-1935
ID: 1935.0729.1CC6.5A12.5FF9.F5D5

Baseline: SITA-Fast   Central 24-2 Threshold Test

Graytone   Pattern Deviation   Graytone   Pattern Deviation
04-11-2005   GHT: Outside Normal Limits   07-14-2005   GHT: Outside Normal Limits

                                    4.5 mm   4.4 mm

FL: 0/11   FN: 0%   FP: 0%   FL: 0/12   FN: 0%   FP: 2%
Fovea: 34 dB   MD: -3.05 dB P < 1%   Fovea: 36 dB   MD: -1.88 dB P < 5%
VFI: 93%   PSD: 7.13 dB P < 0.5% VFI: 94%   PSD: 7.67 dB P < 0.5%

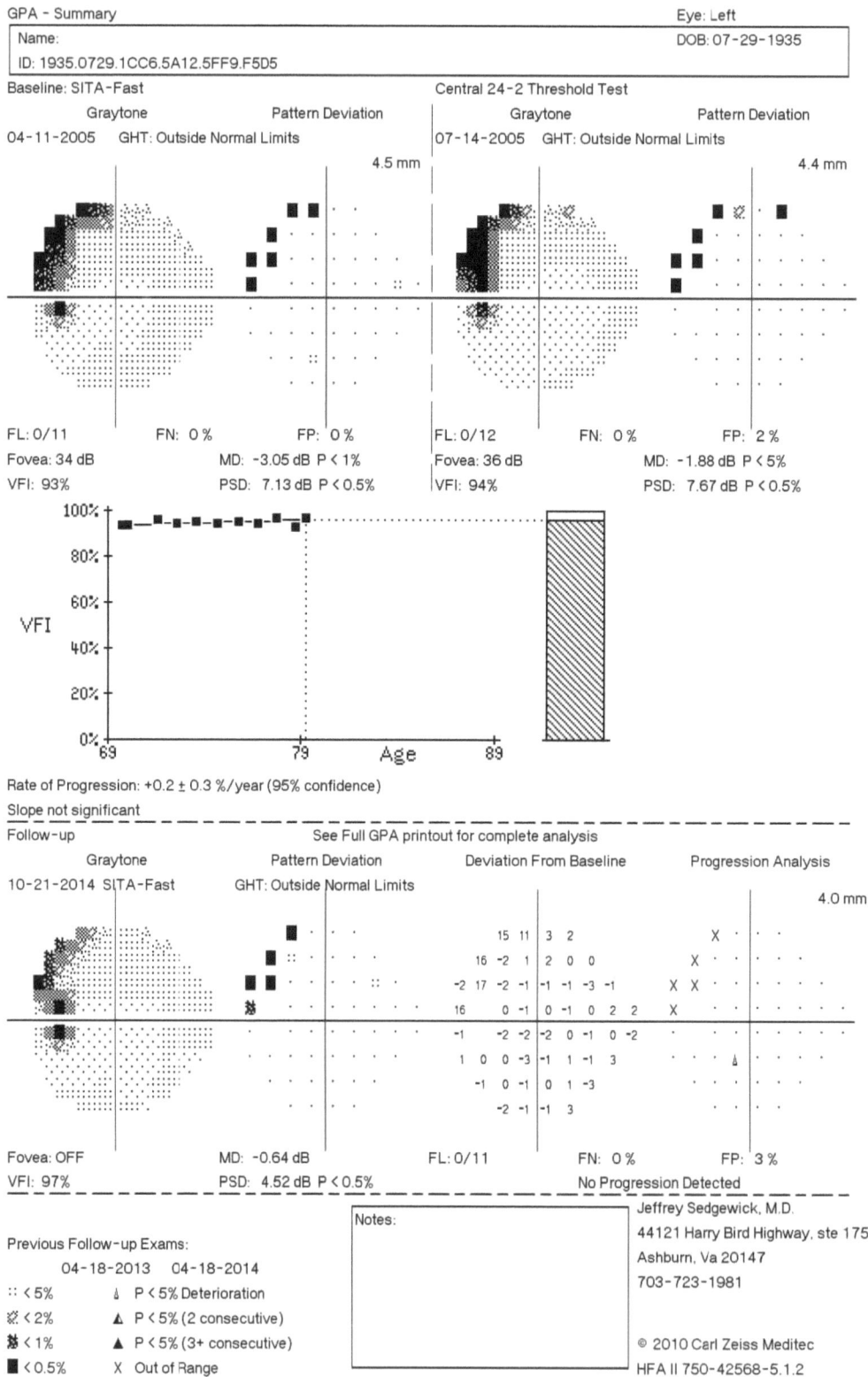

Rate of Progression: +0.2 ± 0.3 %/year (95% confidence)

Slope not significant

Follow-up   See Full GPA printout for complete analysis

Graytone   Pattern Deviation   Deviation From Baseline   Progression Analysis
10-21-2014 SITA-Fast   GHT: Outside Normal Limits

                                                                              4.0 mm

Fovea: OFF   MD: -0.64 dB   FL: 0/11   FN: 0%   FP: 3%
VFI: 97%   PSD: 4.52 dB P < 0.5%   No Progression Detected

Previous Follow-up Exams:
        04-18-2013   04-18-2014
:: < 5%   ↓ P < 5% Deterioration
※ < 2%   ▲ P < 5% (2 consecutive)
※ < 1%   ▲ P < 5% (3+ consecutive)
■ < 0.5%   X Out of Range

Notes:

Jeffrey Sedgewick, M.D.
44121 Harry Bird Highway, ste 175
Ashburn, Va 20147
703-723-1981

© 2010 Carl Zeiss Meditec
HFA II 750-42568-5.1.2

Here is another patient being treated for POAG showing a significant slope of progression (P < 5%) of their HVF OD and a slight non-significant progression OS. The right eye is shown first:

GPA - Summary   Eye: Right
Name: DOB: 04-28-1945
ID: 1945.0428.053D.96B4.CBF1.B767

Baseline: SITA-Fast   Central 24-2 Threshold Test
    Graytone   Pattern Deviation   Graytone   Pattern Deviation
04-10-2009 GHT: Borderline   07-09-2010 GHT: Borderline
                              6.2 mm   6.3 mm

FL: 0/11   FN: 1 %   FP: 7 %   FL: 2/11   FN: 5 %   FP: 11 %
Fovea 36 dB   MD: -1.19 dB   Fovea 37 dB   MD: -1.03 dB
VFI: 98%   PSD: 1.48 dB   VFI: 97%   PSD: 1.77 dB P < 10%

VFI graph — 100%, 80%, 60%, 40%, 20%, 0% vs Age 63, 73, 83   5 years

Rate of Progression: -0.4 ± 0.3 %/year (95% confidence)
Slope significant at P < 5%

Follow-up   See Full GPA printout for complete analysis
    Graytone   Pattern Deviation   Deviation From Baseline   Progression Analysis
09-04-2014 SITA-Fast   GHT: Outside Normal Limits   5.2 mm

Fovea OFF   MD: -2.28 dB P < 5%   FL: 2/11   FN: 12 %   FP: 6 %
VFI: 95%   PSD: 1.82 dB P < 10%   No Progression Detected

Notes:

Previous Follow-up Exams:
        10-08-2012   11-19-2013
∷ < 5%   ⬩ P < 5% Deterioration
⬚ < 2%   ⬚ P < 5% (2 consecutive)
⬛ < 1%   ⬛ P < 5% (3+ consecutive)
■ < 0.5%   X Out of Range

Jeffrey Sedgewick, M.D.
44121 Harry Bird Highway, ste 175
Ashburn, Va 20147
703-723-1981

© 2010 Carl Zeiss Meditec
HFA II 750-42568-5.1.2

GPA - Summary   Eye: Left

| Name: | DOB: 04-28-1945 |

ID: 1945.0428.053D.96B4.CBF1.B767

Baseline: SITA-Fast   Central 24-2 Threshold Test

| Graytone | Pattern Deviation | Graytone | Pattern Deviation |

04-10-2009   GHT: Outside Normal Limits   07-09-2010   GHT: Outside Normal Limits

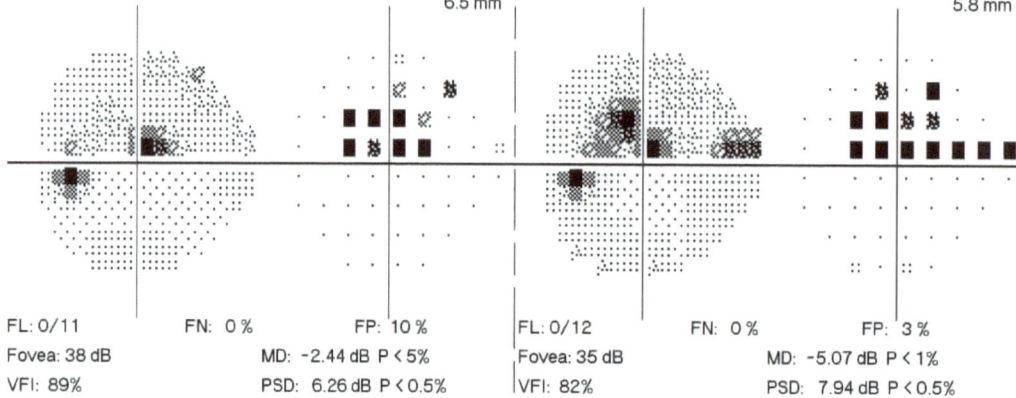

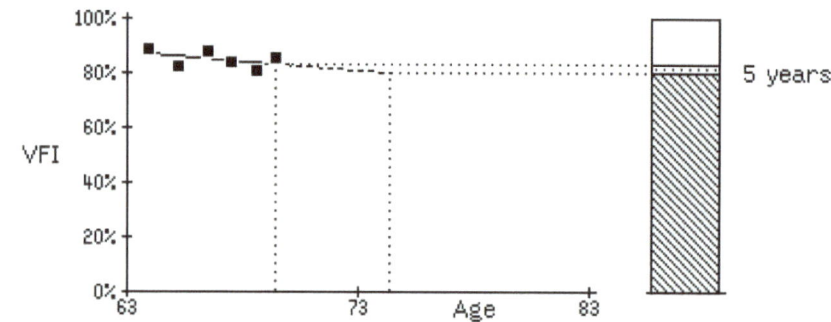

6.5 mm   5.8 mm

FL: 0/11   FN: 0 %   FP: 10 %   FL: 0/12   FN: 0 %   FP: 3 %

Fovea: 38 dB   MD: -2.44 dB P < 5%   Fovea: 35 dB   MD: -5.07 dB P < 1%

VFI: 89%   PSD: 6.26 dB P < 0.5%   VFI: 82%   PSD: 7.94 dB P < 0.5%

VFI

5 years

Rate of Progression: -0.6 ± 1.9 %/year (95% confidence)

Slope not significant

Follow-up   See Full GPA printout for complete analysis

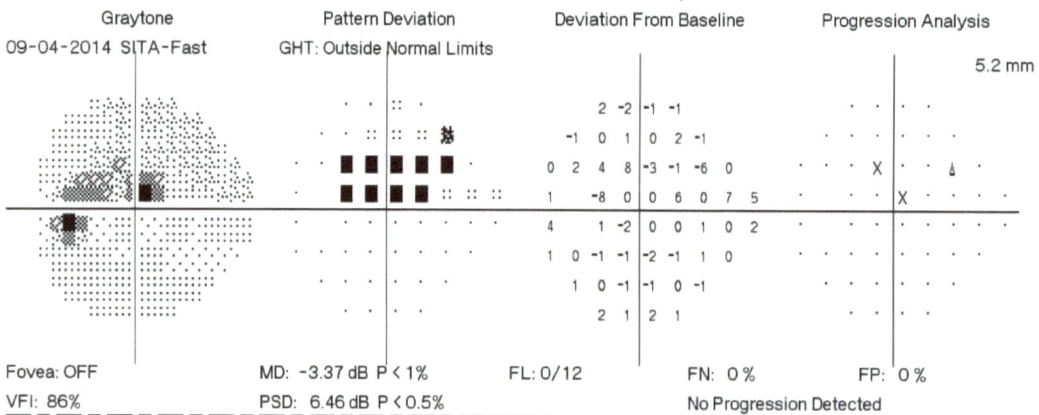

| Graytone | Pattern Deviation | Deviation From Baseline | Progression Analysis |

09-04-2014 SITA-Fast   GHT: Outside Normal Limits

5.2 mm

```
                                        2  -2 -1 -1                      .   .   .
                                    -1   0  1  0  2 -1                .   .   .   .
                                 0   2   4  8 -3 -1 -6  0         .   .   X  .   A  .
                                 1  -8   0  0  6  0  7  5         .   .   X  .   .   .
                                 4   1  -2  0  0  1  0  2         .   .   .   .
                                 1   0  -1 -1 -2 -1  1  0
                                 1   0  -1 -1 -1  0 -1
                                     2   1  2  1                  .   .   .   .
```

Fovea: OFF   MD: -3.37 dB P < 1%   FL: 0/12   FN: 0 %   FP: 0 %

VFI: 86%   PSD: 6.46 dB P < 0.5%   No Progression Detected

| Notes: | Jeffrey Sedgewick, M.D. |

Previous Follow-up Exams:

         10-08-2012   11-19-2013

:: < 5%   ↓ P < 5% Deterioration

☼ < 2%   ▲ P < 5% (2 consecutive)

⅗ < 1%   ▲ P < 5% (3+ consecutive)

■ < 0.5%   X Out of Range

44121 Harry Bird Highway, ste 175
Ashburn, Va 20147
703-723-1981

© 2010 Carl Zeiss Meditec
HFA II 750-42568-5.1.2

62

Another patient showing mild to moderate loss with treatment before the patient was seen by me (I don't have those tests). Notice how the VFI starts at about the 70% level. We usually don't expect future VFI's to improve; we just don't want the VFI to show further loss on future exams.

GPA - Summary   Eye: Left

Name: DOB: 06-23-1949
ID: 1949.0623.7A19.B5BF.ACF5.8A49

Baseline: SITA-Fast   Central 30-2, 24-2 Threshold Test

Graytone   Pattern Deviation   Graytone   Pattern Deviation

10-30-2009   GHT: Outside Normal Limits   11-05-2010   GHT: Outside Normal Limits

7.7 mm   6.6 mm

FL: 0/16   FN: 0%   FP: 2%   FL: 0/15   FN: 0%   FP: 0%
Fovea: 29 dB ■   MD: -14.26 dB P < 0.5%   Fovea: 35 dB   MD: -12.99 dB P < 0.5%
VFI: 72%   PSD: 14.00 dB P < 0.5%   VFI: 73%   PSD: 14.23 dB P < 0.5%

VFI

Rate of Progression: +0.0 ± 0.7 %/year (95% confidence)

Slope not significant

Follow-up   See Full GPA printout for complete analysis

Graytone   Pattern Deviation   Deviation From Baseline   Progression Analysis

06-13-2014 SITA-Fast   GHT: Outside Normal Limits   6.8 mm

Fovea: 35 dB   MD: -12.73 dB P < 0.5%   FL: 1/13   FN: 0%   FP: 3%
VFI: 72%   PSD: 14.13 dB P < 0.5%   No Progression Detected

Notes:

Previous Follow-up Exams:
        11-26-2012   12-06-2013
∷ < 5%   ▵ P < 5% Deterioration
⚠ < 2%   ▲ P < 5% (2 consecutive)
⚠ < 1%   ▲ P < 5% (3+ consecutive)
■ < 0.5%   X Out of Range

Jeffrey Sedgewick, M.D.
44121 Harry Bird Highway, ste 175
Ashburn, Va 20147
703-723-1981

© 2010 Carl Zeiss Meditec
HFA II 750-42568-5.1.2

Here is a middle aged, mixed mechanism glaucoma patient who was non-compliant with their drops for financial reasons and underwent bilateral tube placement showing an *improved* VFI after her tube, contradicting what I just said. Her cups are .95 to .99 OU. The slope is significant at P < 5% but for *improvement*, not degradation.

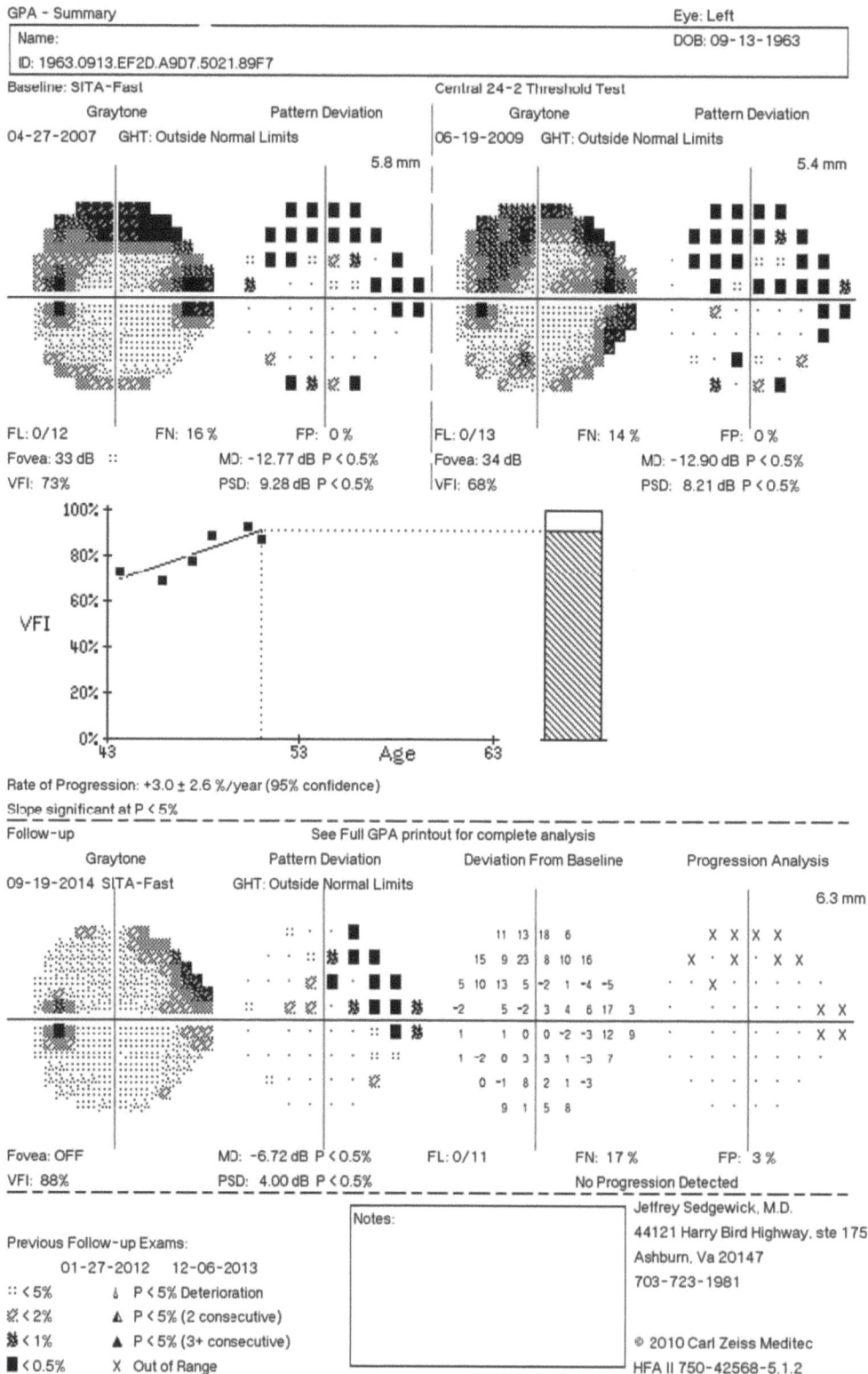

GPA - Summary — Eye: Left

Name: — DOB: 09-13-1963

ID: 1963.0913.EF2D.A9D7.5021.89F7

Baseline: SITA-Fast — Central 24-2 Threshold Test

Graytone — Pattern Deviation — Graytone — Pattern Deviation

04-27-2007 GHT: Outside Normal Limits — 06-19-2009 GHT: Outside Normal Limits

5.8 mm — 5.4 mm

FL: 0/12   FN: 16 %   FP: 0 %   FL: 0/13   FN: 14 %   FP: 0 %

Fovea: 33 dB   MD: -12.77 dB P < 0.5%   Fovea: 34 dB   MD: -12.90 dB P < 0.5%

VFI: 73%   PSD: 9.28 dB P < 0.5%   VFI: 68%   PSD: 8.21 dB P < 0.5%

Rate of Progression: +3.0 ± 2.6 %/year (95% confidence)

Slope significant at P < 5%

Follow-up — See Full GPA printout for complete analysis

Graytone — Pattern Deviation — Deviation From Baseline — Progression Analysis

09-19-2014 SITA-Fast — GHT: Outside Normal Limits

6.3 mm

Fovea: OFF   MD: -6.72 dB P < 0.5%   FL: 0/11   FN: 17 %   FP: 3 %

VFI: 88%   PSD: 4.00 dB P < 0.5%   No Progression Detected

Previous Follow-up Exams:

01-27-2012   12-06-2013

:: < 5%   ↓ P < 5% Deterioration

⬚ < 2%   ▲ P < 5% (2 consecutive)

⬚ < 1%   ▲ P < 5% (3+ consecutive)

■ < 0.5%   X Out of Range

Notes:

Jeffrey Sedgewick, M.D.
44121 Harry Bird Highway. ste 175
Ashburn, Va 20147
703-723-1981

© 2010 Carl Zeiss Meditec
HFA II 750-42568-5.1.2

Here is a 54 yo that I diagnosed with POAG and started on Xalatan OS and SLT. They had a Ta of 18/21, a Tac of 19/21, a c/d of 0.75 OU and a strong FHx of glaucoma. I didn't start drops until the nasal defect was confirmed on the second HVF 24-2, which was worse the first HVF 24-2. Their VFI improved significantly on drops! It seems to me that if a patient's HVF improves with treatment, it is the younger patients that do so.

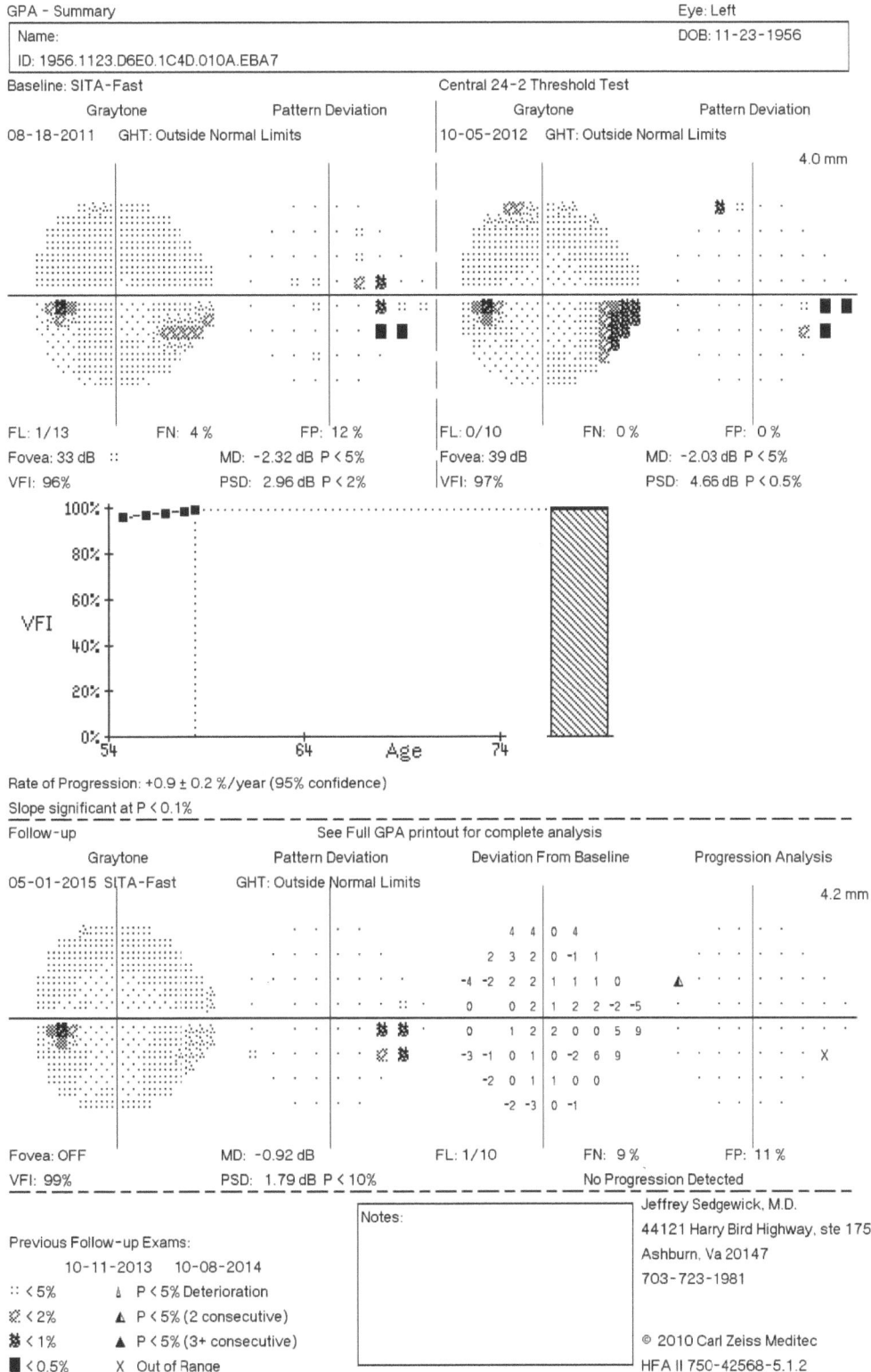

GPA - Summary   Eye: Left

| Name: | DOB: 11-23-1956 |
| ID: 1956.1123.D6E0.1C4D.010A.EBA7 | |

Baseline: SITA-Fast   Central 24-2 Threshold Test

Graytone   Pattern Deviation   Graytone   Pattern Deviation

08-18-2011   GHT: Outside Normal Limits   10-05-2012   GHT: Outside Normal Limits

4.0 mm

FL: 1/13   FN: 4 %   FP: 12 %   FL: 0/10   FN: 0 %   FP: 0 %

Fovea: 33 dB   MD: -2.32 dB P < 5%   Fovea: 39 dB   MD: -2.03 dB P < 5%

VFI: 96%   PSD: 2.96 dB P < 2%   VFI: 97%   PSD: 4.66 dB P < 0.5%

VFI (chart: 100%, 80%, 60%, 40%, 20%, 0%; Age axis: 54, 64, 74)

Rate of Progression: +0.9 ± 0.2 %/year (95% confidence)

Slope significant at P < 0.1%

Follow-up   See Full GPA printout for complete analysis

Graytone   Pattern Deviation   Deviation From Baseline   Progression Analysis

05-01-2015 SITA-Fast   GHT: Outside Normal Limits

4.2 mm

```
                                        4   4 | 0   4
                                    2   3   2 | 0  -1   1
                                -4  -2   2   2 | 1   1   1   0
                                 0       0   2 | 1   2   2  -2  -5
                                 0       1   2 | 2   0   0   5   9
                                -3  -1   0   1 | 0  -2   6   9
                                -2   0   1 | 1   0   0
                                -2  -3 | 0  -1
```

Fovea: OFF   MD: -0.92 dB   FL: 1/10   FN: 9 %   FP: 11 %

VFI: 99%   PSD: 1.79 dB P < 10%   No Progression Detected

Notes:

Jeffrey Sedgewick, M.D.
44121 Harry Bird Highway, ste 175
Ashburn, Va 20147
703-723-1981

Previous Follow-up Exams:
       10-11-2013   10-08-2014

:: < 5%   ⅄ P < 5% Deterioration
⚇ < 2%   ⏶ P < 5% (2 consecutive)
⚉ < 1%   ⏶ P < 5% (3+ consecutive)
■ < 0.5%   X Out of Range

© 2010 Carl Zeiss Meditec

HFA II 750-42568-5.1.2

Here is a middle aged patient with POAG, compliant with drops and good IOPs. I have increased their regimen based on the low level of the starting VFI and the fact that it is probably showing a progression in loss despite the program saying "slope not significant". The more damage that has been done, the more protective of the nerve you need to be by making the target IOP lower:

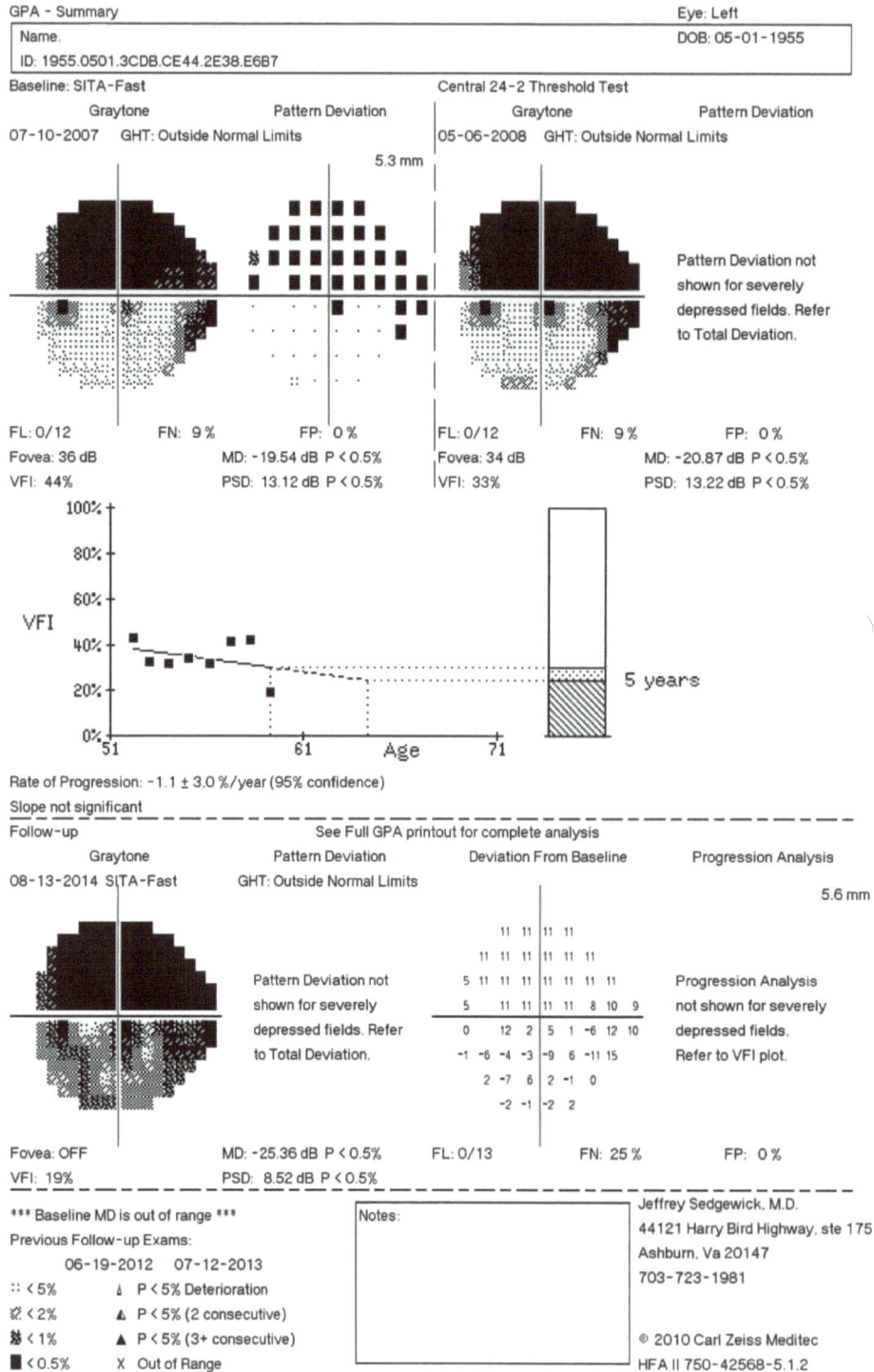

Here are two more patients showing probable progression despite treatment. I have a hard time believing the first patient's VFI stating "slope not significant". I would watch this patient closely and have a short trigger on getting his IOP lower sooner if not now. The second patient has a significant decrease in his slope at P < 1% and needs further IOP reduction:

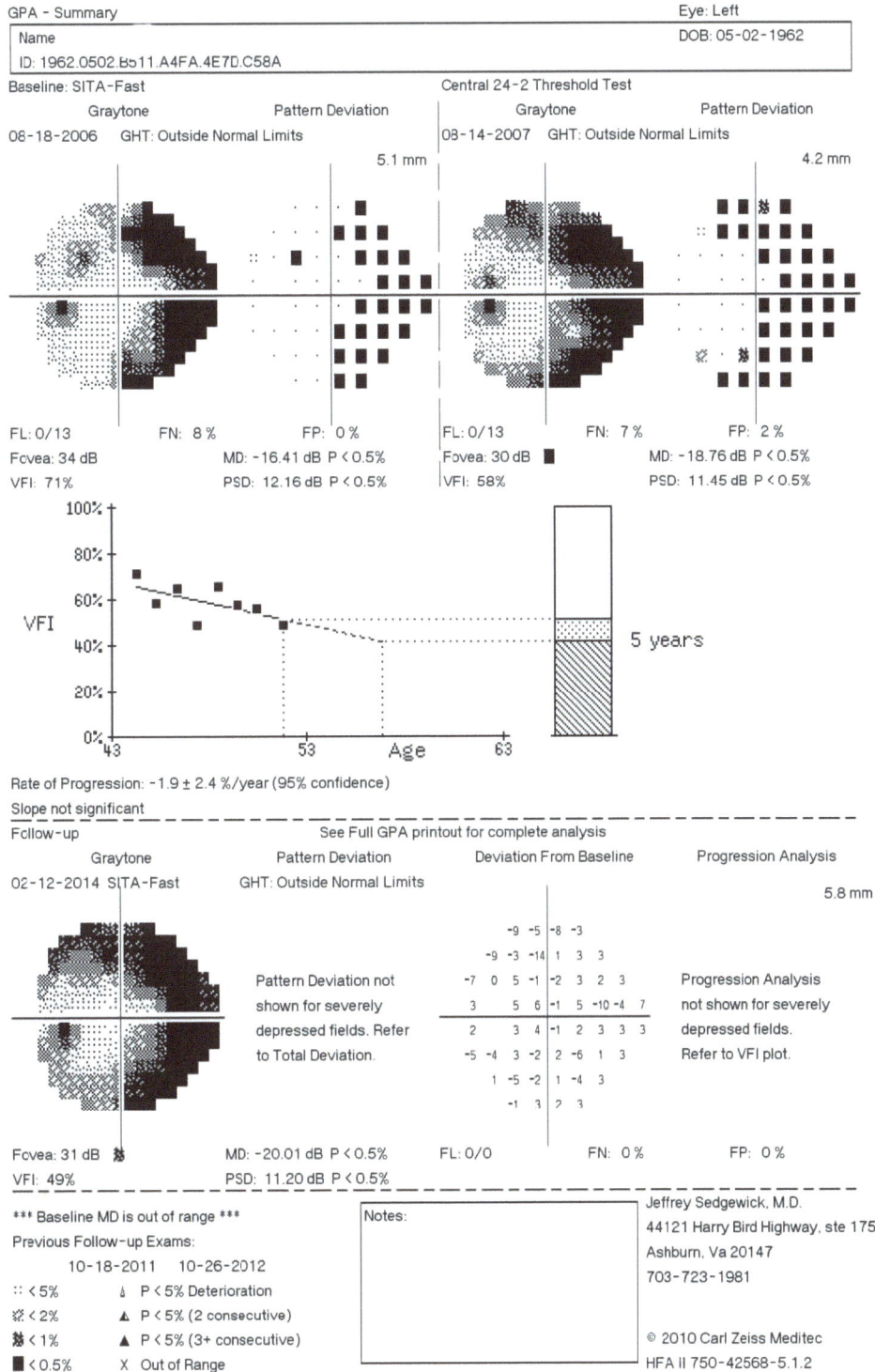

GPA - Summary | Eye: Left

Name | DOB: 05-02-1962
ID: 1962.0502.B511.A4FA.4E7D.C58A

Baseline: SITA-Fast | Central 24-2 Threshold Test

Graytone | Pattern Deviation | Graytone | Pattern Deviation

08-18-2006 GHT: Outside Normal Limits | 08-14-2007 GHT: Outside Normal Limits

5.1 mm | 4.2 mm

FL: 0/13 | FN: 8 % | FP: 0 % | FL: 0/13 | FN: 7 % | FP: 2 %
Fovea: 34 dB | MD: -16.41 dB P < 0.5% | Fovea: 30 dB | MD: -18.76 dB P < 0.5%
VFI: 71% | PSD: 12.16 dB P < 0.5% | VFI: 58% | PSD: 11.45 dB P < 0.5%

Rate of Progression: -1.9 ± 2.4 %/year (95% confidence)
Slope not significant

See Full GPA printout for complete analysis

Follow-up

Graytone | Pattern Deviation | Deviation From Baseline | Progression Analysis

02-12-2014 SITA-Fast | GHT: Outside Normal Limits

5.8 mm

Pattern Deviation not shown for severely depressed fields. Refer to Total Deviation.

```
            -9  -5 -8 -3
         -9  -3 -14  1  3  3
      -7  0  5 -1 -2  3  2  3
         3  5  6 -1  5 -10 -4  7
         2     3  4 -1  2  3  3  3
      -5 -4  3 -2  2 -6  1  3
         1 -5 -2  1 -4  3
            -1  3  2  3
```

Progression Analysis not shown for severely depressed fields. Refer to VFI plot.

Fovea: 31 dB | MD: -20.01 dB P < 0.5% | FL: 0/0 | FN: 0 % | FP: 0 %
VFI: 49% | PSD: 11.20 dB P < 0.5%

*** Baseline MD is out of range ***
Previous Follow-up Exams:
        10-18-2011   10-26-2012
∷ < 5%   ⚠ P < 5% Deterioration
⚠ < 2%   ▲ P < 5% (2 consecutive)
⚠ < 1%   ▲ P < 5% (3+ consecutive)
■ < 0.5%   X Out of Range

Notes:

Jeffrey Sedgewick, M.D.
44121 Harry Bird Highway, ste 175
Ashburn, Va 20147
703-723-1981

© 2010 Carl Zeiss Meditec
HFA II 750-42568-5.1.2

GPA - Summary   Eye: Right

Name   DOB: 04-05-1958
ID: 1958.0405.FCEF.B52D.E2E4.58D6

Baseline: SITA-Fast   Central 24-2 Threshold Test

Graytone	Pattern Deviation	Graytone	Pattern Deviation

01-23-2009   GHT: Outside Normal Limits   11-19-2010   GHT: Outside Normal Limits

*** Low Test Reliability ***   4.1 mm   4.6 mm

FL: 3/13 xx   FN: 0 %   FP: 13 %   FL: 0/13   FN: 0 %   FP: 9 %
Fovea: 37 dB   MD: -8.59 dB P < 0.5%   Fovea: 37 dB   MD: -9.95 dB P < 0.5%
VFI: 82%   PSD: 8.74 dB P < 0.5%   VFI: 79%   PSD: 9.49 dB P < 0.5%

VFI

5 years

Age

Rate of Progression: -2.4 ± 1.0 %/year (95% confidence)
Slope significant at P < 1%

- -

Follow-up   See Full GPA printout for complete analysis

Graytone	Pattern Deviation	Deviation From Baseline	Progression Analysis

06-16-2014 SITA-Fast   GHT: Outside Normal Limits   4.5 mm

	0 -13	0 -4
	-9 -15 -9	1 5 -7
	-5 -1 -9 -3	2 -8 -3 1
	-6 -13 -1 -5 0	0 -8 0
	1 -9 -8 -13 0	1 -9 -2
	-1 0 -1 1	2 -11 -11 -4
	2 1	2 1 0
	0 -2	2 -1

Fovea: 36 dB   MD: -13.13 dB P < 0.5%   FL: 2/14   FN: 0 %   FP: 5 %
VFI: 70%   PSD: 11.47 dB P < 0.5%   Possible Progression

- -

Previous Follow-up Exams: Notes:

        12-06-2011   02-26-2013

:: < 5%   ⌄ P < 5% Deterioration
⌀ < 2%   ▲ P < 5% (2 consecutive)
⊠ < 1%   ▲ P < 5% (3+ consecutive)
■ < 0.5%   X Out of Range

Jeffrey Sedgewick, M.D.
44121 Harry Bird Highway, ste 175
Ashburn, Va 20147
703-723-1981

© 2010 Carl Zeiss Meditec
HFA II 750-42568-5.1.2

Here is a patient of mine with a significant negative slope at p < 5% but, I think, due to the rim artifact on the latest test:

Here is another POAG suspect patient with a clearer picture of a rim artifact depressing the VFI results:

In addition to the type of OAG, age, intensity of treatment as well as other unknown reasons why glaucoma progresses at different rates despite adequate IOP management, the following is, I think, a significant factor that influences the rate of loss:[94]

- In early disease, dB function changes relatively slowly compared to structure

- In later disease, dB function changes relatively faster compared to structure

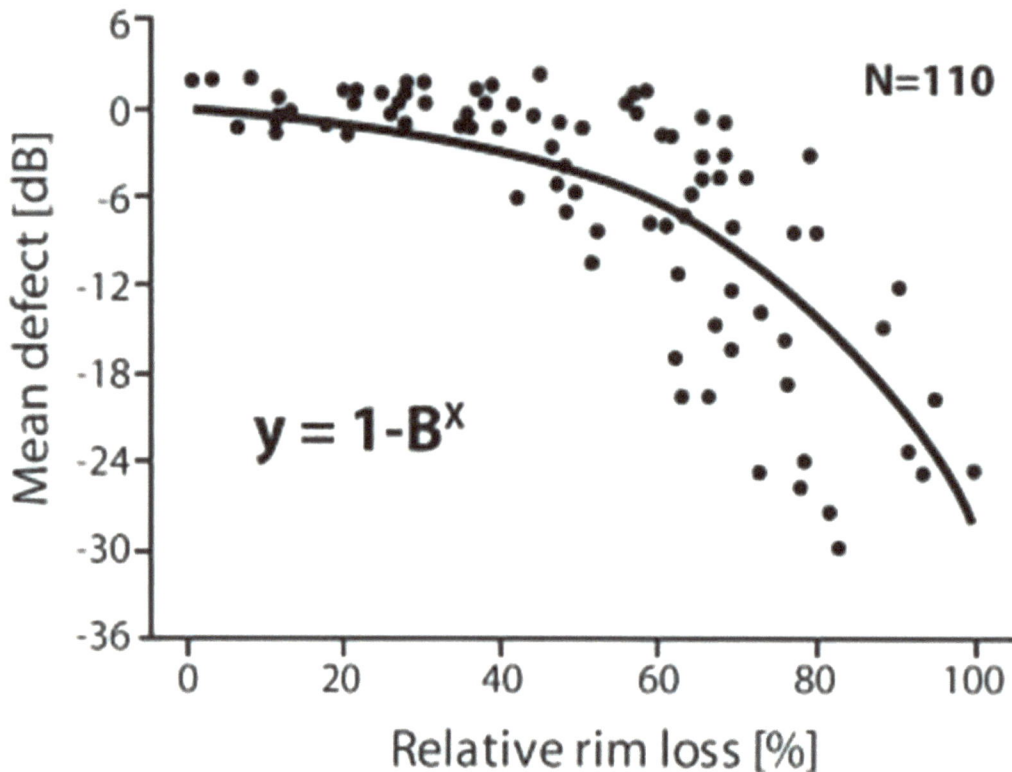

$$y = 1-B^X$$

This figure shows that as the rim tissue thins, the yearly rate of dB loss accelerates.[94] The addition of the VFI will hopefully give us another tool to adjust our target IOP in glaucoma patients or in the risk of developing OAG for our glaucoma suspect patients. Recent studies have shown the importance of 10-2 testing in early glaucoma detection, something that is a novel idea.[164] Currently, the VFI has to have the same type of test each time in order to do its analysis, so you cannot do a 10-2 alternating with a 24-2 and have both tests analyzed by the VFI. I do anticipate that a new HVF 24-2 will be developed at some point, one including more central points, a sort of hybrid of the 10-2 and the 24-2 with this new test being included in the VFI analysis.

6. **Compare** the VF for both eyes and look for optic *tract* defects versus optic *nerve* defects and compare the current test to prior tests, eliminating artifacts and looking for changes, remembering that variability in glaucoma VF defects are common. Bilateral temporal losses should immediately raise your concern for pituitary gland diseases and the need for an MRI of the brain. While the result analysis tools can be used, your instinct after reading many thousands of visual fields is probably

the most valuable tool that you have. Always correlate the VF losses with the appearance of the optic nerve. If a VF defect does not correspond to the location of a notch or disk hemorrhage, or if the VF defect seems to respect the vertical midline or you have extensive VF loss with a normal optic nerve exam, consider non-glaucoma causes of VF loss. These include AION, optic neuritis, optic atrophy, vein or artery occlusions, among others. Re-check to see if an APD is there or if the patient has a history of a CVA. The PMHx is critical at this point.

I always treat both eyes of a non-ischemic and ischemic CRVO for glaucoma in order to reduce the chance of an ischemic CRVO developing in that eye or in the other eye in the future.

G. SLIT LAMP/ BIO:

You should place the slit lamp on a high enough magnification so that the cornea fills up most of the view. I guarantee that you will miss Krukenberg's spindle, cells in the AC, map/dot/fingerprint dystrophy as well as a host of other diseases if you do not use a high enough magnification. This is the most common error I see ODs make, they have the slit lamp magnification on too low of a setting. Also, since you need a higher magnification when looking at the macula and optic nerve, the 78D lens should be used instead of the 90D. This is a another common error and I can tell instantly how well an eye doctor performs a slit lamp exam by what magnification the slit lamp is on when examining the macula as well as their choice of the slit lamp retinal lens.

First, look at the cornea of the right eye for any abnormalities. Significant negatives include Krukenberg's spindle, corneal dystrophies such as map/dot/fingerprint (which is fairly common and missed a lot), SEIs in SCL wearers, cortical cataracts, cells in the AC and vitreous, growths on the conjunctiva (ones that have a whitish surface to them are suspicious of malignancy and need a referral) among other items. Use a long stemmed Q-tip to bring the lower lid down while the patient looks up, looking for growths and nevuses. Push on the lid margin looking for meibomianitis (very common). Use the Q-tip to pull the upper lid up while the patient looks down, looking for growths and nevuses. Push on the lid margin and examine it as in the lower lid. Evert the upper lid in all FBS patients and corneal FB patients looking for FBs stuck up there. If the AC is narrow, gonioscopy must be done to ensure that dilating the patient is safe. If gonioscopy reveals an angle without trabecular tissue visible and any PAS, refer the patient for a possible LPI before dilating the patient. Explain to the patient that it is not safe to dilate without further testing as dilating might cause them to go into angle closure glaucoma with IOPs in the 60s and potential blindness in a week or two. Don't worry about scaring them. They might not go to the referral doctor and later claim that they would have gone if you had only told them that they could go blind if they didn't.

Estimate the density of cataracts and correlate that with the BCVA that the patient has with those cataracts. If the patient sees significantly worse than you estimate for their cataracts, you need to entertain other causes of reduced BCVA. Here is a diagram of the two main types of IOL's used in cataract extraction, an anterior chamber and a posterior chamber IOL. (©AAO 2014)

Anterior chamber
type lens

Anterior
hyaloid
face

Posterior chamber
type lens

Posterior
capsule

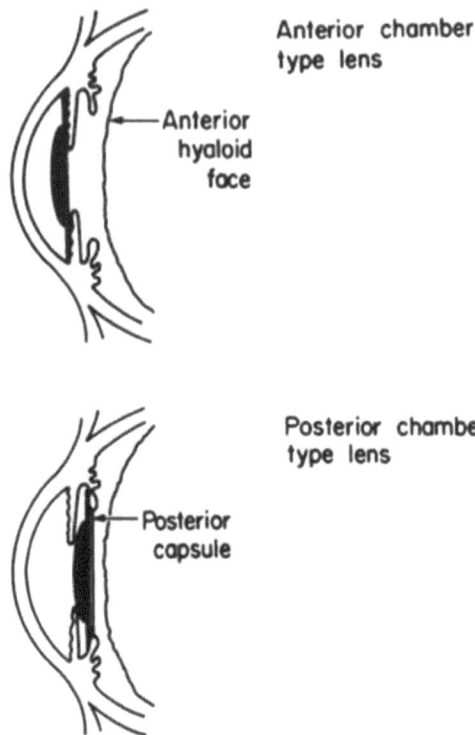

In the dilated patient, use your 78D lens (not the 90D lens right?) to look at the optic nerve first. Note if the optic nerve is flat or not and if collaterals are present. Look at the rim tissue. Narrow rims tend to go in concert with a large C/D but not always. A C/D of 0.5 with normal rims is a lot different than a C/D of 0.5 and very narrow rims. Estimate the ratio of the C/D vertically and note the presence of any disk hemorrhages, notches, or nerve fiber layer wedge defects (NFLD). You have to mentally review this check list in your head *on every patient* or you will miss seeing most of them (as in the OHTS)

Glaucoma specialist examining patients in the ocular hypertensive treatment study (OHTS) missed 84% of the disk hemorrhages on the initial exam. They were subsequently found during photo review of the disks[74]. This does not surprise me as I have had the same thing happen to me right in front of a patient who saw a disk hemorrhage on a photo review that I had not seen earlier. I didn't let on that I had missed it but it was embarrassing to me to say the least.

If three cups fit into the disk, the ratio is 0.35. Mentally split the ON in half. If the cup is the same size as the rim tissue in each half, then the C/D ratio is 0.5. If there is twice as much cup as rim tissue in each half, the C/D is 0.75. C/Ds of 0.5 and bigger are a risk factor for glaucoma.[35] This is covered in more detail later.

Next, look at the macula within and just beyond the arcades, noting any deposits, scars, macular holes (with or without a positive Watzke-Allen test), nevus or any other abnormality. Look along the arcades above and below the macula looking for emboli and anything else abnormal. *A BCVA of less than PH 20/20 has to be explained!* Don't forget the Worth 4 dot test either.

BIO:

You need less magnification and a wider field of view when looking at the peripheral retina so we use the 28D BIO lens, rather than the 20D lens. Scleral depression is also easier with the 28D lens as it has a smaller diameter and is easier to hold. With the 28D lens, you start at the arcades and look towards the ora again looking for any abnormalities such as nevi/ lattice/tears etc. It is important to start at the arcades and continue to the ora so that you are sure that you have seen the entire retina. (We went to the arcades with the 78D lens, right?) We will cover some of the more common retinal diseases later.

Scleral depression is performed in all PVDs or R/O RD patients unless you can visualize the ora serrata. To do scleral depression, numb the eyes. Have the patient look opposite to the area of retina that you want to see. Place the depressor above the tarsal plate in the upper lid or below the tarsal plate in the lower lid. Then, have the patient look in the same direction as the depressor. Push gently on the globe as far back from the limbus as you can. Your objective is to push the far peripheral retina and ora into view and look for any retinal tears. I find that it is difficult to depress the lower lid through the lid, so I will sometimes depress directly on the conjunctiva instead.

Sometimes, it is hard to figure out which part of the retina you are looking at with the BIO. Think of it this way. The area of the retina you are looking at is the same area that the patient is looking towards. If the patient is looking superior and nasal, you are looking at the superior and nasal area of the retina. This is because the opposite side of the patient's retina presents itself to you since the center of the eye does not move. It isn't hard to get it right if you just remember that. On a photo record of a retina, a photo of the OD will be presented with a normal orientation with the ON nasal to the macula. When looking with the 78D and the slit lamp, the image is reversed and inverted, just like it is with the BIO.

DISEASES

I think that giving examples is the best way to teach disease. This is how it is taught in ophthalmology residency programs. You get 10-20 diseased patients a day, 5 days a week for the first year. A staff member is present to look over your shoulder and point out what to look for and how to manage the disease. Within 12 months, you have examined thousands of diseased patients and your comfort level is quite high with just about whatever walks into your exam room. You have to know the basic science first though, so let's cover the more common diseases, as concisely as I can make it:

1. Glaucoma: What is Glaucoma?

If you ask 10 glaucoma specialists what glaucoma is, you'll probably get 11 different responses. In a medical optometry/ophthalmology practice, you don't have the luxury of pontification. You have to explain to the patient in plain English why you are concerned about the health of their eye, what you are going to do to address it and how you are going to stop them from going blind. Some newly diagnosed glaucoma suspect patients are afraid of going blind in the next 6 months so you need to give the patient confidence through your explanations that that isn't going to happen. You spend this time because you care about the patient and because you want to withstand a lawsuit. So, with this in mind, let's stick to the KISS principle when it comes to defining what glaucoma is.

Glaucoma is a process whereby the optic nerve is usually slowly destroyed due to the IOP being too high for that particular eye.

"IOP plays a major role in the development of glaucomatous optic neuropathy in most individuals and is considered the most significant risk factor." [36] "However, up to one-third of glaucomatous injury in North America is associated with low IOP."[37] The inclusion of corneal Pachymetry in our glaucoma work ups *should* have cleared a lot of this confusion up but, it is still not a settled issue.

Let us also simplify our definition of who has glaucoma: if a patient is legitimately taking glaucoma drops or has had glaucoma surgery performed on them, then *they have glaucoma*. If you are concerned that a patient might develop glaucoma in the future but they are not currently taking drops or have not had glaucoma surgery performed on them, then they are a *glaucoma suspect*. I am aware that other sources break the definition down differently but I think that my definition is a simple way to understand glaucoma and explain what glaucoma is to patients.

The goal of glaucoma treatment is to detect patients at high risk for, or ones that have, glaucomatous damage as early as possible and then to reduce the IOP to lessen or eliminate any subsequent damage.

What is the Problem in Glaucoma?

The aqueous is produced by the ciliary processes inside the eye, behind the iris. The aqueous flows through the pupil and drains out of the eye mainly through the trabecular meshwork (TM) into the canal of Schlemm (red arrow) but also through the uveal scleral outflow tract as well (green arrow). Both © 2014 AAO.

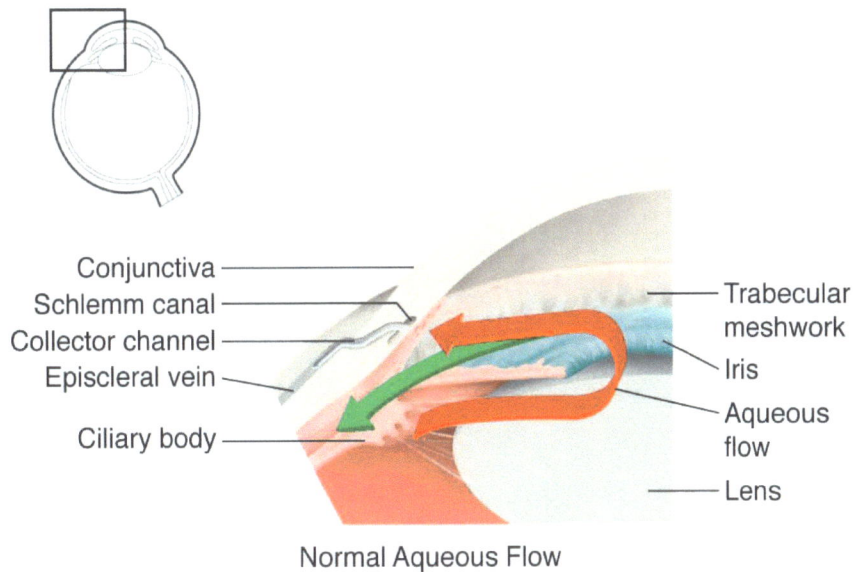

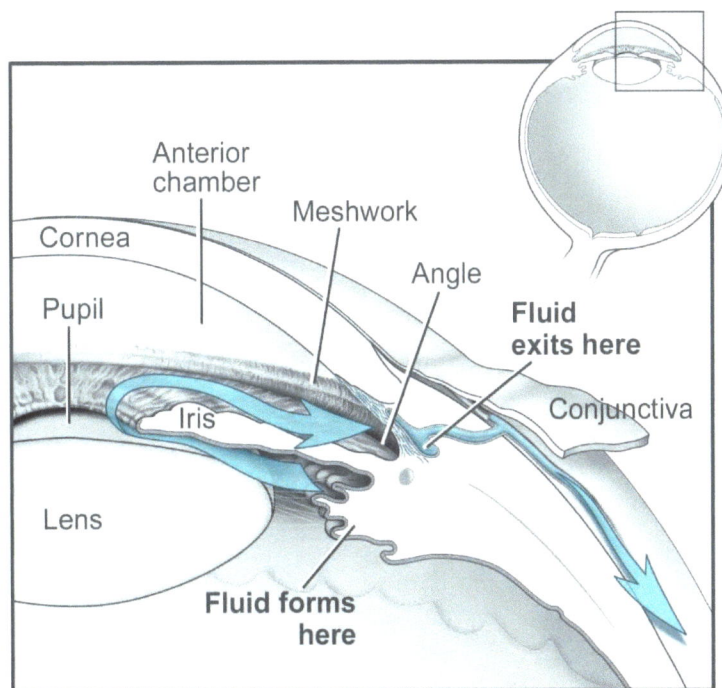

Normal Aqueous Flow

The TM is between the peripheral edge of the iris and the base of the cornea. It is believed that resistance to outflow at the trabecular meshwork level for OAG, and the iris blocking the TM in ACG, is responsible for the higher IOP typically seen in glaucoma.[38] This resistance to outflow results in the aqueous not being able to exit the eye. As a result, the aqueous backs up inside the eye, leading to a higher IOP which pushes on and slowly kills the optic nerve. The following 2 diagrams show this (both © AAO 2014).

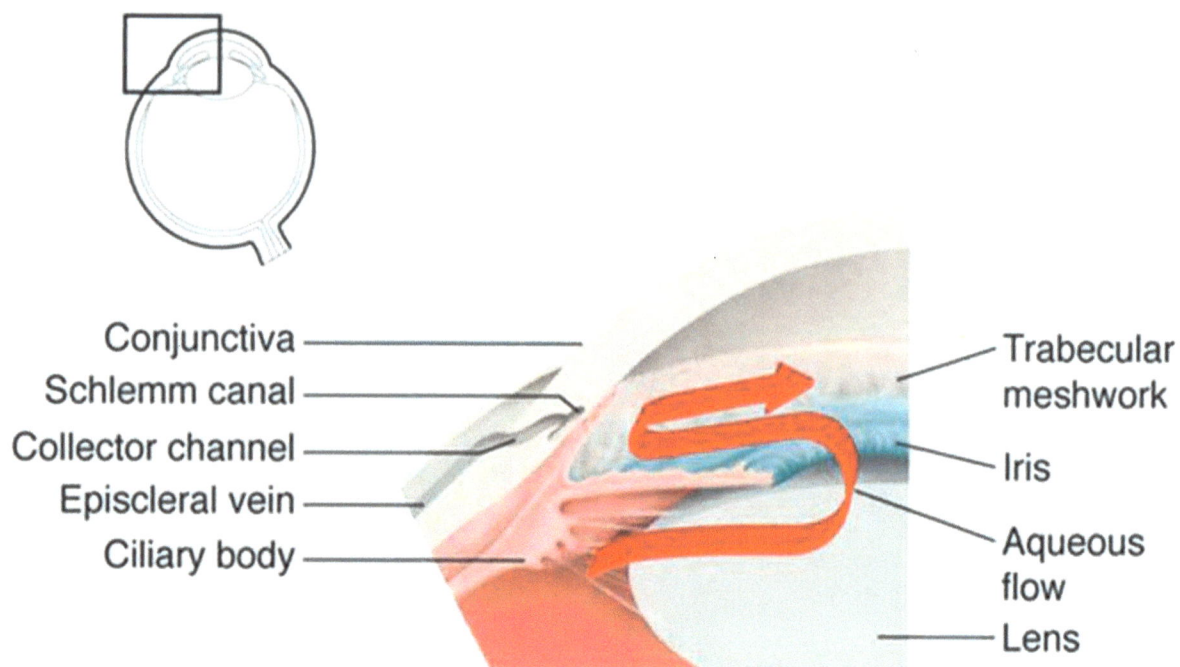

Conjunctiva — — — — — — — — — — — Trabecular meshwork
Schlemm canal — — — — — — — — — Iris
Collector channel — — — — — — — — Aqueous flow
Episcleral vein — — — — — — — — Lens
Ciliary body — — — — — —

Primary Open-Angle Glaucoma

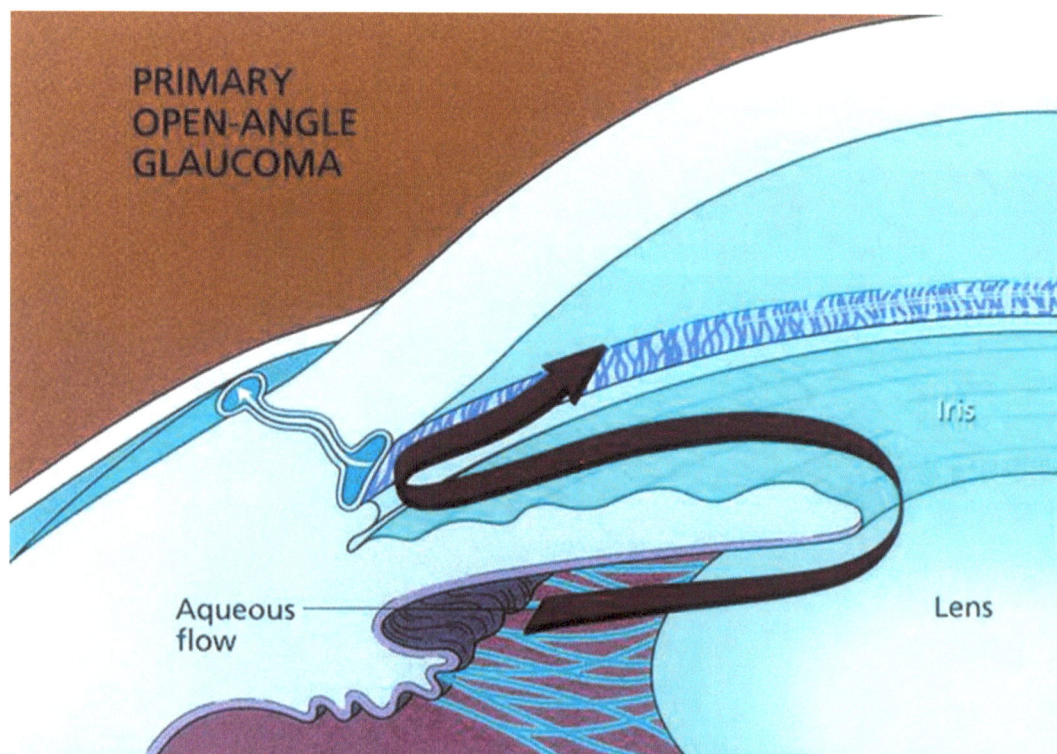

PRIMARY OPEN-ANGLE GLAUCOMA

Iris

Aqueous flow

Lens

This whole process of killing the optic nerve due to the IOP being too high for that particular eye is called glaucoma. Even in normal tension glaucoma (NTG) which we will cover later, lowering of the IOP is still the mainstay of treatment (this is the case since most NTG patients have a thinner than average cornea[59] and have a higher "true" IOP).

Usually a high NCT reading, a narrow angle, an increased C/D ratio, or significant risk factors discovered on a routine eye exam begins the concern about a patient being at risk for glaucoma. It is important to realize that even though a glaucoma suspect may not have glaucoma now, whatever gave rise to them being a suspect in the first place has not gone away with a negative initial W/U and that most patients remain a glaucoma suspect for the rest of their lives, with their risk *increasing* as they become older.

The two major categories of glaucoma: open and closed angle glaucoma.

**If the iris does not close off or occlude the trabecular meshwork, it is called an open angle.
If the iris is pushed up against the meshwork and closes it off, it is called a closed angle.**

Open Angle Glaucoma

Open angle glaucoma without an identifiable cause is called primary open angle glaucoma (**POAG**). POAG is the most common form of glaucoma in the US. You cannot diagnose an open angle until gonioscopy has been done. When the patient has an identifiable cause of open angle glaucoma it is called secondary open angle glaucoma (**SOAG**).

Examples of SOAG include:

1) **Pigmentary Dispersion Syndrome/Glaucoma (PDS/PDG):**

This is described in the academic literature as a "rare" type of SOAG. Don't believe it. I see a lot of PDS and you will too. The best theory, in my opinion, as to how PDS/PDG occurs is as follows. For some reason, there is a higher aqueous pressure in front of the iris than behind it. Because of this, the iris bows backwards against the lens and the zonules. Our best guess as to why this pressure differential occurs is that blood pressure pulses from the retinal arterioles set up pressure waves that get transmitted through the pupil, reflect off of the cornea and push the iris backwards.

In any case, this backwards configuration, easily seen on gonioscopy as "queer configuration," causes the problem. Every time the iris dilates and constricts, pigment from the back of the iris is rubbed off by the zonules. These pigment particles float free in the aqueous and are deposited on the back surface of the central cornea where it forms pigmented lines known as Krukenberg's Spindles (which you will miss if you are not on high magnification). Kruckenburg's spindle (© AAO 2014):

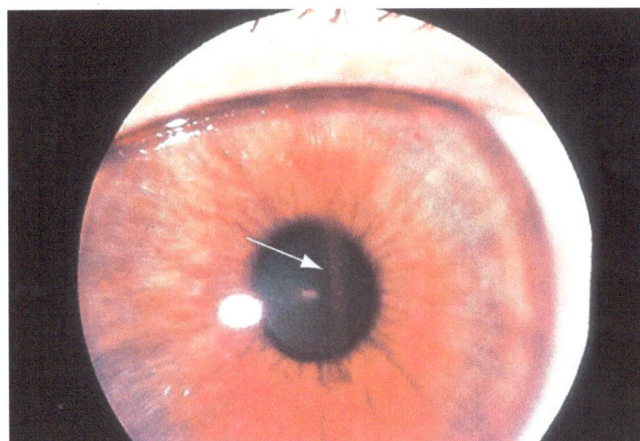

The pigment also gets deposited in the TM, plugging it up. If enough pigment gets rubbed off, transillumination defects (TI) occur in the iris which are seen as radial lines in the iris. Here is a photo showing TI defects in a radial pattern as well as heavy TM pigmentation and the queer configuration on B-scan (both © AAO 2014):

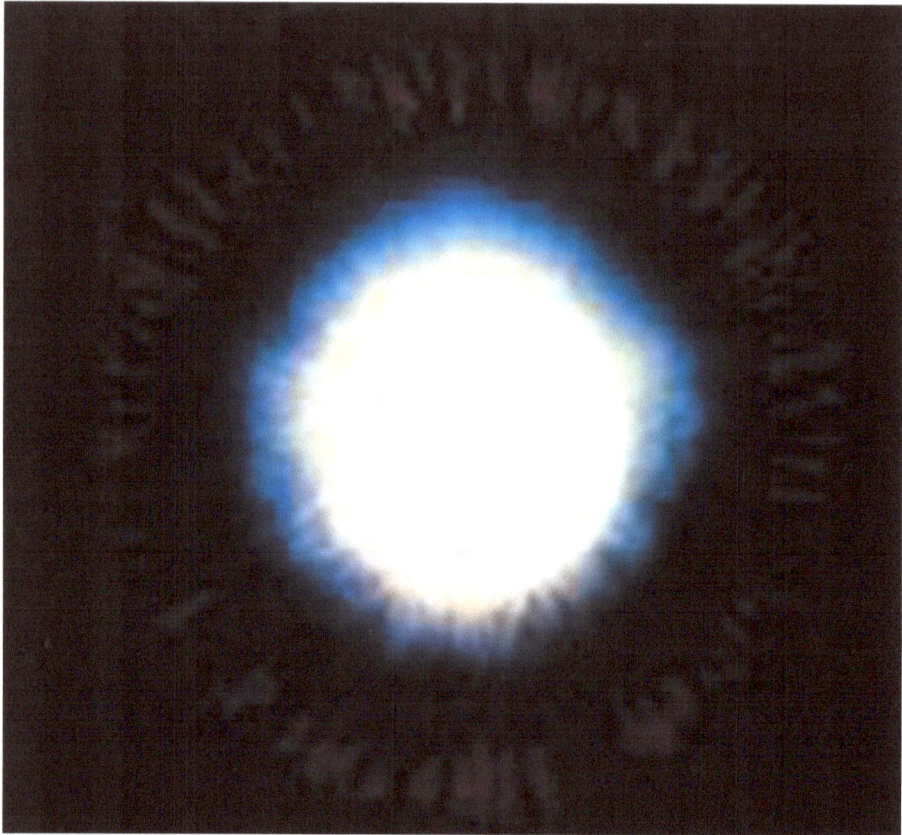

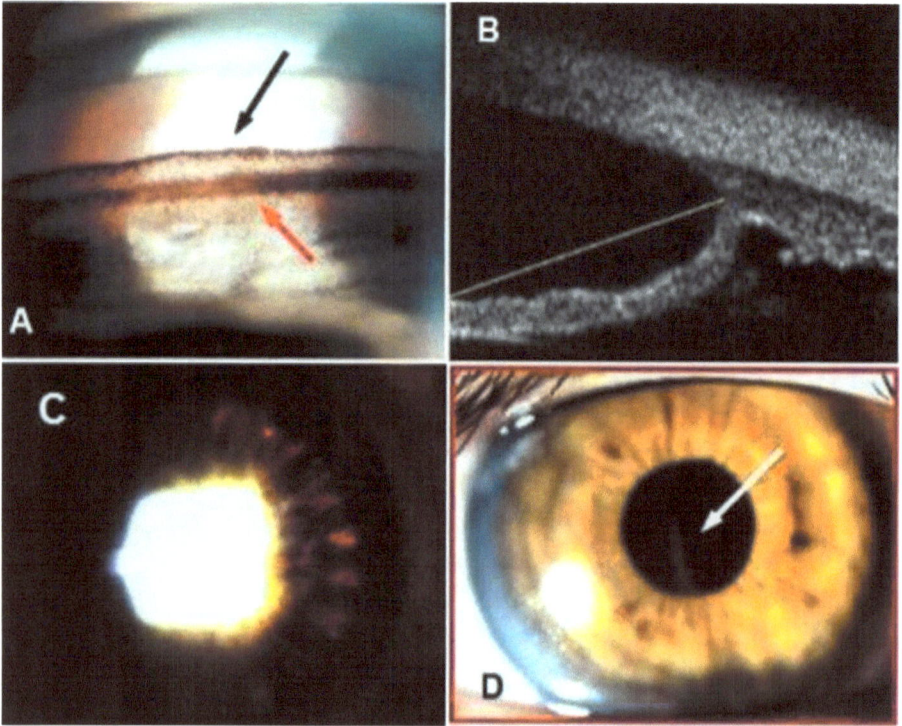

Exercise causes our pupils to dilate and constrict, which then releases a lot more of the pigment, causing the IOP to go to very high levels for a short period of time. This can force aqueous into the cornea, causing a cloudy cornea. This is one of the classic symptoms of PDS, cloudy vision after exercising; something I have never seen in a PDS/PDG patient, despite having looked for it.

Treatment of PDS includes pilocarpine which constricts the pupil, causing a mild iris bombe and a lifting of the iris off of the zonules. Pilocarpine needs to be taken for the rest of the patient's life and can cause mild headaches. I prefer another treatment. If some of the tests show early damage (once damage is evident, it is called PDG) or if the IOP goes above 20, an LPI provides a shunt to eliminate the pressure difference on either side of the iris and reduce, if not eliminate, the backwards bowing of the iris, providing a potentially permanent reduction in the release of pigment from the iris. A 3 year study (median follow up 35.9 months) of LPI vs no LPI in PDS patients with IOP >21 did not show a benefit for the placement of an LPI in the conversion of PDS to PDG (n=105 in both arms of the study).[39] I think that with more years of follow up, there may be a proven benefit for an LPI in PDG. SLT in PDS is a very good option to get the IOP reduced. The more pigment the TM has in it, the more it seems to respond to SLT. Standard POAG IOP lowering drops are used in the treatment of PDG.

One more word about PDS: the academic literature is full of descriptions of PDS that I have not found to be reflective of my private practice. PDS is described as a rare entity that confers a 50% chance of developing glaucoma and an LPI is rarely mentioned as a treatment modality. I have had about 50 cases of PDS in my clinic alone in the last 8 years and I find that less than 10% of these patients develop damage of the optic nerve. I have checked many of my patients after exercise and I have not seen a single spike in IOP. I also find that an LPI will significantly reduce the backwards bowing in the majority of patients and is a very low risk, in-office procedure that I offer frequently. Why the difference in my experience and the academic literature? Well, technology has changed. In-office LPIs were not available more than 15-20 years ago. Most of the research is from academic centers and they tend to get the more severe cases of a disease referred to them. This means that their research is based on a biased data pool. I think that this is true for PDS as well. Recently, the treatment of corneal ulcers was found to be very different at academic centers than in private practice which was explained due to some of these same factors.

The Olmstead County study of a community based population of PDS patients supports my impression.[40] All patients seen for ocular issues from 1976 to 1999 in Olmstead County (where the Mayo Clinic is, which sees almost all of the people in this county, forming a large set of data available for research) resulted in 113 total cases of PDS being followed over a 24 year period. The risk of conversion to PDG was 15% at 15 years after diagnosis with PDS. The most significant risk factor for the conversion to PDG was an initial IOP of 21 or higher, which gave a 46% chance of conversion after 15 years. An IOP less than 21 gave a risk of conversion of 2% over a 15 year period. This is much lower than most other articles from academic institutions have published and is closer to what my experience is.

2) Pseudo exfoliation (PXF or PXE):

PXF is found in more than 50% of OAG patients in Scandinavian populations[41] but occurs in about 50% of Ethiopians in the D.C. area. Ethiopians with PXE also have a high incidence of subluxation of the lens as well, forming SACG. You will see PXF in your practice. The zonules are loose in PXF which makes cataract extraction (CE) more difficult. This is a photo of PXE showing the anterior lens material (© AAO 2014):

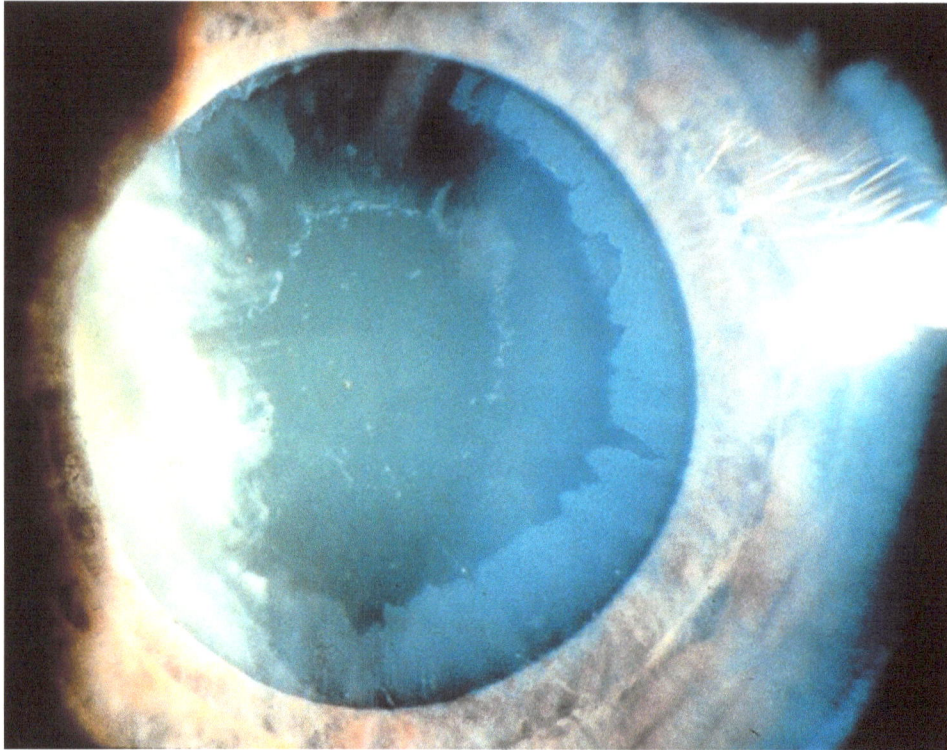

3) Uveitic (Possner-Schlossman and Fuch's Heterochromic):

Both of these entities present with high IOP, mild KPs or no KPs, and 1+ cells in the anterior chamber. They are treated as in POAG. Treat the uveitis with intense steroids, even in a patient who is a steroid responder. Treat the high IOP with glaucoma drops and use the brand name Pred Forte (PF) for the uveitis since the brand name PF is absorbed better into the eye. Taper the steroid ASAP as the uveitis subsides. These two entities are mainly uveitic conditions with temporary, secondary glaucoma.

4) Steroid induced:

The increase in IOP may occur at any time during long-term use with a much higher chance of this in POAG patients.[42] It usually takes 2-4 weeks after starting steroids to develop.[43] I check the IOP 10 days after patients start PO or topical steroids and occasionally every 4-6 weeks after that. The risk is highest with intra-vitreal steroid use, then topical steroids, followed by PO/systemic steroids and rarely, inhaled use of steroids.

5) Angle Recession:

This results from a tear or split in the CBB from trauma. The glaucoma usually occurs 10 to 20 years after the trauma.[44] These patients are followed yearly after the trauma for *life*.

6) Increased episceral pressure:

This can occur from a carotid-cavernous fistulas (significant head trauma with LOC, a basal skull fracture, an ocular bruit and enlarged conjunctival veins) or Sturge-Weber Syndrome (you see this on the face, especially on the upper lid).

7) Ghost Cell glaucoma:

Old RBCs from a hyphema or a vitreal bleed are seen plugging up the TM. Here is a photo of a patient with ghost cells (© AAO 2014):

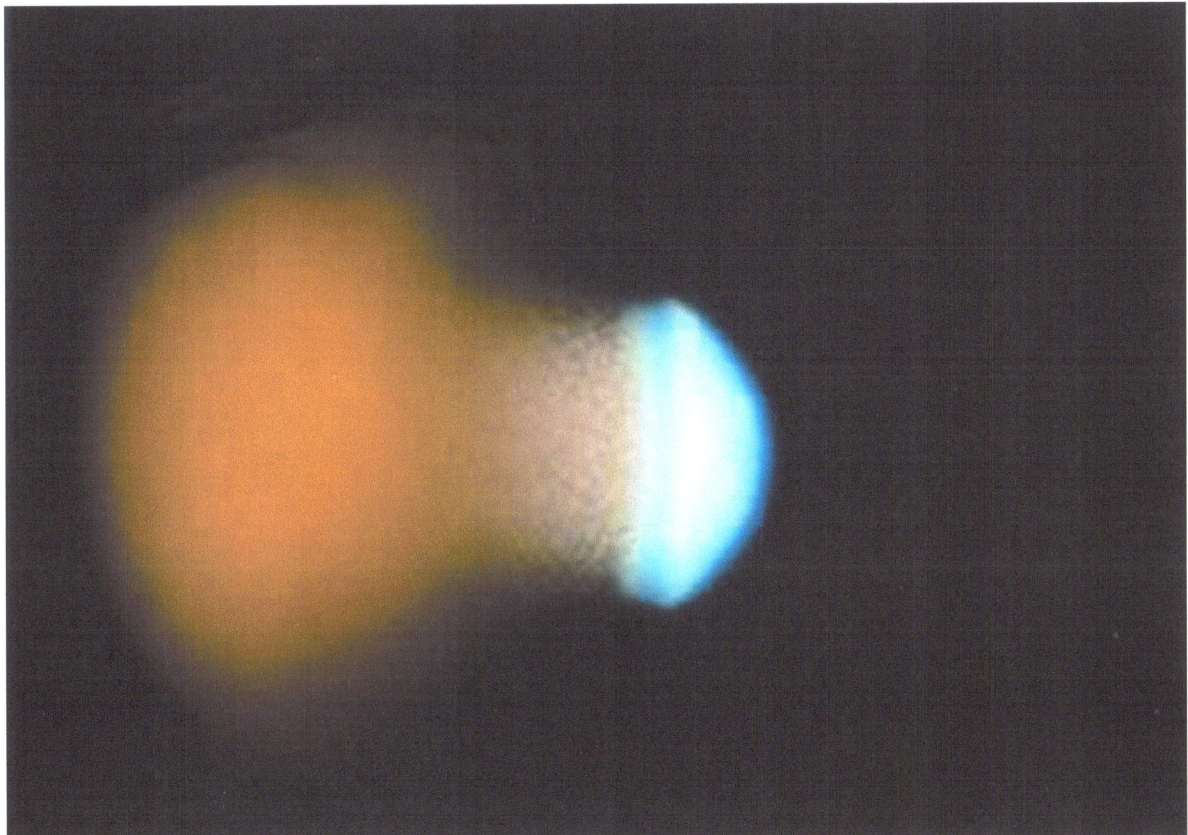

WHAT ARE THE RISK FACTORS FOR POAG?

The major risk factors for POAG are:

1) Intraocular pressure. The Goldmann tonometer is more accurate than the NCT and should be used on all glaucoma and glaucoma suspect patients.[53] The NCT is okay as a screening test for checking IOP but not on glaucoma patients.

The Zeiss manual on the Goldmann tonometer shows the correct end position for the mires when measuring IOP. They are to have the inner circles just touch each other as the endpoint for the measurement of IOP. However, for patients with a significant pulse excursion, the endpoint is described as "the edges cross over each other with the pulsation of the eye" (Zeiss.com). I have the excursion be halfway beyond and halfway into the inner mires touch as my endpoint for the IOP measurement.

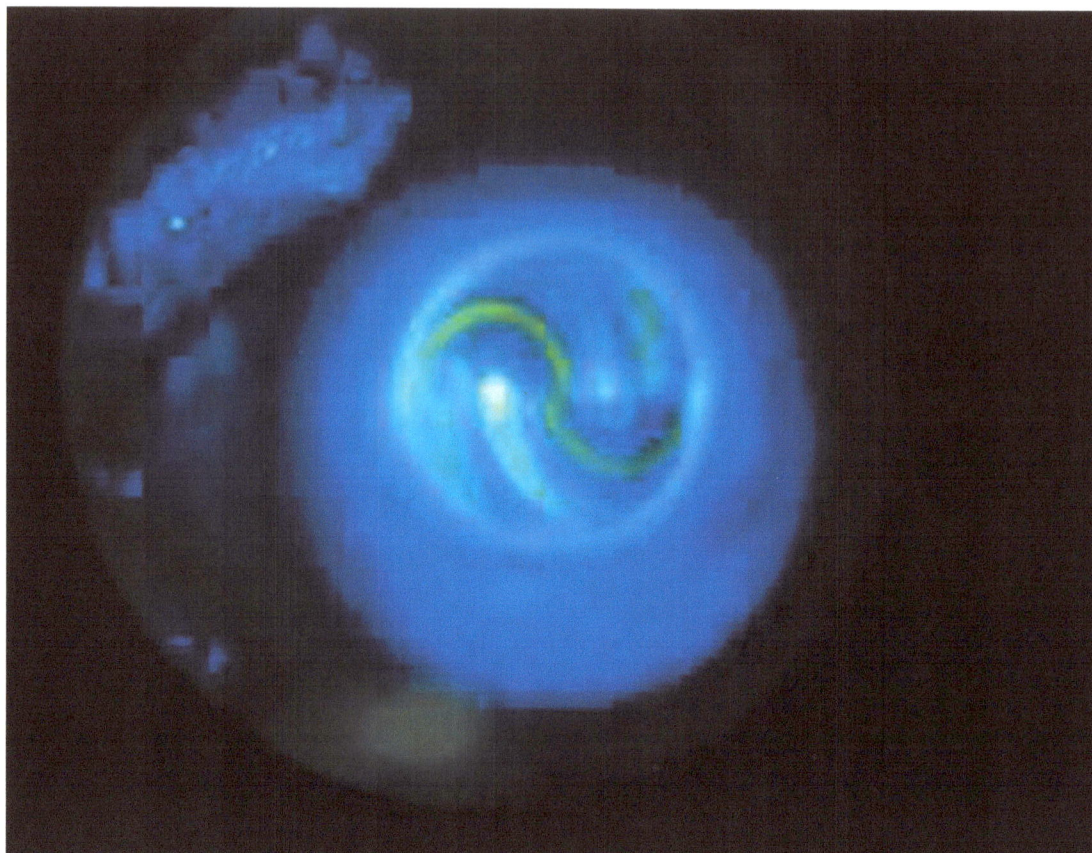

I use an IOP of 21 or higher, corrected or uncorrected for central corneal thickness (CCT), as a risk factor for glaucoma. "Because glaucoma can develop at any IOP level within the normal range of pressures observed in the general population, IOP is a continuous risk factor for the development of glaucoma, and any cut-off between 'normal' and 'abnormal' IOP is arbitrary."[54] So, let's use a cutoff of 20 for the upper limit of normal and an IOP of 21 and higher as suspicious for glaucoma. In my book, and traditionally in ophthalmology, IOPs of 30 and higher are diagnostic of glaucoma. i.e., you start the patient on drops based on the IOP being that high alone, not taking into account other test results. It is debatable whether the corrected or uncorrected IOPs are used (see below). A number of articles show that the risk of glaucoma goes up with higher IOPs.[55,56] The following graphs show that there isn't one IOP that is safe and one that is not. In the los Angeles latino Eye Study (LALES), it is stated that " ...**an IOP range of 19 to 20 mmHg was identified as the turning point where there was the greatest increase in OAG prevalence**".[55] The following graph uses uncorrected Goldmann IOP and the *prevalence* of glaucoma at baseline in the LALES:

- *Prevalence* **is the percent of a population that have a disease at one point in time.**
- *Incidence* **is the percentage of a population that develop a disease (new cases) over a given amount of time.**
- **The best studies are prospective in design, double blinded, multicentered, and involve a large number of patients in each wing of the study.**

LALES *prevalence* risk of POAG at baseline exam IOP, 5,970 people screened with no prior history of glaucoma, follow up = none (prevalence data), n = 167 patients diagnosed with OAG at the time of screening, using uncorrected IOP[55]

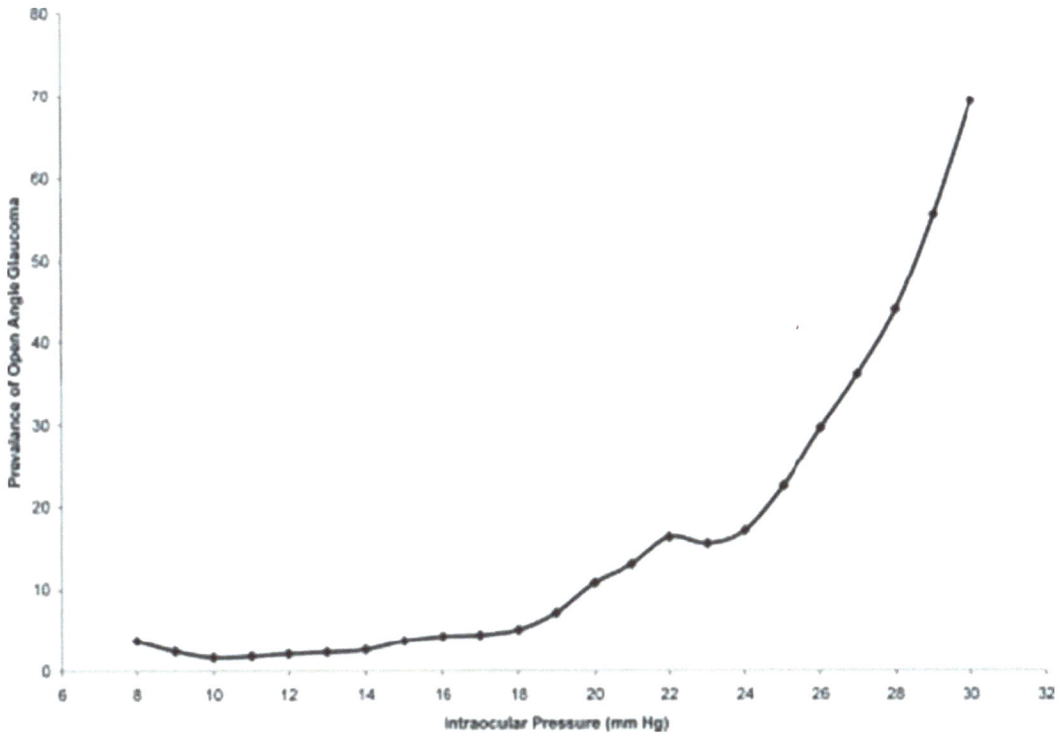

The graph below shows the *incidence* of patients developing OAG in the LALES.[57]

LALES *incidence* risk of POAG in 3,772 people followed for 4 years, 87 developed glaucoma. The graph below shows baseline uncorrected IOP with incidence of conversion to glaucoma adjusted for covariate factors.[57] The top graph on the next page is the same data after factoring in CCT. [57]

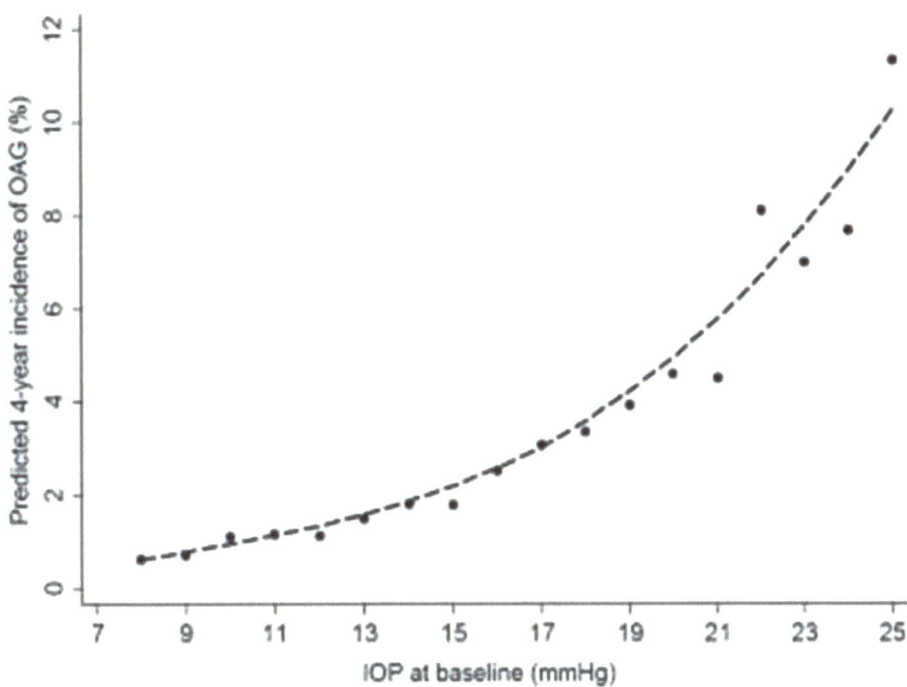

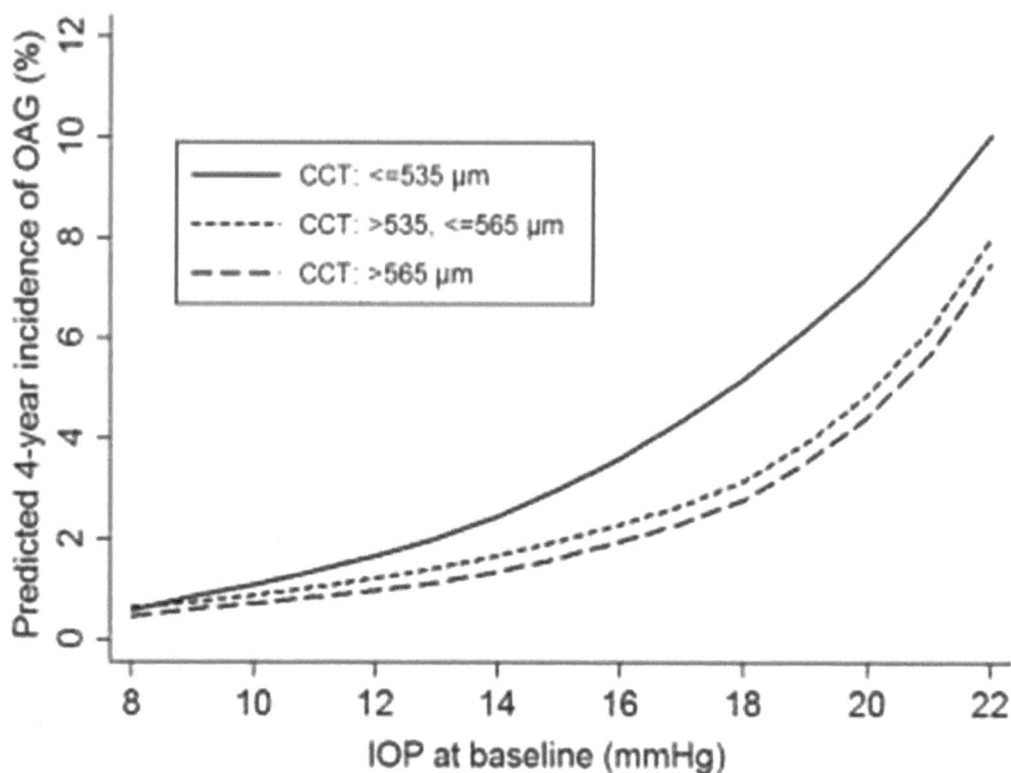

Notice that *at all IOPs*, the risk of developing OAG is *higher with a thinner cornea* and *less with a thicker cornea*.

Barbados: *incidence* risk of developing POAG over 9 years of follow-up at the given baseline uncorrected IOP. 3,222 people screened, n = 125 patients developing OAG[56]

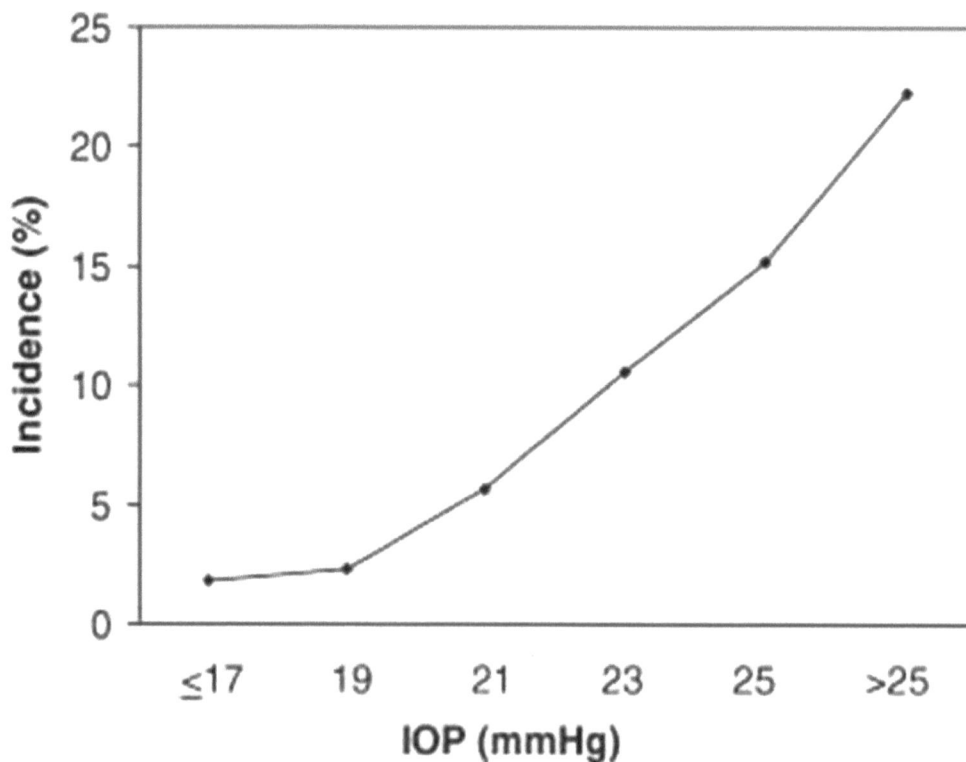

As you can see, there isn't a cutoff between a normal IOP and an abnormal IOP similar to blood pressure measurements. Again, the LALES showed an "exponential increase" in the risk of developing glaucoma starting at an IOP of 19 to 20, using either corrected for corneal central thickness (CCT) or uncorrected IOP measurements[55], so I use 20 as my upper limit of normal and 21 as a risk factor for glaucoma. I also prefer to use the corrected IOP in all of my decisions but I will consider a patient with an uncorrected IOP higher than 20 as a glaucoma suspect even if the corrected IOP is below 21 and everything else is normal. I do inform the patient that their risk is low since their corrected for CCT IOP is normal but I explain that a number of researchers don't believe in using corrected IOP measurements, even though I do. I follow these low risk patients once a year.

This brings up a very contentious issue in optometry and ophthalmology:

Use corrected or non-corrected tono?

The controversy over why some low tension patients get glaucoma and why some high tension patients don't has been going on for decades. This is probably the biggest reason why IOP was thought to be only one of a number of risk factors for the development of glaucoma. The addition of corneal thickness measurements to our exam regimen should have cleared a lot of this up, but, unfortunately, it hasn't. A group of researchers are calling a thin cornea a "risk factor"[58] for the development of glaucoma, not a mis-measurement of the true IOP.

When the tonometer tip pushes against the cornea, the cornea *pushes back* with the force of the "corneal rigidity" plus the IOP. If you assume a given corneal rigidity and calibrate or eliminate that amount from the calibration of the "push back" force, whatever is left is the IOP.

CORNEAL RIGIDITY + IOP = TONOMETER READING
Push back force   = push forward force or tonometer reading

Goldmann assumed that all healthy corneas were about 550 um thick in his calculations. We know today that healthy corneas vary in thickness by a wide margin, and this leads us to the errors that we see today when we measure IOP. A cornea thicker than 550 um has more corneal rigidity push back force than Goldmann assumed, which increases the push back force that the tonometer has to equalize with its push forward force. The tonometer assumes that the increase in push back force is due to a higher IOP and not the corneal rigidity, giving rise to a falsely elevated IOP reading. A cornea thinner than 550 um has less rigidity push back force than goldmann assumed, reducing the total push back force that the tonometer has to equalize, which the tonometer attributes to a low IOP. This gives a falsely low IOP reading.

Many ocular hypertensives have thicker than average corneas, giving a falsely high IOP reading with the true "corrected" for CCT IOP being lower.[58] In the Ocular Hypertensive Treatment Study (OHTS, details below) using *uncorrected* IOP readings, 90% of the subjects had IOP $\geq$ 22, which makes sense as it is the ocular *hypertensive* study after all. Whereas after using the Ehlers correction calculation for CCT, only 48% had an IOP $\geq$ 22.[58] **Almost half of the ocular hypertensives in the OHTS were not hypertensives after IOP readings were corrected for CCT!** This helps explain why ocular hypertensives don't tend to get glaucoma; their true IOP is much lower than measured.

[The Ocular Hypertensive Treatment Study (**OHTS**), a prospective, multicenter, randomized study that randomized a total of 1,408 patients with ocular hypertension to two groups, half received OAG treatment and half did not. Patients completed 60 months of follow up. Treatment reduced the risk of developing OAG by over 50% (9.5% vs 4.4%, P < 0.0001).[65] This completed phase 1 of the OHTS. After receiving 60 months of follow up, both groups were given treatment for OAG. They were followed for an additional time, for a total of 13 years, to determine if the non-treatment group had a penalty for delaying treatment. They did not (P < 0.77). [107] This was phase 2 of the OHTS].

Corrected IOPs were not used in the Collaborative Normal Tension Glaucoma Study (**NTGS**, details later) either, where patients with glaucoma and low measured IOP were studied. The average corneal thickness for patients with NTG is 510 to 520 um, much thinner than average.[59] So, low-tension patients in the NTGS probably had a thinner cornea than average, resulting in a higher true IOP than what was measured. This helps explain why they get glaucoma; their true IOP is much higher than measured.

Also, the (NTGS) found that lowering IOP at least 30% lowered the risk of VF loss at 5 years from 35% to 12%, confirming that IOP has an important role in treating patients with Normal Tension Glaucoma.[60] This strongly infers that a true, higher IOP is what is being treated.

Questions regarding this issue

If a thin cornea is just a risk factor for the development of glaucoma, why then why would a thicker than average cornea *reduce* the chances of developing glaucoma at all IOPs (see the LALES graph below)?

If a thin cornea is just a risk factor and not a mis-measurement of IOP, then all LASIK patients with a thin cornea post-operatively would then be at risk for glaucoma.

The researchers in the LALES (n=5,970) showed a higher risk of glaucoma with patients who had thin corneas *at all IOPs* and a lower than average risk for patients who had thicker than average corneas at all IOPs, with this difference mostly collapsed towards the "normal" risk when the CCT correction was removed.[55]

Graph from LALES measuring the *prevalence* risk of glaucoma versus CCT broken into 3 strata, no follow up (prevalence data), 5,970 screened, n = 167 patients had undiagnosed glaucoma at the baseline exam[55]

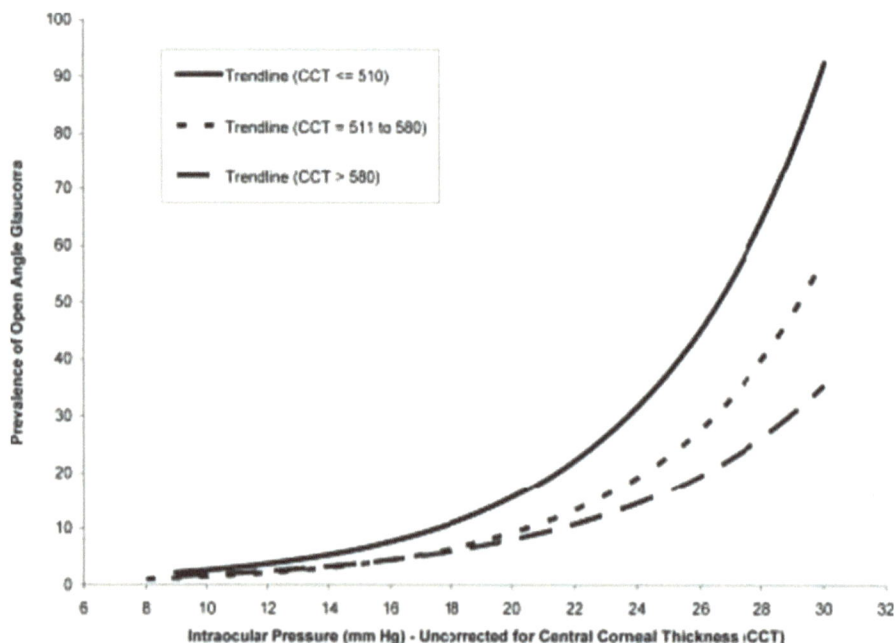

The Ocular Hypertensive Treatment Study (OHTS) concluded that corneal thickness (really thinness) was the strongest predictor of glaucomatous damage.[62] The researchers of the OHTS state that "It is likely that the predictive power of corneal thickness is due to its effect on the measured IOP."[70] But, the study then goes on to say: "However, we cannot exclude the possibility that corneal thickness is related to other factors affecting susceptibility to glaucomatous damage."[70]

Both the OHTS and the LALES label a thin cornea as a "risk factor" rather than saying that a thick or thin cornea was a mis-measurement of the true IOP. "Whether this increased risk of glaucoma is due to underestimating actual IOP in patients with low corneal thickness or whether low corneal thickness is a risk factor independent of IOP measurement has not been completely determined."[63] In my general ophthalmology opinion, I do think that the corrected IOP is more accurate than the uncorrected IOP measurement and should be the main number used in all IOP measurements. You can see how trapping a hair while doing Goldmann tonometry can artificially raise the IOP measurement as well as how the measured IOP is lower after LASIK. Again, if we are to follow some researchers' advice, **then all LASIK patients with thin corneas are at risk for glaucoma**. I use the corrected tonometer numbers in all of my analysis of glaucoma patients. Medeiros from UCSD says " ...the conclusion that CCT is a true independent risk factor for glaucoma is not validated at this time and requires further investigations."[64] He goes on to say that, while present correction models for CCT may not be as accurate as they need to be, interpreting CCT as an independent risk factor is a mistake and infers that a correction of CCT is a valid (my word) clinical tool.

In the LALES, they "found that the same level of baseline IOP measured by Goldmann applination tonometry predicted a higher rate of developing OAG among individuals with thinner CCT than among those with thicker CCT." [61] Unfortunately, the researchers then go on to say that this shows that a thin cornea is an independent risk factor rather than an under measurement of the true IOP. They say that this independent risk factor may be related to a corneal resistance factor or corneal curvature or other biomechanical properties of the cornea. I use to think that this change in risk was simply because some corneas are just thinner or thicker than the 550 um Goldmann assumed when he calibrated his tonometer and that this throws off the IOP reading. Do to more recent research, Corneal Hysteresis has become another independent risk factor for the development of glaucoma.

In the OHTS, the risk of glaucoma damage was reduced from 9.4% to 4.5% with IOP lowering drops over the 5 year follow up period,[65] a 58% reduction. The question then arises, why not treat all patients with IOP > 20? Because we would have to treat about 10 patients with ocular hypertension to catch one that would develop glaucoma over a 5 year period. This is felt by many to be an unnecessary expense, morbidity and waste.

2) Corneal Hysteresis (CH) Recently, corneal hysteresis has been differentiated from CCT as an *independent* risk factor for the development of glaucoma with both being separate and independent from each other. In a *prospective* study, Medeiros followed 114 eyes with OAG for an average of 4.0 years.[204] The multivariable analysis showed a significant correlation between a worsening of VFI values (the negative values on the graph lines below) and a lower CH value, with an even stronger relationship than CCT (CH accounts for 17.4% of the correlation versus 5.2% for CCT). The following complex figure shows this (©Elsevier 2013). The numbers on the lines represent VFI losses in Db per year. Follow the -2 Db loss per year line. You can see that for lower CH values (bad), you get a -2 Db loss in VFI per year at lower IOP values.

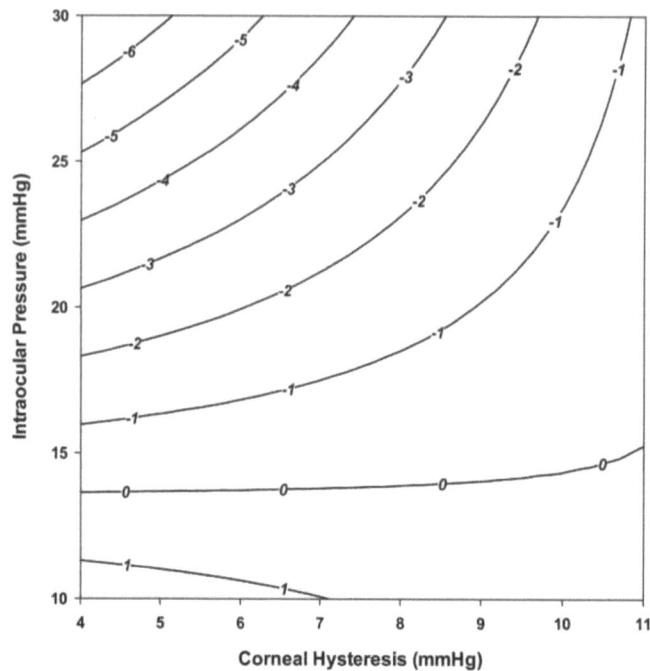

Intraocular Pressure (mmHg) vs. Corneal Hysteresis (mmHg)

De Moraes found a similar result when he prospectively followed 153 OAG eyes of 153 patients for an average of 5.3 years. CH (p < 0.01) and peak IOP (p < 0.01) was significantly associated with a worsening of the rate of VF change. (Age was close at p < 0.03). [207]

Park retrospectively measured 82 eyes of 82 normal tension glaucoma patients for CCT, CH, and RNFL thickness.[205] Park compared the eyes that progressed on HVF tests (n = 46) to those that didn't progress (n = 36) over the study period mean of 77 months. Upon multivariable analysis, a lower CH score (p < 0.01), but not CCT (p = 0.35), was significantly associated with the group with the progressive loss on VF. (Average RNFL thickness was close at p = 0.04). So, CH seems to be an independent risk factor for the development of glaucoma damage.

Research also shows that CH may change based on a lowering of the IOP. Sun studied 40 patients with *unilateral* chronic angle closure (CAG) that had their CCT, IOP, and CH measured between the two eyes and in the same eye prospectively before and after treatment.[208] While there was no statistical difference in CCT before and after treatment in the same eye with CAG nor between the two eyes of the same patient before or after treatment, CH significantly increased (p < 0.001) after treatment within 2 weeks and stayed there for the 4 week follow up period. This study shows that while CCT does not change following lowering of the IOP somehow CH, as measured with the ORA, does.

The figure below on the left (© Elsevier) shows the *pretreatment* correlation between IOP and CH values: [208] You can see the significant negative correlation (p < 0.001) between pretreatment IOP and CH values. As the pre-treatment IOP goes up, the CH value goes down. The figure on the right shows the change in IOP versus the change in CH values both 2 weeks and 4 weeks after a trabeculectomy and peripheral iridectomy were performed in all eyes. A significant correlation was seen post operatively between an increase in CH values and a reduction in IOP (p < 0.001). The larger the IOP decrease was after surgery, the more the CH value increased, both the IOP and CH moving towards a "normal" value. (Remember, the CCT didn't change pre and post treatment. In the OHTS and the LALES study, the CCT of each person's paired eyes shows a high degree of similarity in 93% of people).[58, 209] The *pre*-treatment average CH value for all 40 eyes with CAG was 6.83 +/- 2.08 mmHg, the average 4 week *post*-treatment CH for the same eyes was 9.50 +/- 1.66 mmHg with the fellow, normal eye without chronic angle closure average CH being 10.59 +/- 1.38 mmHg. You can see that, while there remained a significant difference between post treatment CH (9.50 mmHg) and the CH of fellow normal eyes (10.59 mmHg), the difference was much less after treatment (CH increasing from 6.83 mmHg to 9.50 mmHg) and given more time, may have approximated the normal eye's CH value. This shows that CH seems to be a factor, independent of CCT, which can change with IOP lowering treatment.

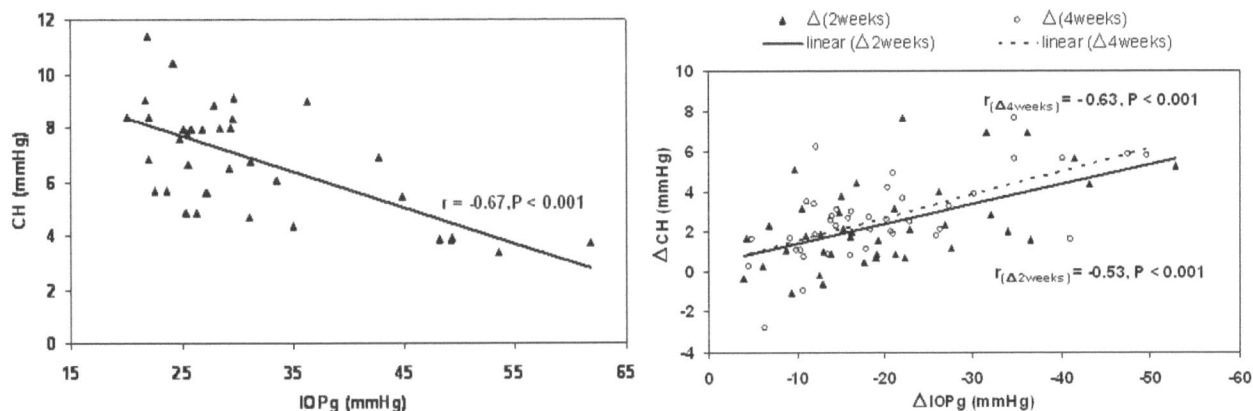

So, CH is a separate risk factor for glaucoma from CCT and lower values for CH are bad. <u>What is CH?</u> It is the viscous and elastic properties of the cornea (and maybe other tissues of the body as well) that dampen forces applied to it.[205] It is measured by an Reichert's **Ocular Response Analyzer** (OSA) by blowing air at the cornea, just like the NCT does. The pressure of the air follows an increasing and then decreasing bell curve pattern as shown on the left diagram below (courtesy of Reichert Inc.) as the green line. The machine measures the pressure it takes to flatten the cornea from the resting convex position to the plano position, seen as the first red peak below. Then, it over flattens the cornea to a concave position and then measures the pressure it takes to allow the concave cornea to resume the flat position again as it come from the concave position, seen as the second red peak below. A stiff cornea takes a lot of pressure to keep the cornea from going flat from the concave position, which gives a low difference between the first and second reading (a low CH value which is bad). An elastic cornea takes less pressure to prevent the cornea from going flat from the concave position, which gives a large difference between the first and second value (a high CH which is good). The difference between the first flat reading and the second flat reading is the corneal hysteresis, as shown in the left diagram below. Again, stiff corneas have little difference between the first and second reading (bad), pliable corneas have a big difference (good).

The distribution of the CH for the Park study is shown to the right with the average being around 10 mmHg. [205] Currently, insurance pays about $15 for the test and the machine costs $15,000.

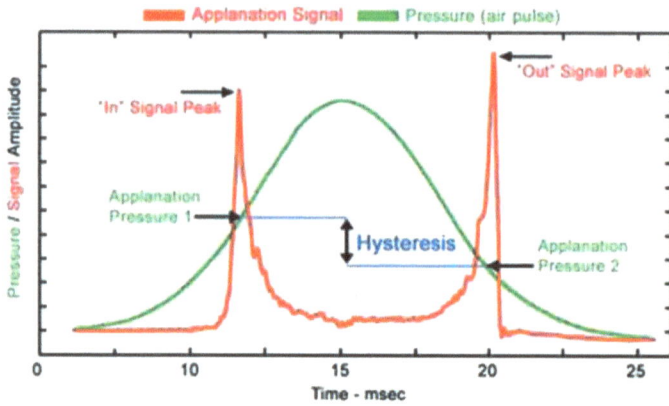

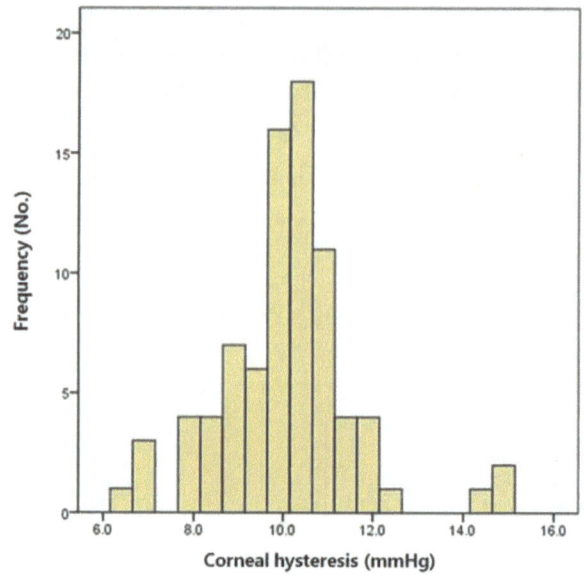

3) Age 65 or higher. As you can see by the studies below, there again isn't one age at which the risk goes from normal to abnormal, similar to IOP. I use 65 years old as a cut-off for a higher risk of glaucoma.

<u>**Prevalence**</u> **of Glaucoma all types, Yazd Iran, 1,990 patients screened, n = 87** [66]

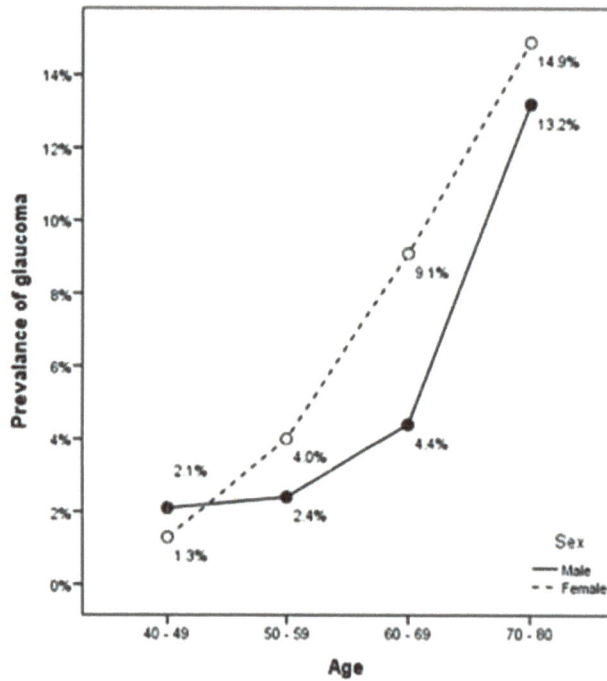

Incidence of Glaucoma from 4 separate studies according to age [67]

◆ Dalby ■ Melbourne ▲ Barbados ● Rotterdam

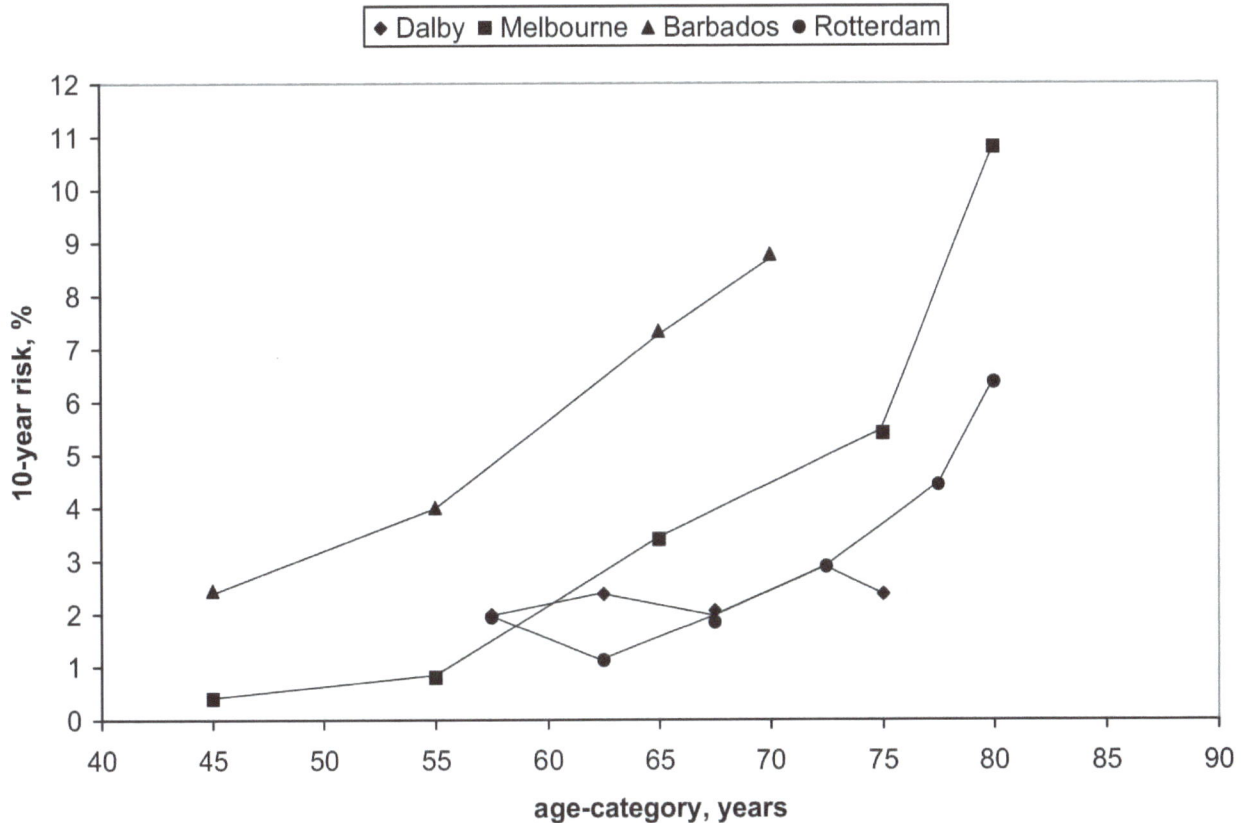

For the Melbourne study, 2,594 people completed the study, n = 62, average follow up 4.5 years
For the Rotterdam study, 2,571 people completed the study, n = 108, mean follow up 9.8 years
For the Barbados study, 3,222 people completed the study, n = 125, average follow up 9 years
For the Dalby study, 1,511 people completed the study, n = 26, average follow up 10 years

Estimated annual incidence of Glaucoma for different age groups with a diagnosis of glaucoma from 1965 to 1980, published in 2001; Olmstead MN, 60,666 people screened, n = 295 eyes with treatment for OAG [68]

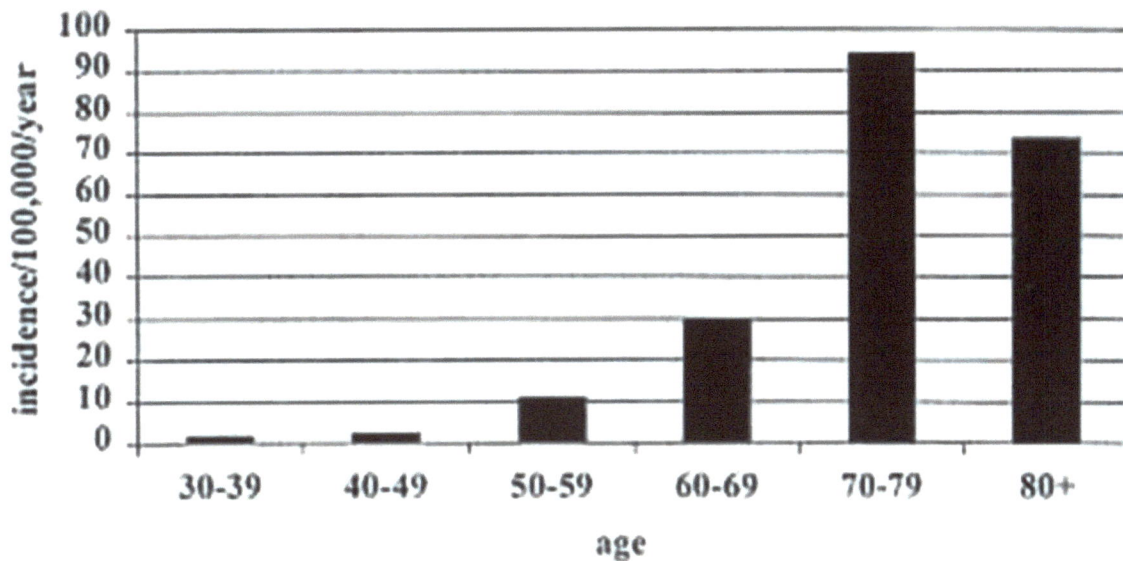

This graph is from the Early Manifest Glaucoma Trial (EMGT), a multicenter study, showing the *prevalence* and severity at diagnosis of OAG (32,918 people screened, n = 406 patients with newly diagnosed OAG, some bilaterally).[165] Notice that the prevalence of OAG increases with age, most eyes had mild glaucoma at the time of diagnosis in all age groups, and that the extent of VF loss is similar in all age groups from age 60 and older.[165]

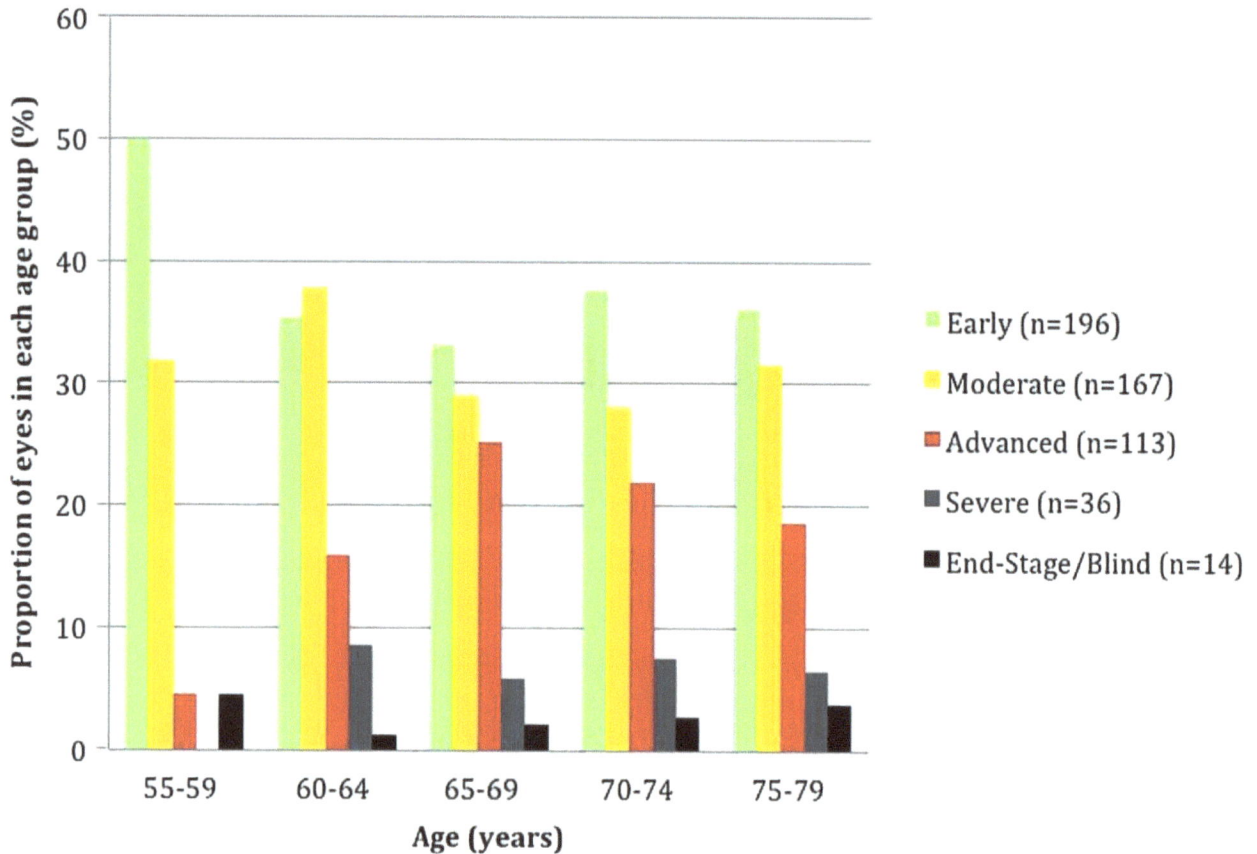

4) **Family history.** A family history has long been accepted as one of the strong risk factors for the development of glaucoma. Family history, however, was not found in the OHTS to be a risk factor for the development of POAG.[69] This was attributed to patients confusing a family history of glaucoma with cataracts or other diseases and the researchers not following up on the medical history of the participants to confirm a reported positive family history of glaucoma.[70] The 9 year follow up in the Barbados study showed a 2.4 fold increase risk in people with a family history of glaucoma.[71]

I consider a positive family history of glaucoma to be a significant risk factor for the development of OAG and CAG.

5) **Cup to Disk Ratio.** A C/D ratio of $\geq$ 0.5 is a risk factor for the development of POAG. [35,70] Only 5% of non-glaucoma patients have a C/D larger than 0.6.[35] A C/D ratio can be misleading though. A large C/D ratio can also mean that the nerve sheath is larger than average and that the 1.2 million ganglion cell axons are spread further apart, giving a falsely large C/D. This is thought to be the case in African-Americans in the OHTS.[70,72] Conversely, a small optic nerve sheath can squeeze the fibers tight together, artificially reducing the size of the cup, masking a glaucoma concern.[35] Most people are born with small to medium-sized cups but some patients are born with a large cup. Even given the above, a large C/D ratio by itself is a major risk

factor for the development of glaucoma.[70] Here is a photo of a small optic nerve with situs inversus on the left and on the right, the opposite disk, an abnormally large disk the morning glory disk:

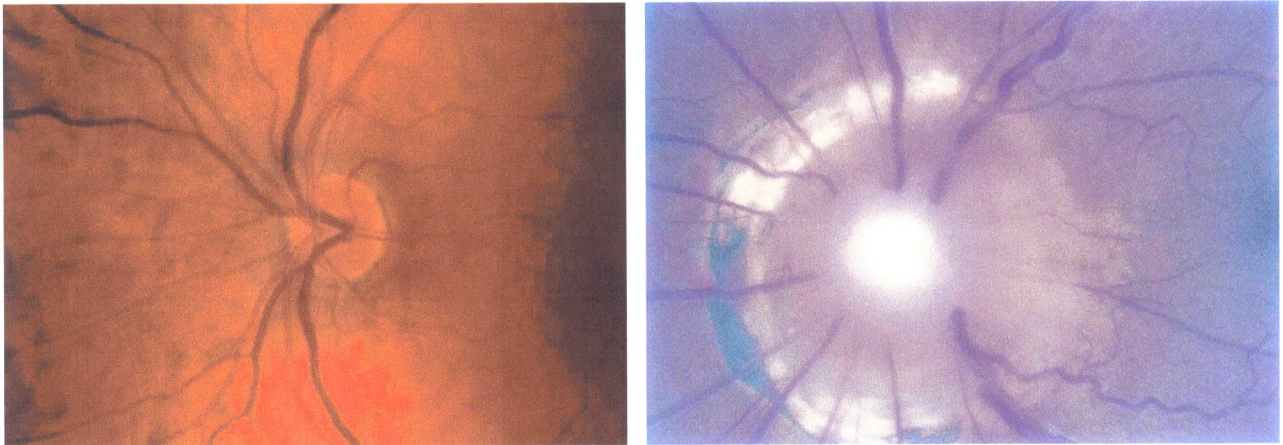

Here is a classic optic nerve hypoplasia nerve in a patient of mine showing a small nerve and peripapillary atrophy:

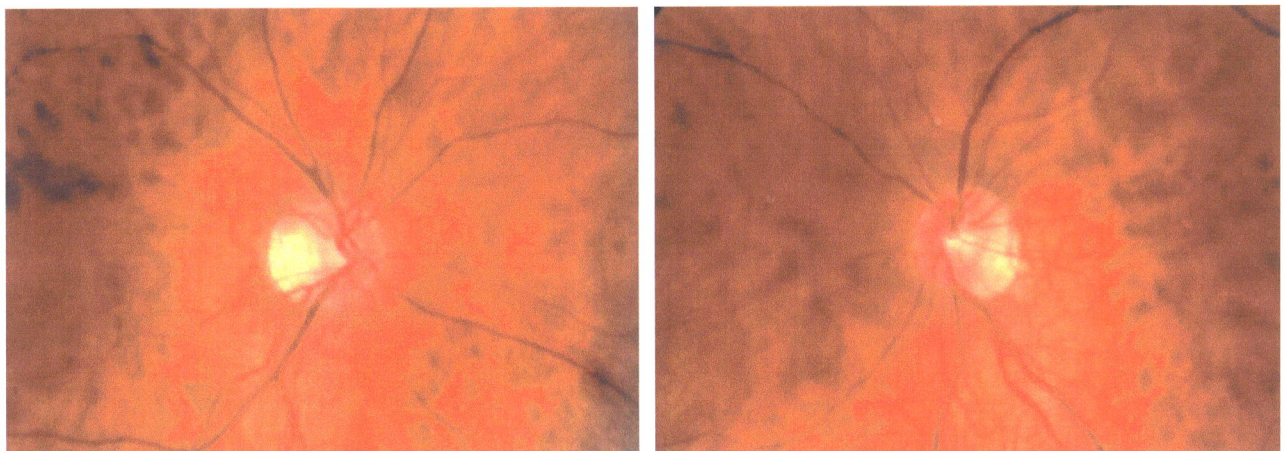

Over time, glaucoma destroys the ganglion nerve fibers and this causes the *C/D ratio to enlarge* and *the rim tissue to thin*, exposing more of the lamina cribosa and *pushing the blood vessels in the disk nasally*. Therefore, these changes in the appearance of the disk over time are of great concern for the development of glaucoma. Changes in the cup usually take years to become noticeable. Taking digital photos of the optic nerve every year is mandatory on all glaucoma and glaucoma suspect patients, even though medicare may only pay for photos once every other year. Some cups have a saucer slope and can make assigning a C/D difficult. I use the contour method, not the color method, of assigning a C/D ratio to an eye. Just do your best and document that it is a saucer slope on the exam form. I assign only the largest C/D ratio for each eye, typically the vertical one. Upon chart review, insurance companies may contest an eye having "optic cupping neuropathy" if the two C/D ratios for an optic nerve are 0.6 vertically and 0.3 horizontally, as they may pick the lower horizontal C/D ratio and deny the patient as a glaucoma suspect.

Here is the *prevalence* of glaucoma for various C/D ratios from the Singapore study (1,232 people screened, n = 61 eyes having glaucoma):[73]

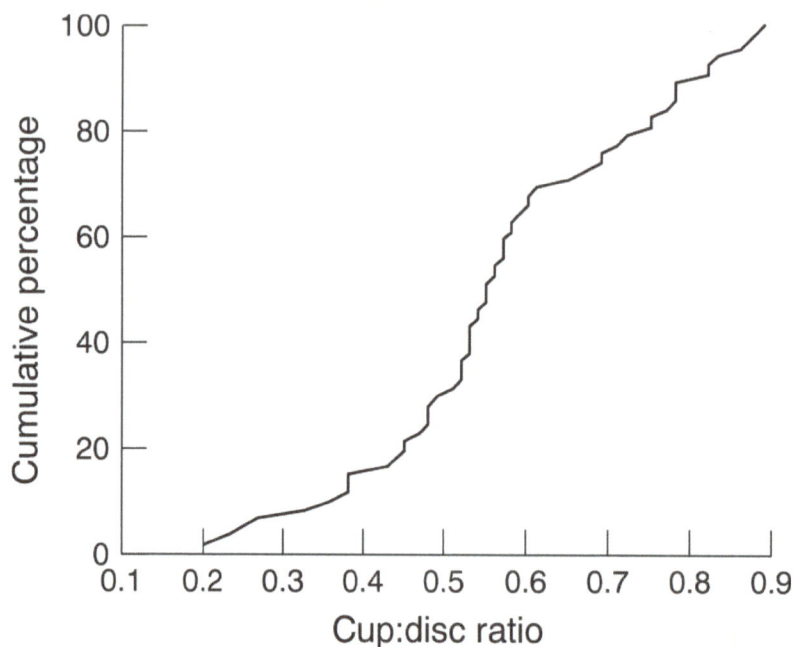

As you can see, there isn't one C/D ratio that is normal and one that isn't, just like age and IOP.

Rim tissue thickness, in my opinion, is probably more important than the C/D ratio. A C/D of 0.6 with a healthy thick rim probably has a lower risk of developing glaucoma than a C/D of 0.4 with a small optic nerve sheath and very thin rims. Paleness inside the cup is normal in glaucoma as the lamina cribosa is white. Paleness of the ganglion cell axons *in the rim tissue* is not normal. Pale rim tissue is optic atrophy and causes of optic atrophy have to be sought out. "With rare exceptions, glaucoma results in increased cupping and pallor within the cup, but not pallor of the remaining rim tissue."[56]

Recent research has studied the areas just proximal to the optic nerve as a way of also determining if glaucoma damage is present. This is the alpha and beta zones shown below. © AAO 2015. The red area is the optic nerve, the green area is the alpha zone and the blue area is the beta zone.

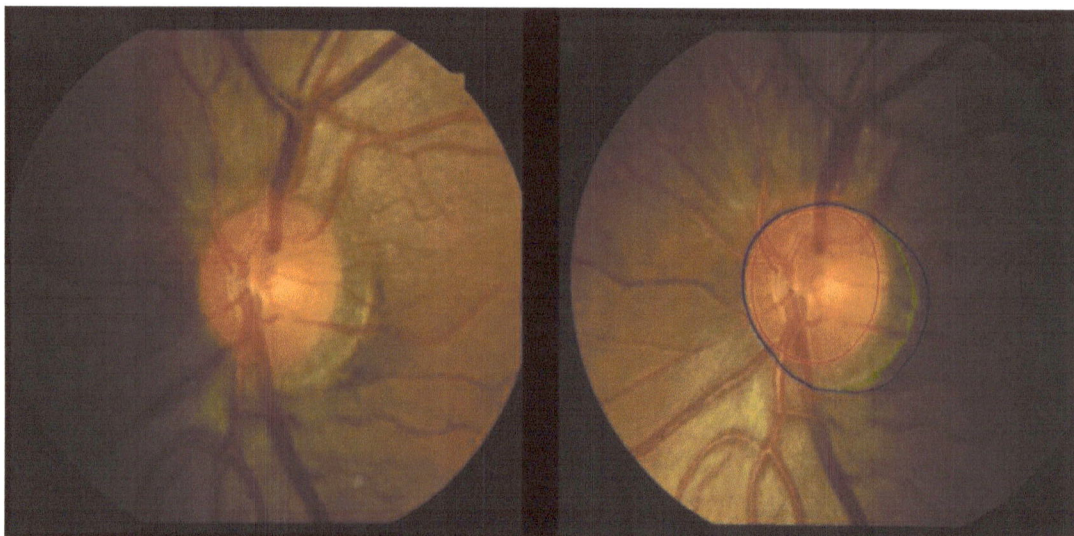

279 eyes that converted to OAG in the OHTS were compared to 279 eyes that did not convert in the same study.[194] No significant difference was found in the alpha and beta zones. Currently, I don't place much importance on the alpha and beta peripapillary areas in OAG detection and treatment.

6) Disk Hemorrhages (DH), Nerve Fiber Layer Defects (NFLD), Notches, and blood vessel (BV) changes. Disk hemorrhages typically cut across the edge of the optic nerve and when present, are very concerning for the presence of glaucoma. Disk hemorrhages suggest active disease.[75] Up to a third of glaucoma patients will have a disk hemorrhage at some point.[76] In a study examining patients with low tension glaucoma every 4 weeks for the presence of a DH and then every week until the DH was gone, 92% of DH lasted 4 or more weeks with a range of 2 to 35 weeks for resolution of the hemorrhage.[51] Think of the presence of a DH as the optic nerve crying out under stress. You do need to R/O diabetes. The Hemoglobin A1C is probably the best test for this as it is the prior 3 month average of blood sugars. The hemorrhage from a vein occlusion, PVD or anticoagulation will typically be in the retina or vitreous, and not exactly at and only at the disk margin, like it is in glaucoma. Photograph and document any disk hemorrhages.

The OHTS found, in multivariate analysis (taking into account demographic data and other known risk factors that could skew the results), a **3.7 fold increase (P < 0.001)** in the risk of developing POAG in patients who had a disk hemorrhage (n=3,236 eyes screened, 168 eyes developed glaucoma with a mean follow up 31 months since the discovery of the hemorrhage, 128 eyes had disk hemorrhages. Most of the eyes, 87%, that had a disk hemorrhage did not develop glaucoma).[74] *A 3.7 fold increase in risk is a big risk factor.* The patients in the OHTS were dilated once a year. Again, all of the DH in the study mentioned earlier resolved in less than 1 year. The researchers suggest that this relative risk could have been even higher with more frequent DFEs, since more DH may have been discovered.

Interestingly, they found that 84% of the hemorrhages were missed on live exams (by glaucoma specialists!) and were only discovered during a photo review of the disks. This shows the importance of having a mental checklist that you go through when examining the optic nerve, noting the C/D ratio and looking for NFLD and disk hemorrhages in all of the patients.

The Early Manifest Glaucoma Trial (EMGT) had a different conclusion than the OHTS.

[In the EMGT, 44,243 patients were screened for glaucoma, 255 had newly diagnosed OAG with loss on a HVF. Half were randomized to "treatment" with ALT and Betaxolol or to "no treatment." The results showed a 50% reduction (P = 0.007) in the rate of glaucoma progression in treated eyes (n=227 completed the median follow up of 8 years). 55% of all of the patients developed DH at some point.][106, 197, 198]

Researchers in *this* study found the same number of patients with DH on *any* of the live clinical exams as those found on photo review (140 versus 141). They did miss a significant number of DH on live clinical exams when of all the exams were compared for photo review (P< 0.0001). [198] The authors attributed this difference to the EMGT forms requiring the ophthalmologists to actively check off whether a DH was present or not. This supports my recommendation to have a mental list that you go through when examining a disk in a glaucoma patient. They also concluded that, while the presence of DH shortened the time to OAG progression, **there was no evidence to support the idea that patients with a DH are being undertreated or are in need of a more intensive IOP lowering treatment**.[198] The same number of patients in both the treated group and the non-treated group developed DH (P = 0.93).[198] Here is a graph showing the frequency of DH found on live exam versus photo review as well as the number of eyes that had multiple number of DHs. [198]

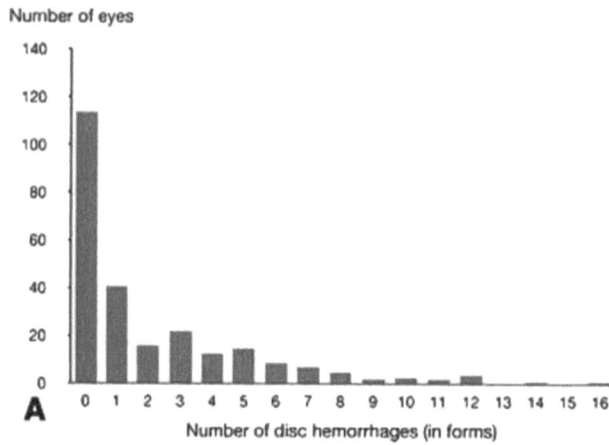

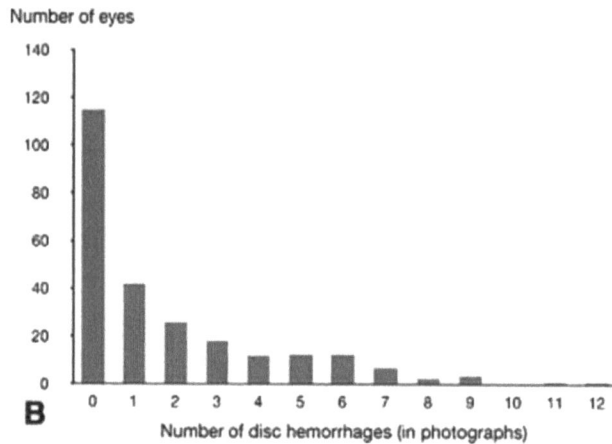

The graph below[74] looks at the OHTS data in another way. It asks the question, "did treatment show a reduced risk of developing disk hemorrhages?" The graph shows the strong trend of more frequent DHs in the untreated group. But while not statistically significant (p = 0.13) over a follow up of 31 months since the discovery of the DH, it leads one to believe that with a longer follow up, the difference may have become statistically significant (my words). Most of the patients who developed a disk hemorrhage did not develop POAG (87%) over the mean follow up of 31 months. So it may be that while it takes a while to develop OAG after a DH occurs, a DH should certainly increase your concern about a patient eventually developing glaucoma.

Budenz et al · Optic Disc Hemorrhages in the OHTS

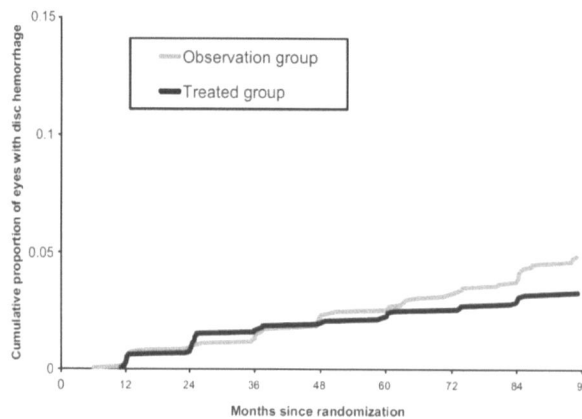

Figure 2. Graph showing cumulative incidence of optic disc hemorrhages in annual photographs of treated and observation groups.

Here are some photos of DHs. **Diabetes needs to be ruled out in all DH patients**:

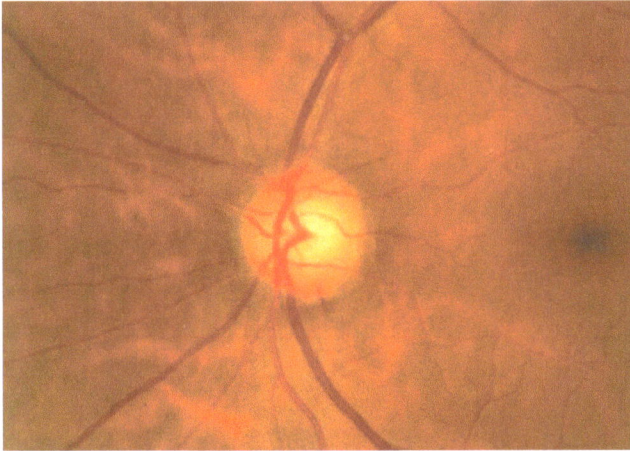

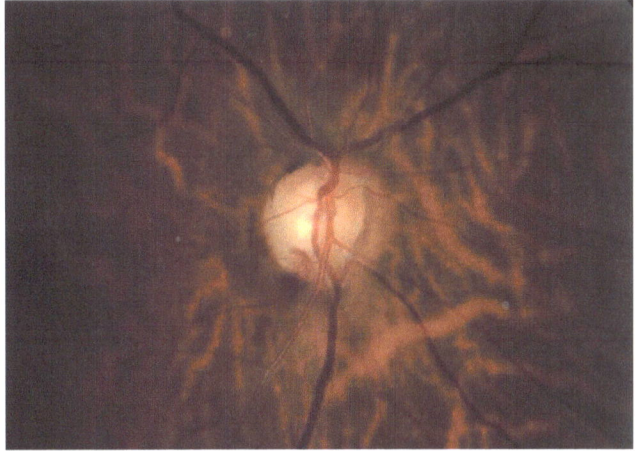

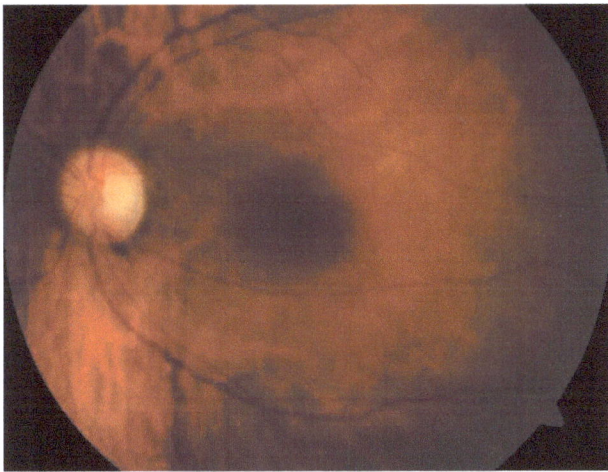

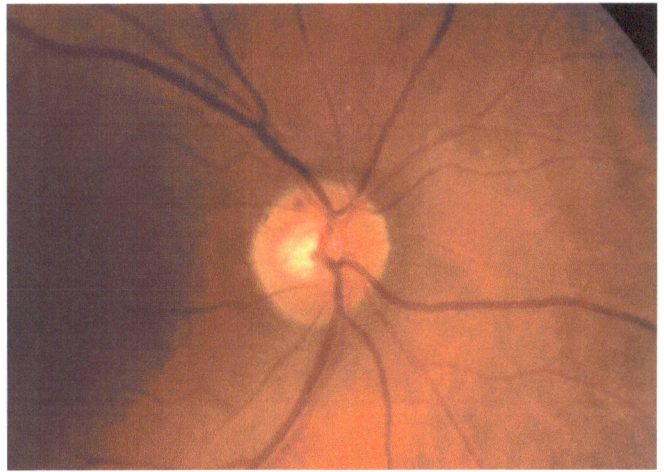

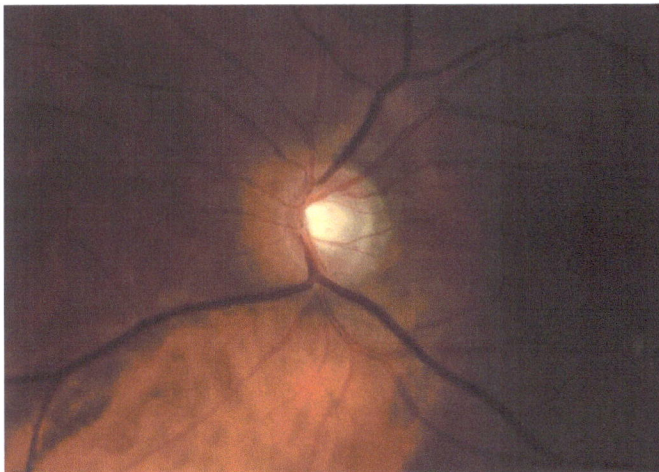

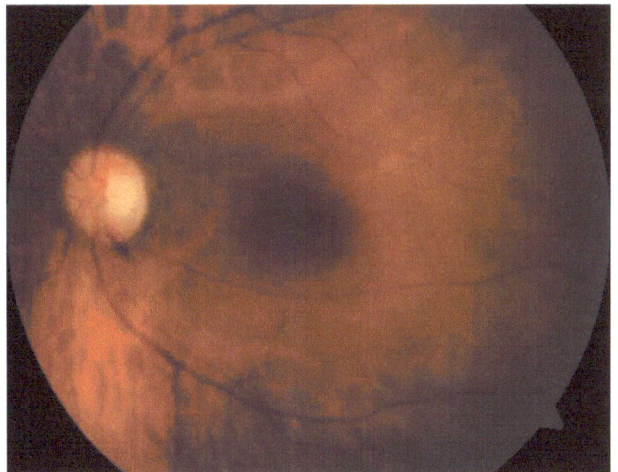

Here are photos from the same patient 1 year apart with an atypical appearance of a DH in the second photo:

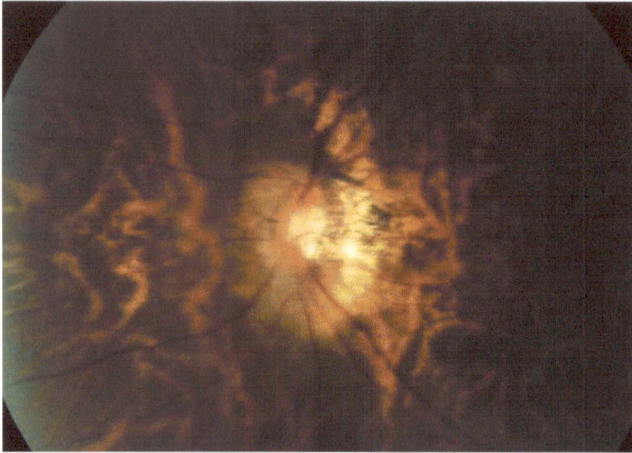

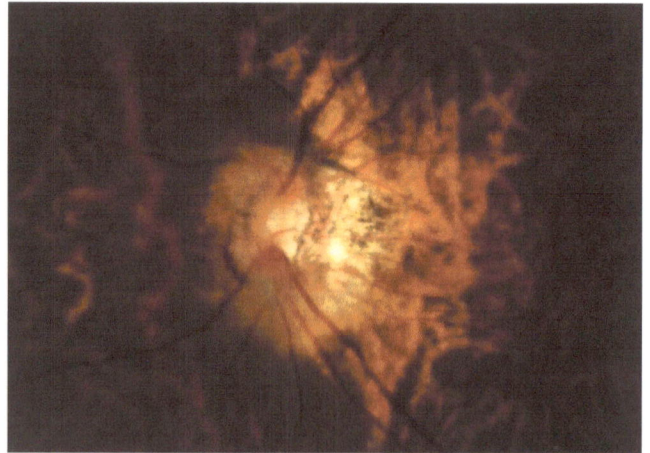

More DH in patients of mine:

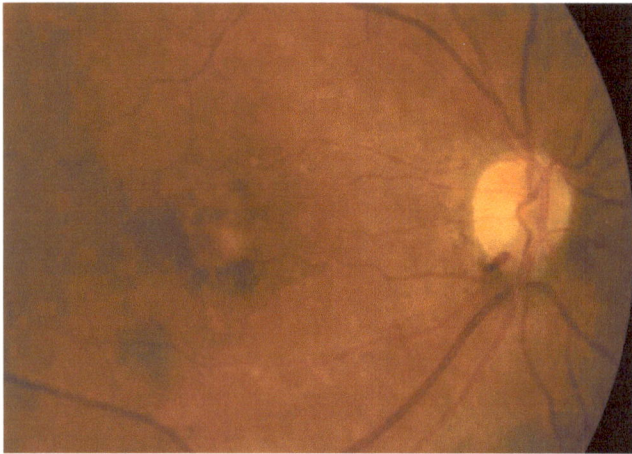

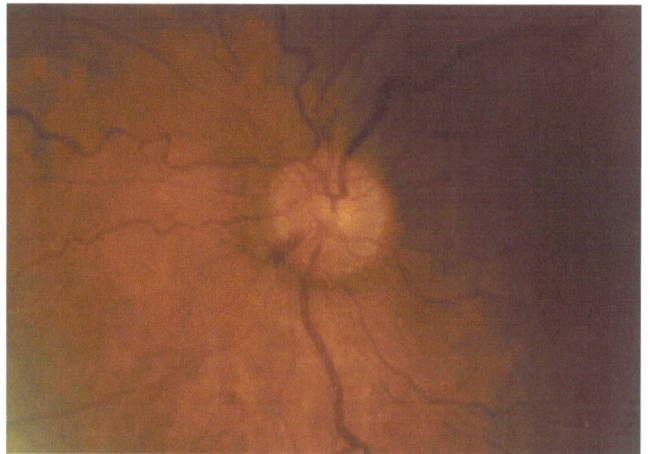

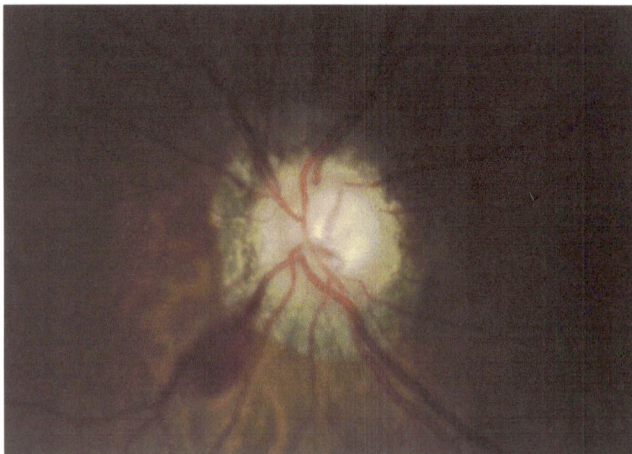

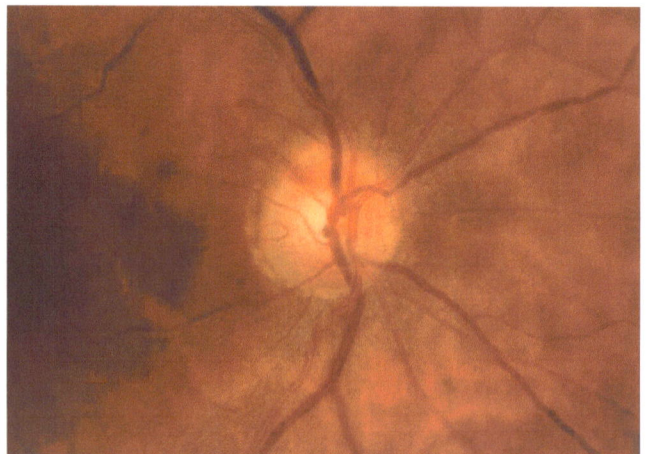

Here is a flame hemorrhage just off of the disk that I missed initially and then caught on the next years photo review on this low risk POAG suspect. This hemorrhage could be related to OAG but it also could be a diabetic hemorrhage (negative A1C in this patient):

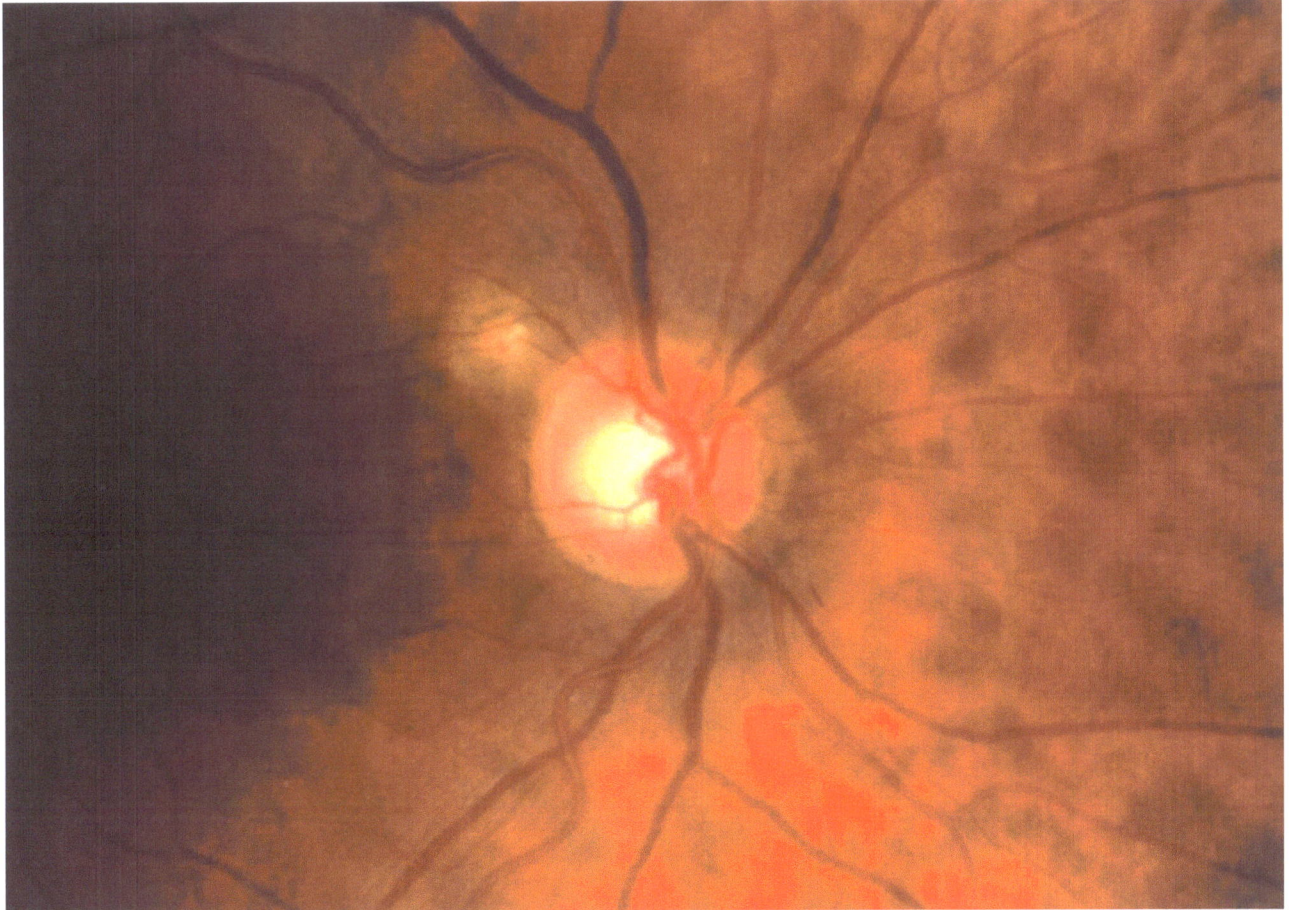

Nerve Fiber Layer Defects (NFLD): These are seen as grey wedges that start at the disk and follow the NFL arcing over the fovea. Document any NFLB defects. Disk hemorrhages may give rise to NFLD over time. Here are some examples of NFLD in my patients. The first photo in each pair shows the NFLD with a corresponding DH. The paired second photo shows the persistent NFLD without the DH a year later:

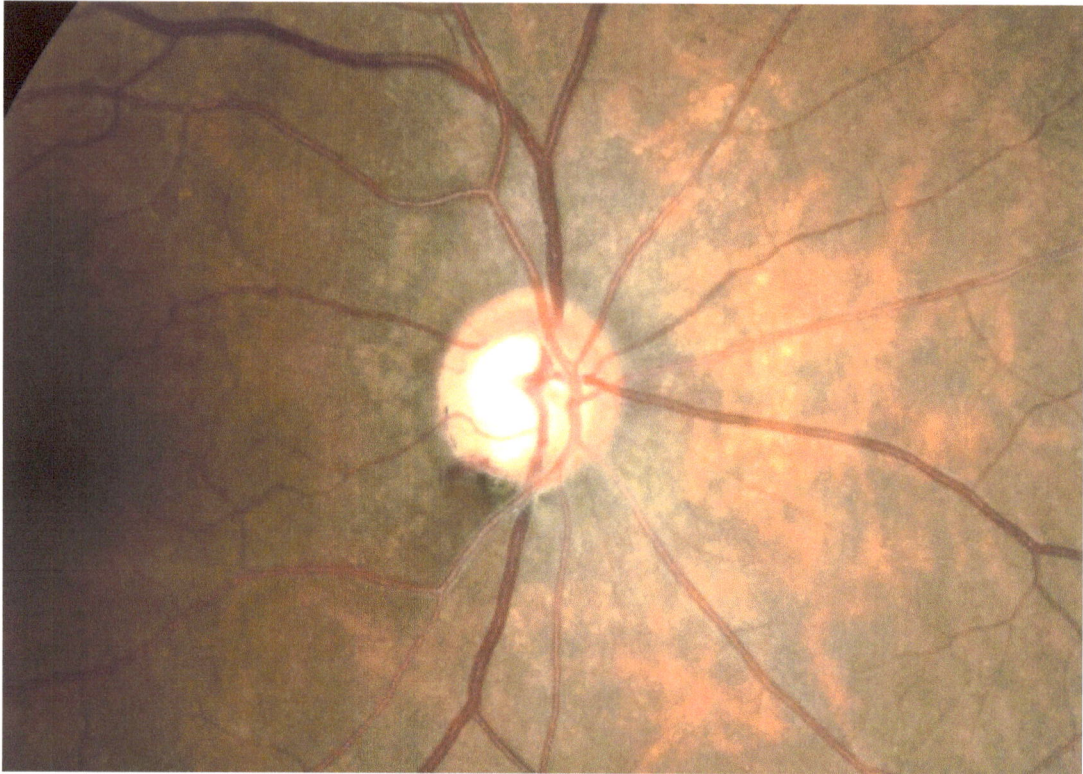

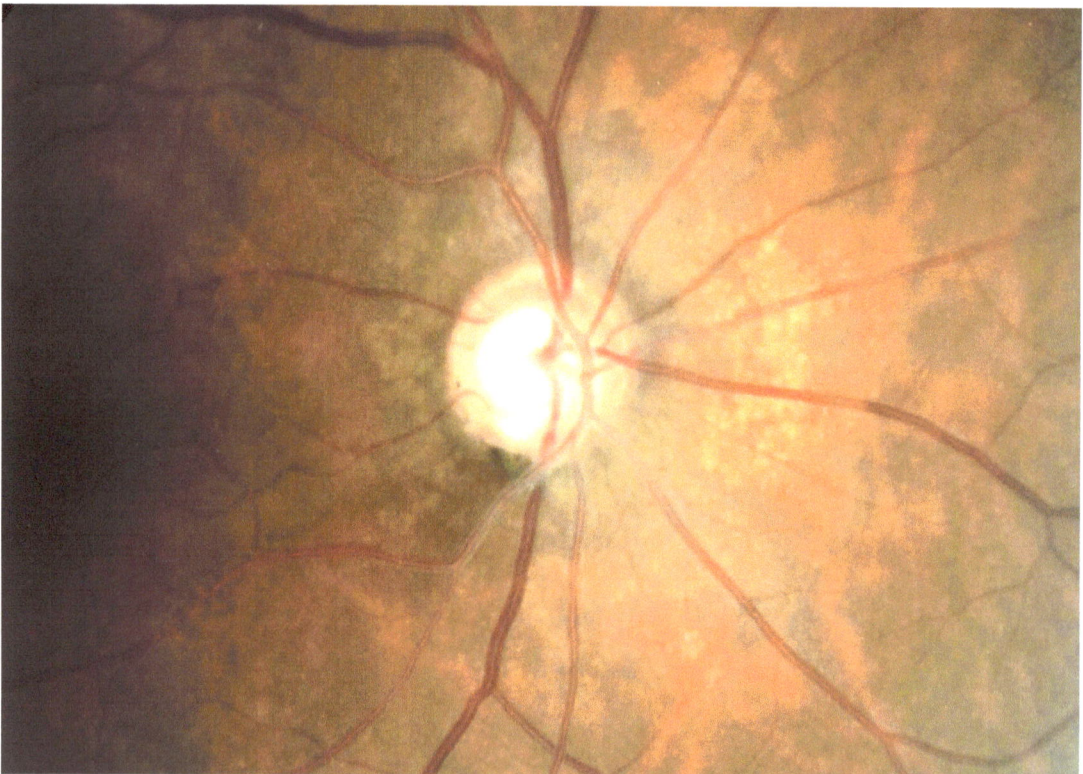

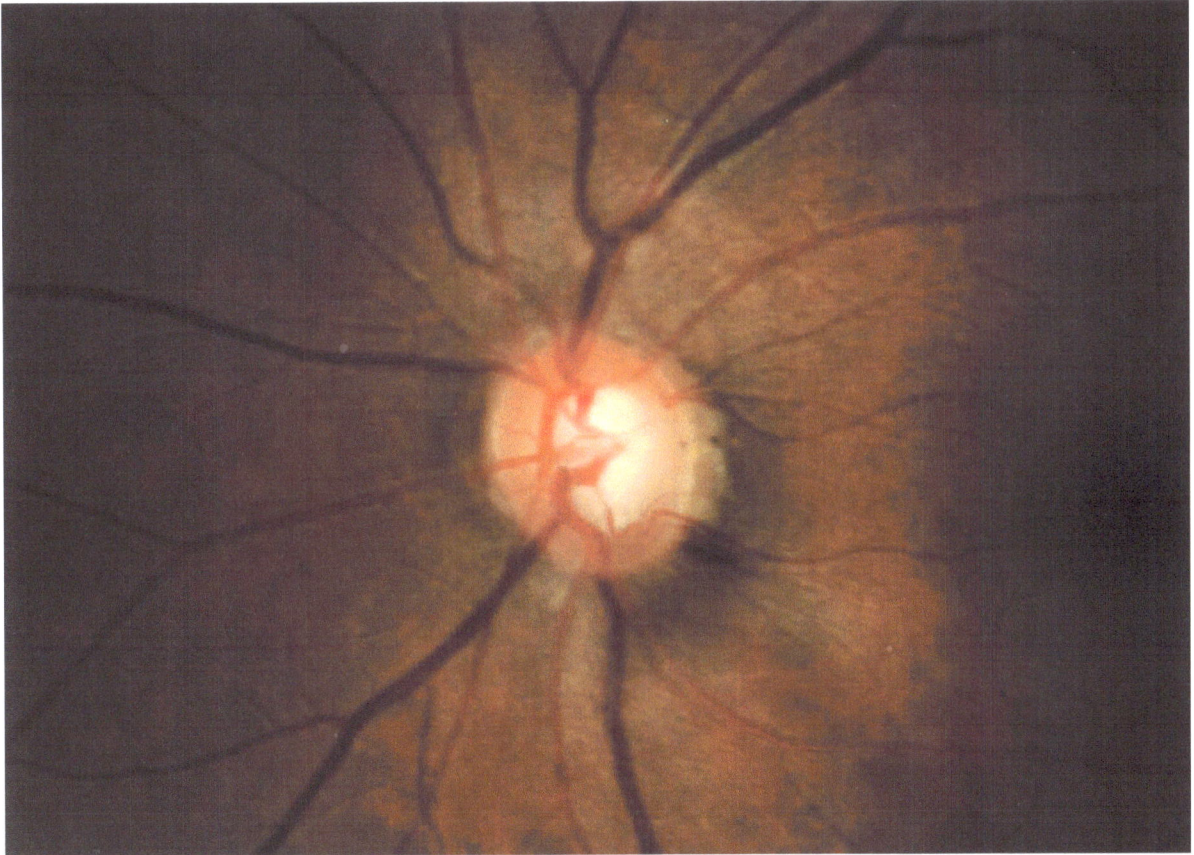

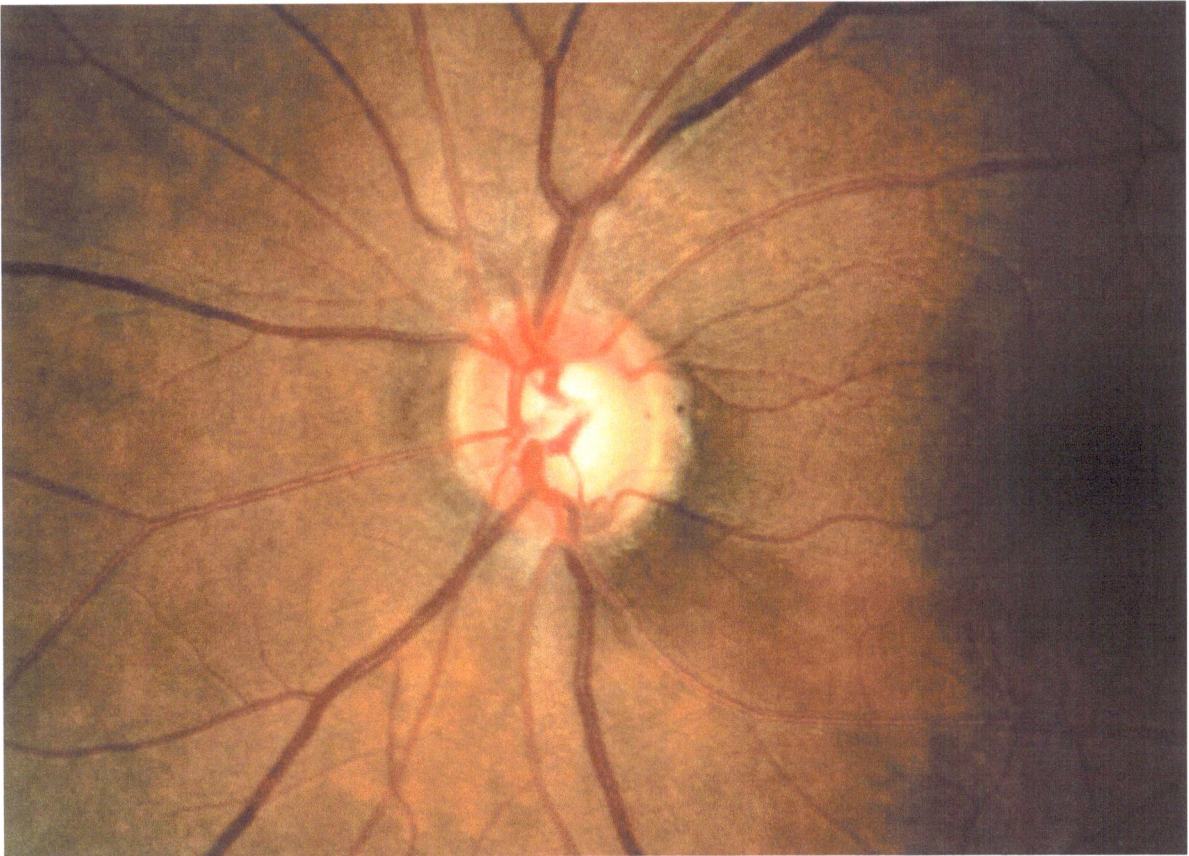

This one had the NFLD first (probably with an unrecognized DH) and then the DH that I saw:

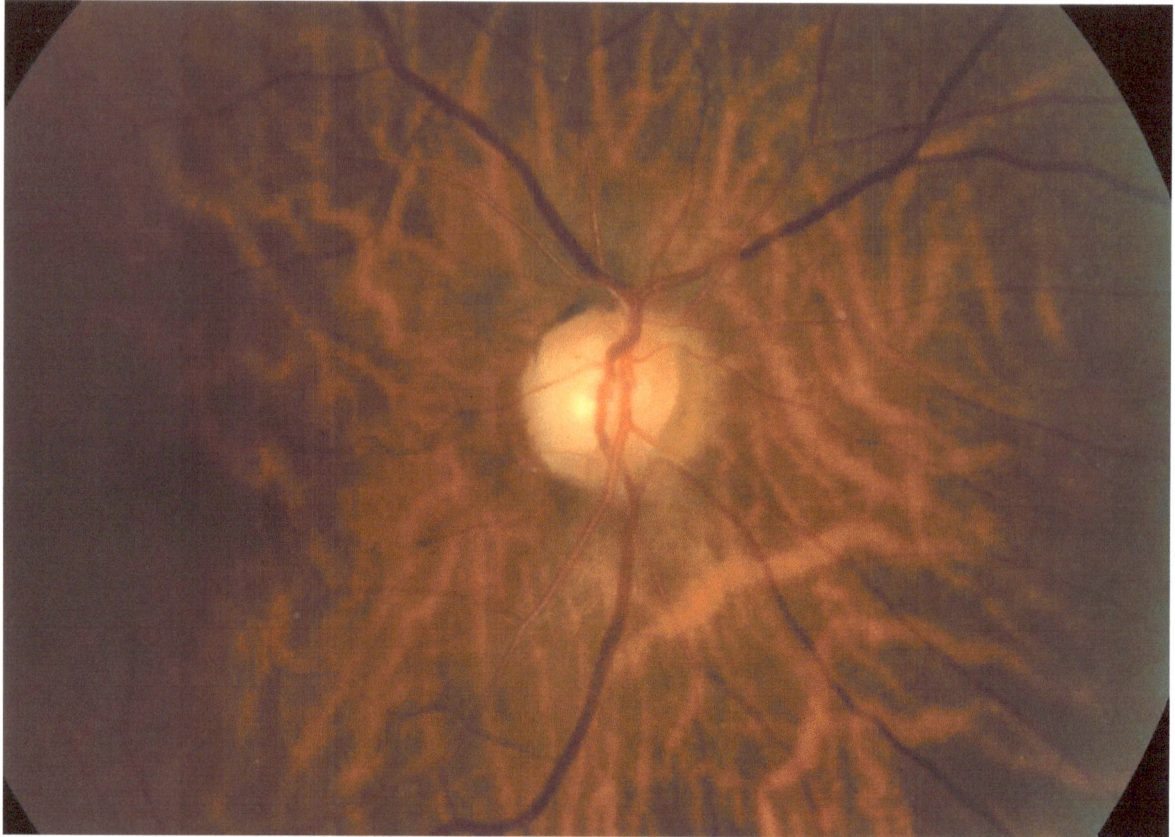

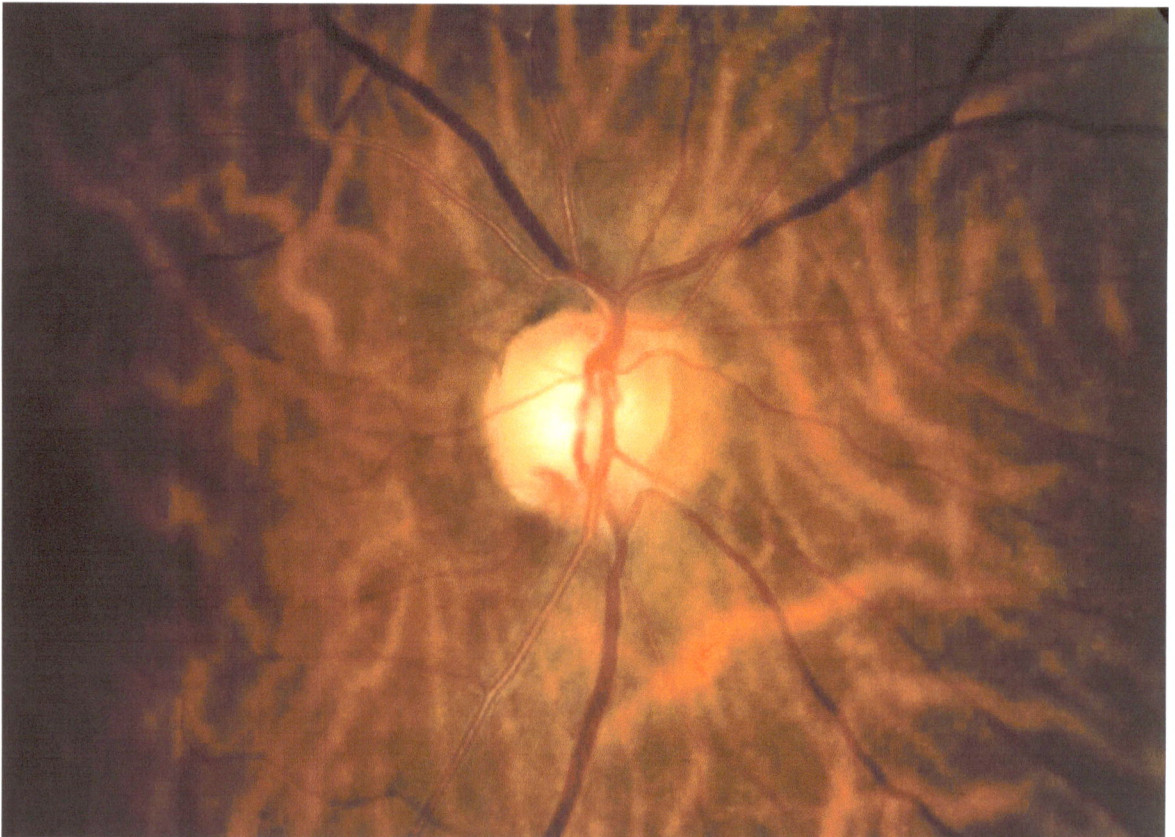

Notches are a focal thinning of the rim tissue, typically in the superior and inferior temporal areas of the rim tissue. They are also *very* concerning for glaucoma. Document any notches and correlate VF defects with their location. A superior or inferior temporal notch with a corresponding VF defect is diagnostic of glaucoma, excluding other causes such as vein occlusions.

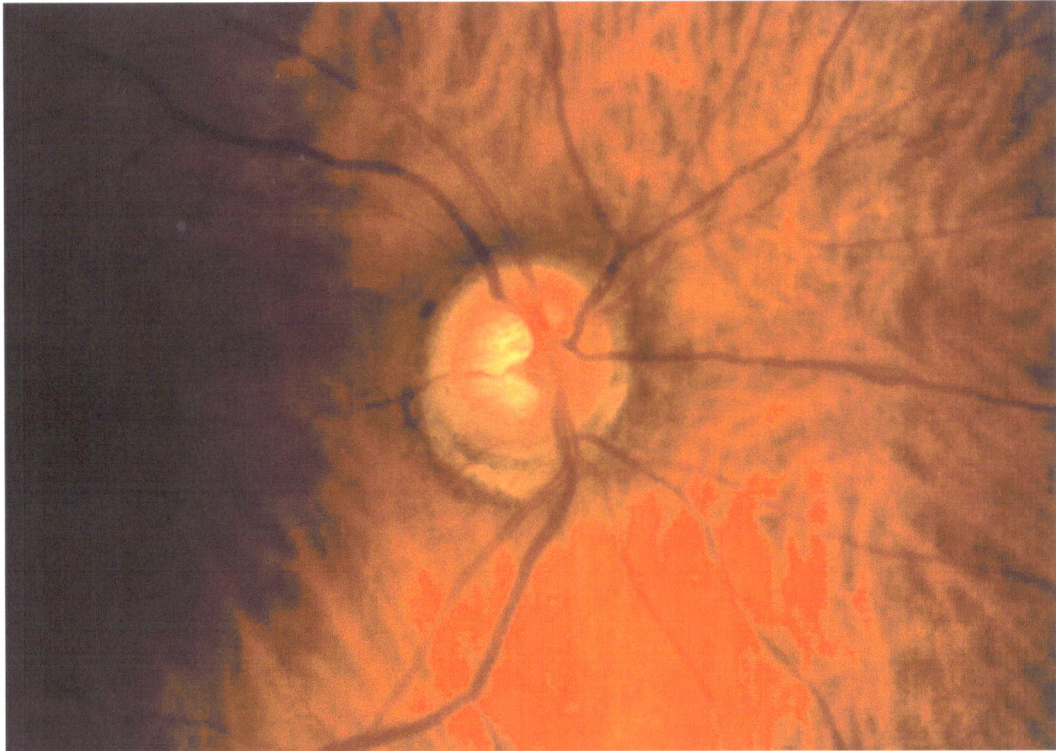

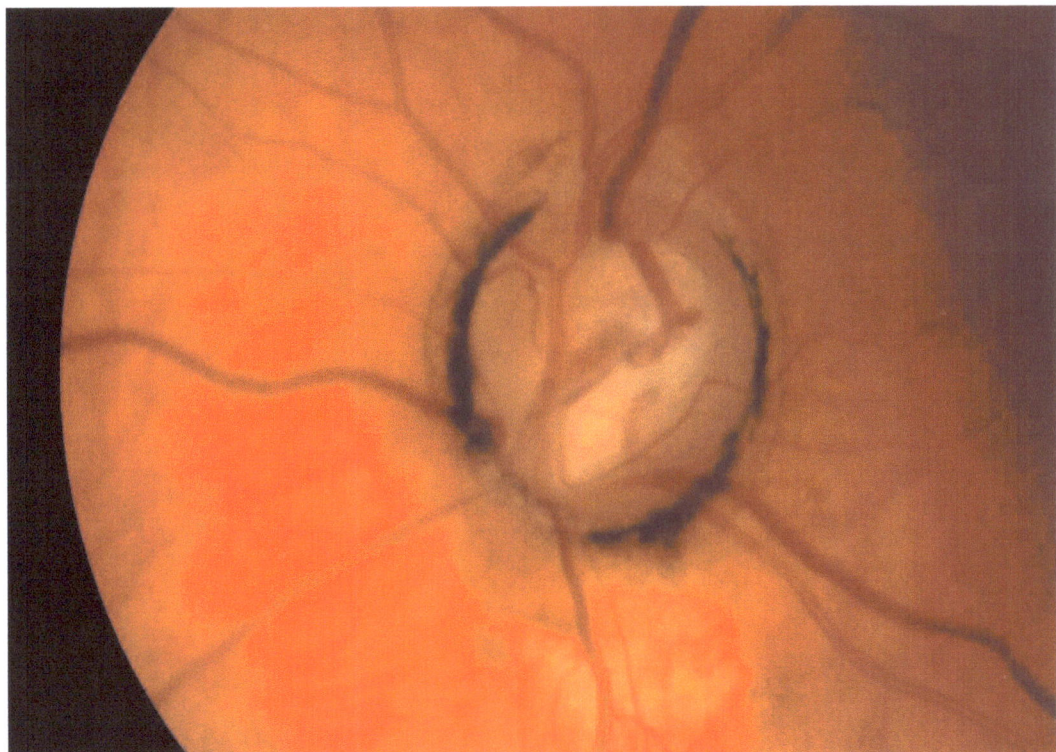

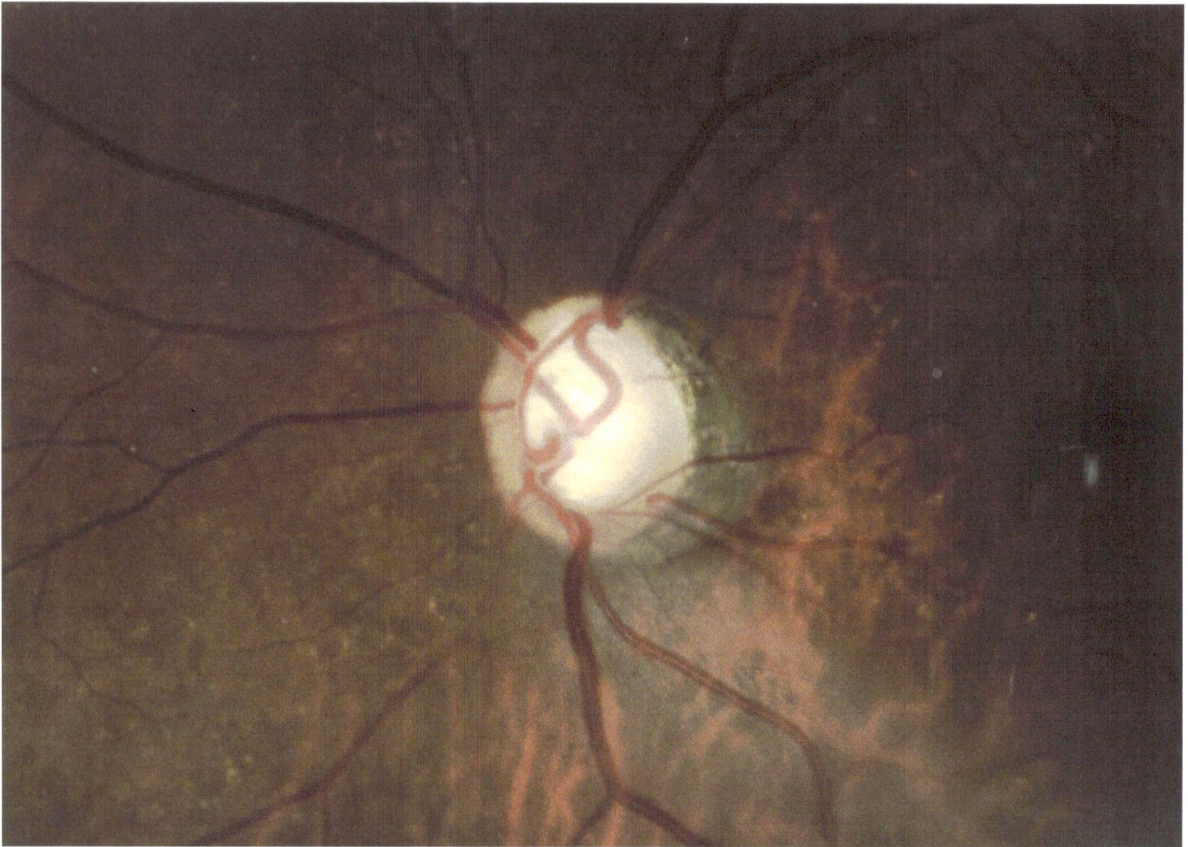

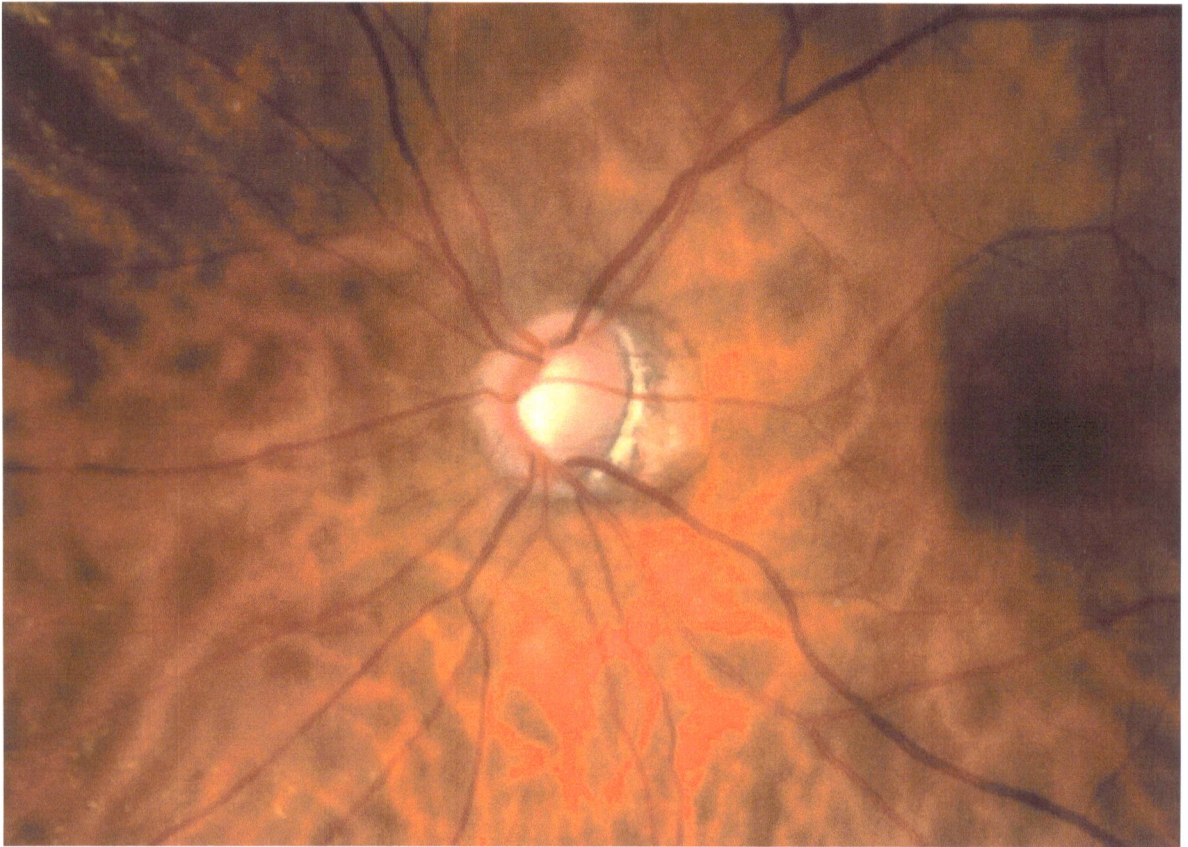

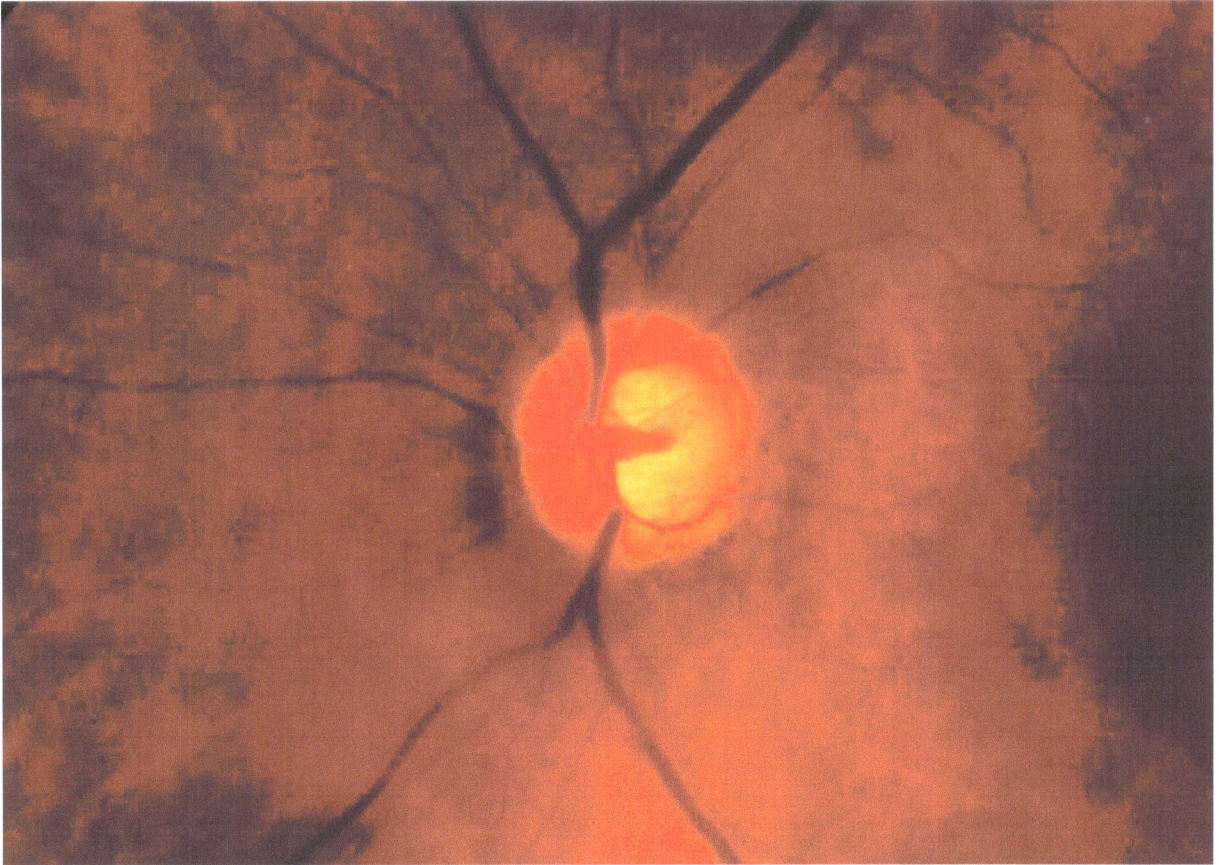

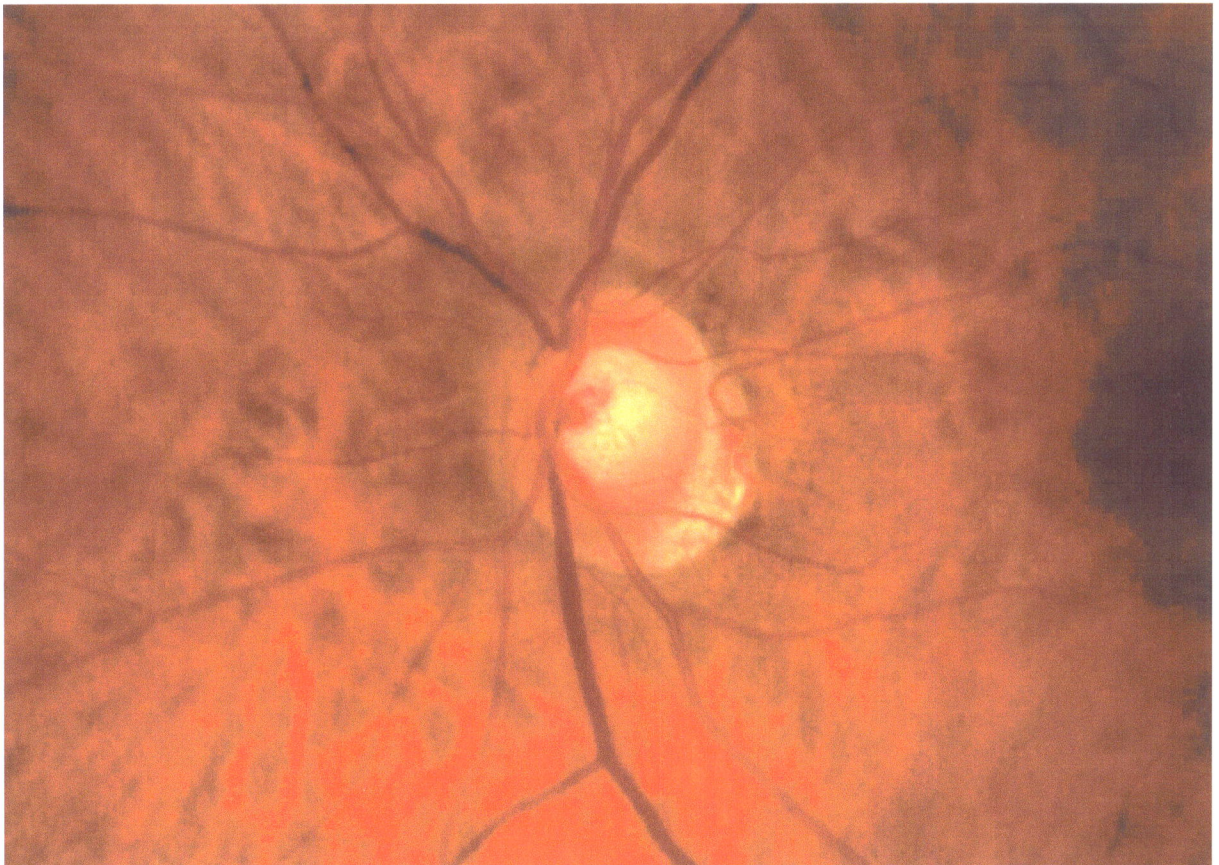

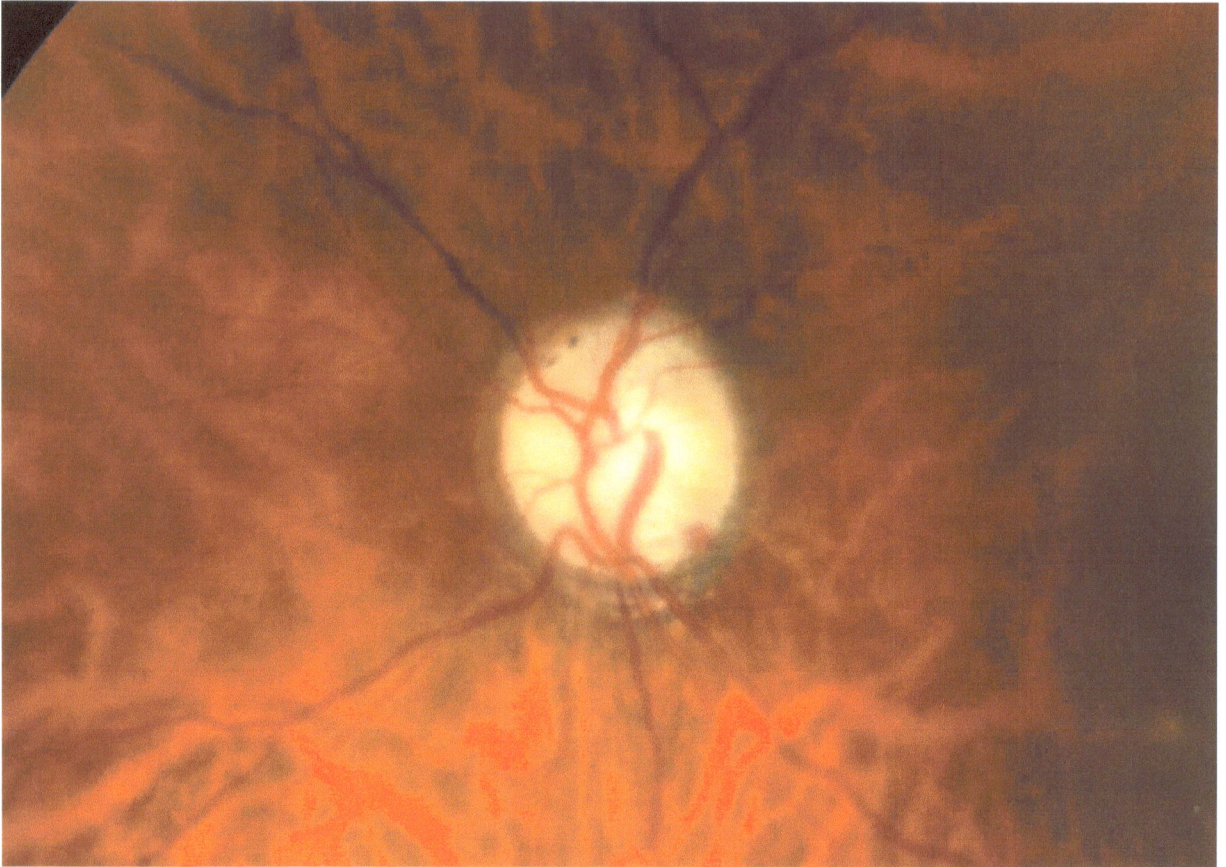
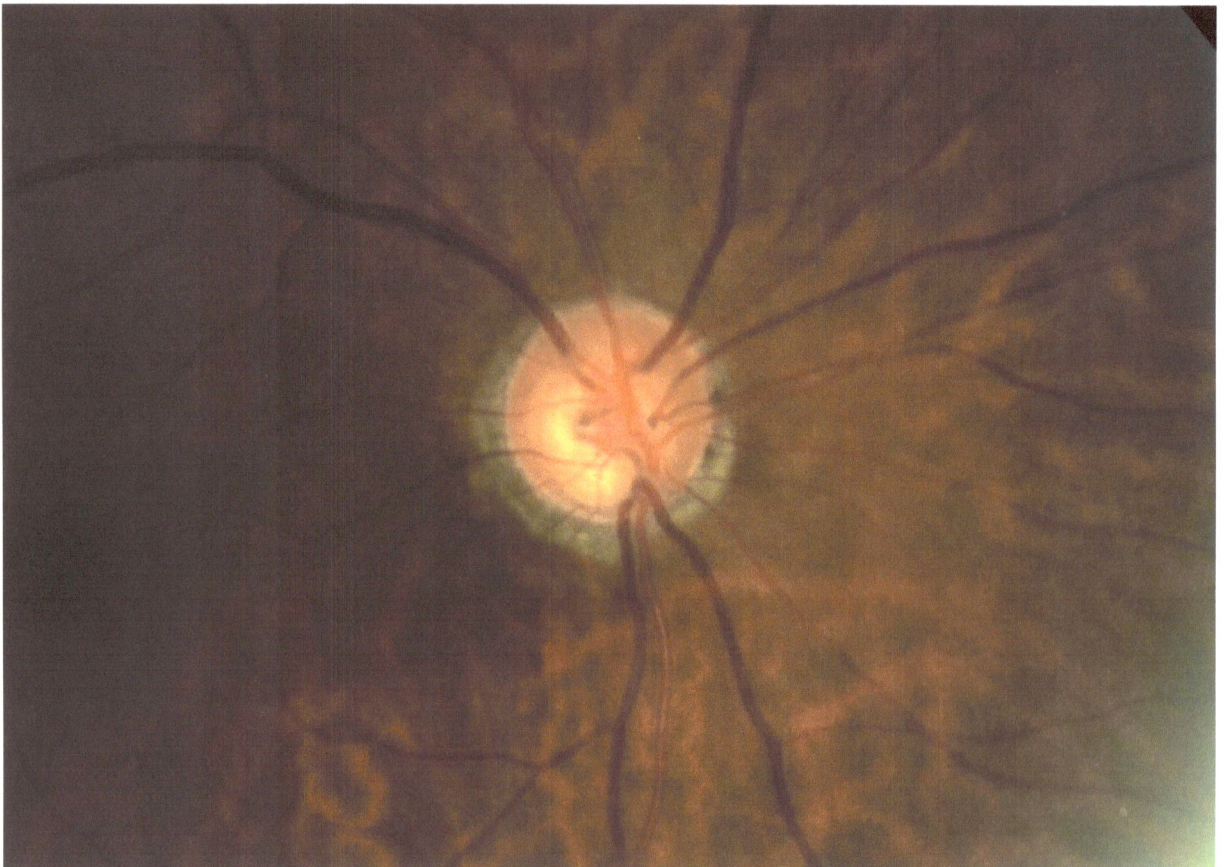

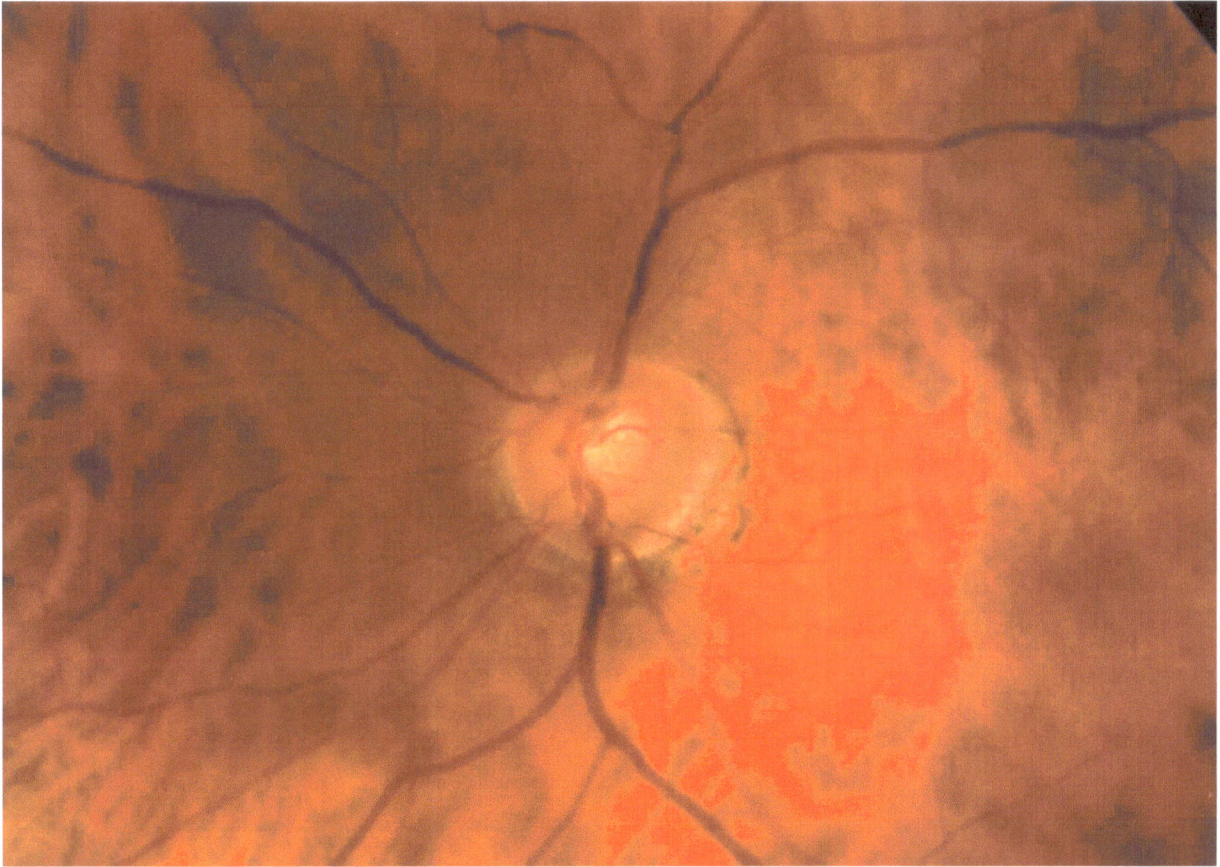
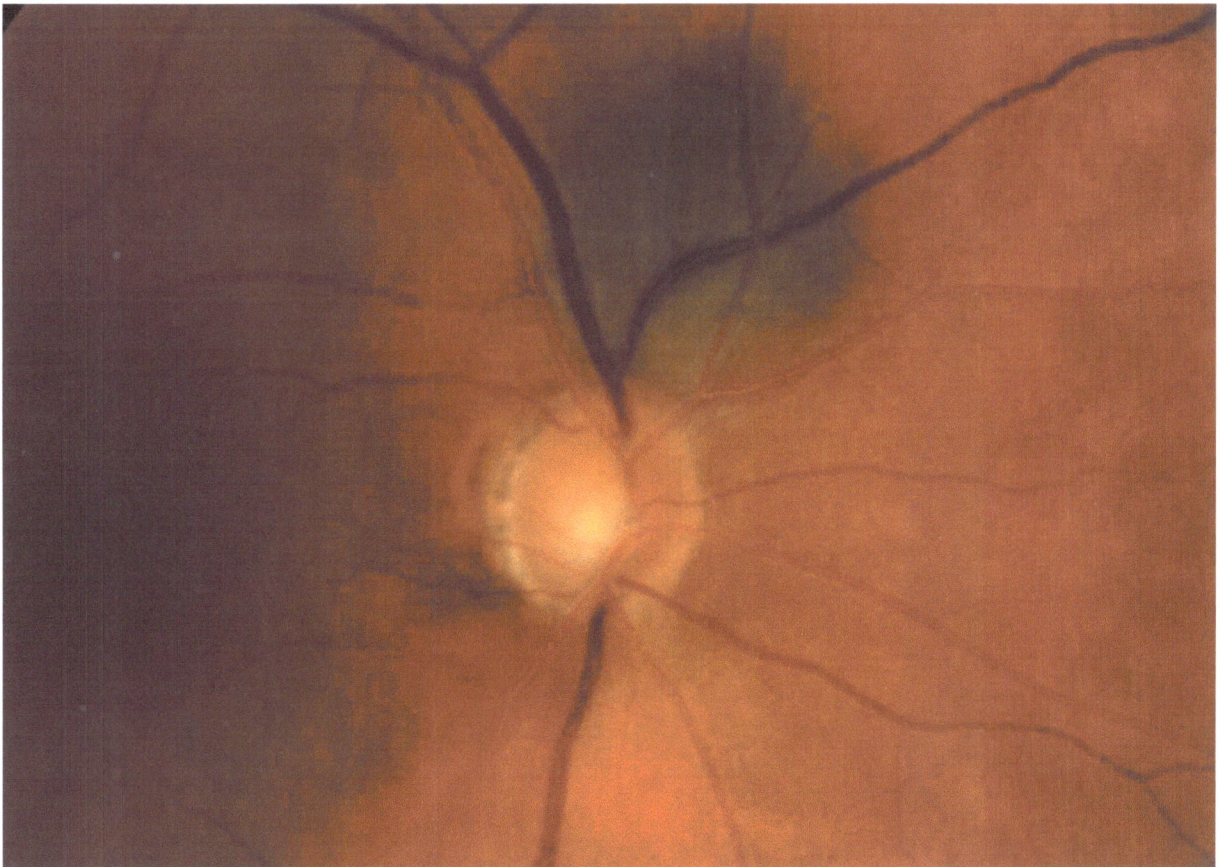

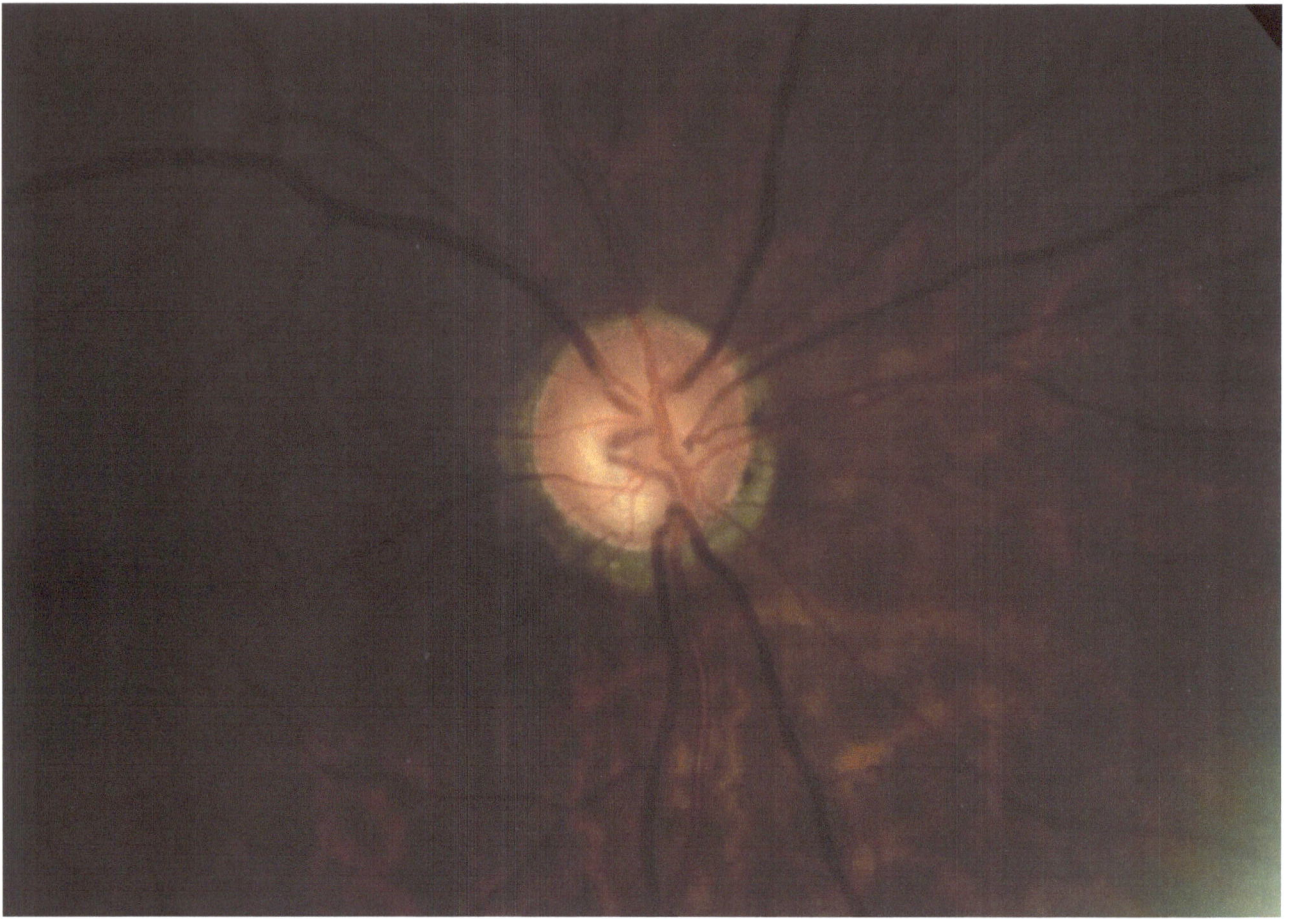

7) Asymmetry:

in IOP between the two eyes
in IOP in the same eye (which is diurnal variation, which will be covered later)
in the C/D between the two eyes

IOP asymmetry between the two eyes: The higher the difference in IOP between the two eyes, the higher the risk for glaucoma to develop in the eye with the higher IOP.[77] The following graph is from a *prevalence* study of patients initially diagnosed with glaucoma by Alice Williams[77] (A total of 652 patients were examined, half with newly diagnosed glaucoma and half without glaucoma) showing the chance of having glaucoma with varying pre-treatment IOP differences between the two eyes of the same patient. A difference of > 6mm confers a 57% chance of having glaucoma and was determined to be a "great risk" for the development of glaucoma.[77] This was a prevalence study, IOP was measured at only one point in time for each patient, so the time frame for OAG development was not able to be calculated.

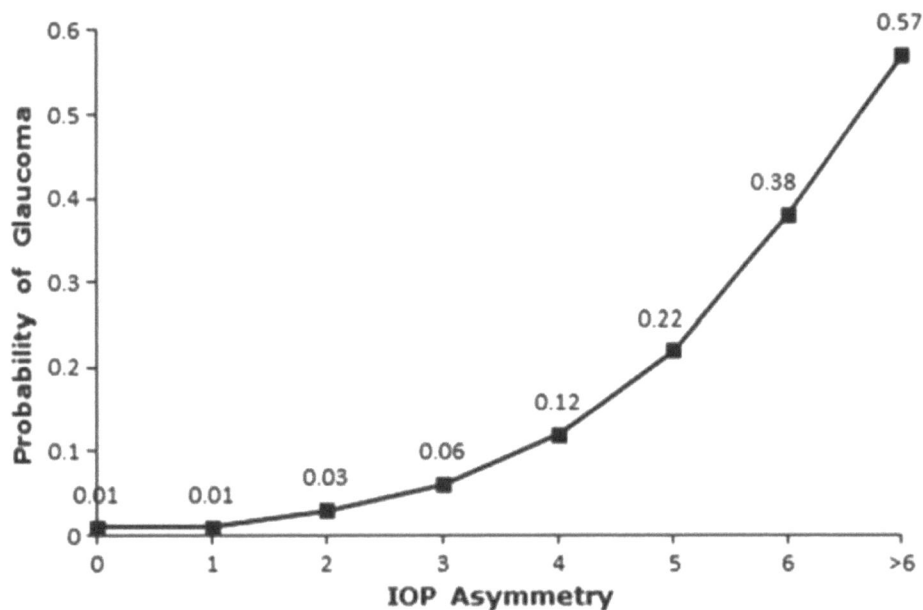

FIGURE 3. Probability of having glaucoma as a function of intraocular pressure (IOP) asymmetry between the fellow eyes. Probability of having glaucoma was calculated with a logistic regression analysis, assuming an overall worldwide prevalence of 2%.

The OHTS data was examined to see if patients with IOP differences between the two eyes increased the risk for the development of glaucoma.[78] 3,236 patients were screened for a median follow up of 6.11 years, n=146 for the number of patients who developed glaucoma during the follow up period. They concluded that "asymmetric IOP appears to be an important predictor of POAG beyond the predictive value of the IOP value itself" with p < 0.0004.[78] A difference of 3 mmHg or more was used as the cut-off in this study. This study also showed that of the 146 eyes that developed glaucoma, 104 (or 71%) had glaucoma in only one eye. So, as a side topic, **POAG tends to develop monocularly, at least initially**.

C/D asymmetry: Less than 1% of people have a C/D difference of more than 0.2.[35] Various studies have not shown a clear significance of C/D asymmetry between the two eyes of patients who develop glaucoma. The Wills Eye Manual places the cut-off of significant C/D asymmetry between the two eyes at > 0.2.[79] The European Glaucoma Prevention Study[80] followed 1,077 people for a median of 59 months and showed a multivariate increased risk of 1.4X for the development of POAG in patients with a C/D asymmetry of $\geq$ 0.4. A C/D asymmetry of 0.3 was used as a 99th percentile of normal in a number of studies[73] with variable significance.

The Blue Mountains Eye Study in Australia examined 3,126 people, or 6,252 eyes, and found a statistically significant difference in developing OAG between patients with C/Ds < 0.3 and $\geq$ 0.3. The $\geq$ 0.3 group had (prevalence data) significantly more OAG than the other group (P< 0.0001).[81] The authors do state though that "the highest positive predictive factor for any asymmetric variable used as a test for OAG was only 17%, and this occurred at a cup-to-disk ratio asymmetry of 0.3 or more, which was found in only 10% of patients with OAG and 1% of patients without OAG."[81] So, they don't recommend diagnosing POAG on a C/D ratio asymmetry of $\geq$ 0.3 but it certainly is another red flag. I use a difference of 0.3 or higher as an increased risk factor for POAG in my clinic.

8) ? Race. The prevalence of POAG is up to 4 times higher in African-Americans than Caucasions.[82] This has led to the theory that African-Americans are at a higher risk of developing glaucoma. The OHTS researchers found that the African-American participants had corneas 23.5 um thinner than non-Africans and, combined with the larger optic nerve sheaths and hence C/Ds found in Africans, eliminated race as a significant risk factor for the development of glaucoma in the multivariate analysis.[70] In support of this, the basic clinical science course glaucoma volume says: "in OHTS, African-American patients were 59% more likely than Caucasions to develop glaucoma in a univariate analysis, but this relationship was not present after corneal thickness and baseline vertical cup-to-disk ratio were factored into the multivariate analysis (African-American patients had lower corneal thickness and larger vertical cup-to-disk ratios on average)"[87]

There are a number of other minor risk factors mentioned in other studies, but none that have been shown to be consistently significant. Interestingly, diabetes showed a protective relationship to the development of glaucoma in the OHTS, i.e. having diabetes mellitus reduced your chances of developing glaucoma.[70]

The more risk factors you have, the higher your risk for developing glaucoma. This includes abnormal test results as well. "Most patients with POAG will retain useful vision for their entire lives. The patients who are at greatest risk of blindness have visual field loss at the time of diagnosis of glaucoma."[83]

A number of risk calculators have been developed to stratify patients risk for developing glaucoma. I don't use them. The OHTS point system[84] does not include disk hemorrhages, family history, the presence of NFLD or of asymmetry in the C/D or IOP either in the same eye or between the eyes. I add up the risk factors that a particular patient has in my mind, roughly in the order that I have presented them to you above, and arrive at a conclusion of treatment or no treatment at that particular point in time. The risk of the patient not returning for follow up is also another wild card that you must take into account in your decision of when to treat. This is the art part of the art and science of medicine. When you schedule the next follow up is also a clinical judgment based on the risk of developing glaucoma for that particular patient.

Angle Closure Glaucoma (Acg)

Angle closure glaucoma (ACG):

ACG is defined as the iris closing off the aqueous from draining through the TM. ACG is broken down into Primary ACG (PACG), which is defined as the presence of iris bombe pushing the iris into the TM, and secondary ACG (SACG), which has a different identifiable cause of the iris being pushed (or drawn) against the TM. ACG is much rarer in Europeans than POAG. ACG has a prevalence of 0.1% in Caucasions.[45] The prevalence of ACG is 30 times higher in the Inuit population.[45] Asian populations have a prevalence somewhere in between whites and the Inuit populations.[45] Family history is important, as one third to one half of first degree relatives can also have occludable angles.[46] PACG is a leading cause of bilateral blindness worldwide.[38] IOPs in ACG can be high enough to cause severe and permanent visual loss within several hours.[47]

PACG Iris bombe: The differentiation between PACG and SACG is important because Iris bombe in PACG is usually *cured* with a laser peripheral iridotomy (LPI). Iris bombe is caused by the aqueous not being able to flow through the pupil due to the iris-lens being in opposition to each other at the pupil edge. This bows the iris forward into the TM, closing it off. It is most commonly found in hyperopes over the age of 55.[45] This block to aqueous flow at the iris/lens interface is maximal in the mid dilated position,[52] just as dilation is wearing off and the patient is arriving home from your dilated fundus exam! This is also why most acute ACG patients present with the iris in the mid-dilated position. Dynamic gonioscopy can be used to break an acute attack of angle closure. If an angle is still too narrow after an LPI has been performed, think of plateau Iris configuration. Chronic uveitis patients can have extensive posterior synechia also resulting in ACG with iris bombe. The treatment is still an LPI. Here is a photo showing the development of angle closure due to iris bombe (© AAO 2014):

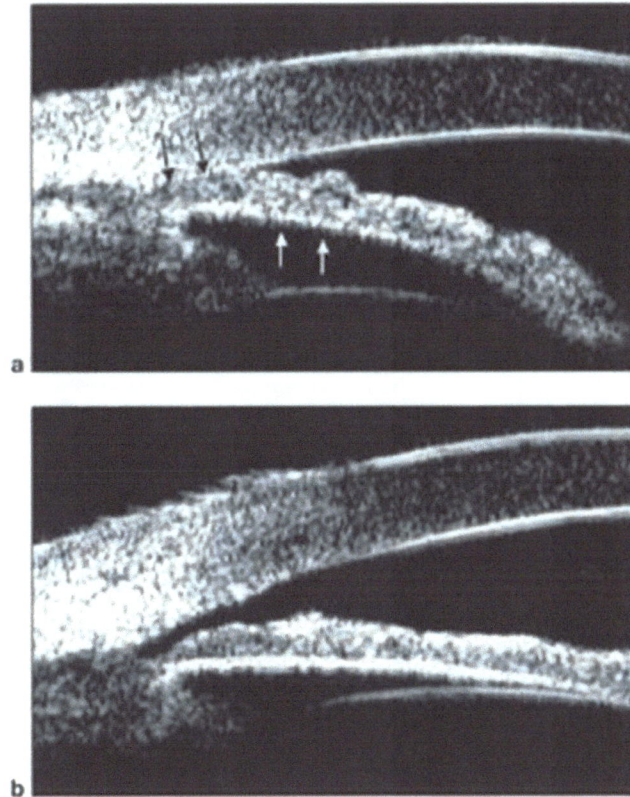

Here is a slit lamp photo of iris bombe (© AAO 2014):

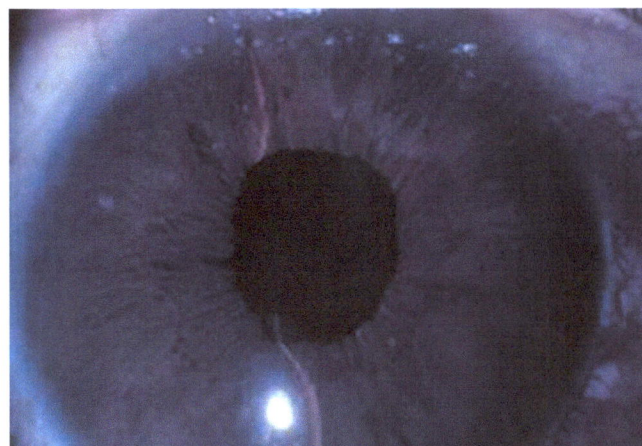

Here is a photo of an eye in angle closure. Notice the mid dilated pupil position and the mildly hazy cornea (© AAO 2014):

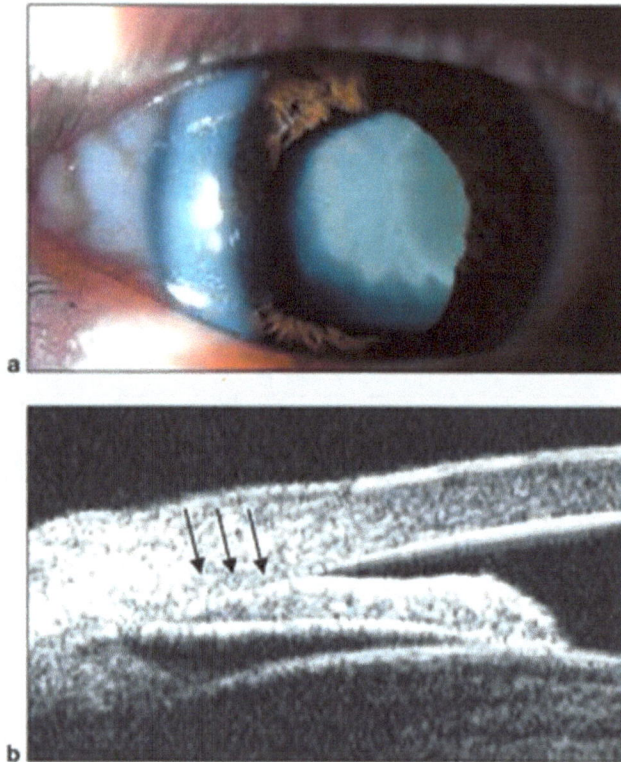

Here is a diagram of the aqueous outflow in an angle closure glaucoma patient showing the importance of pupillary block in primary ACG (© AAO 2014):

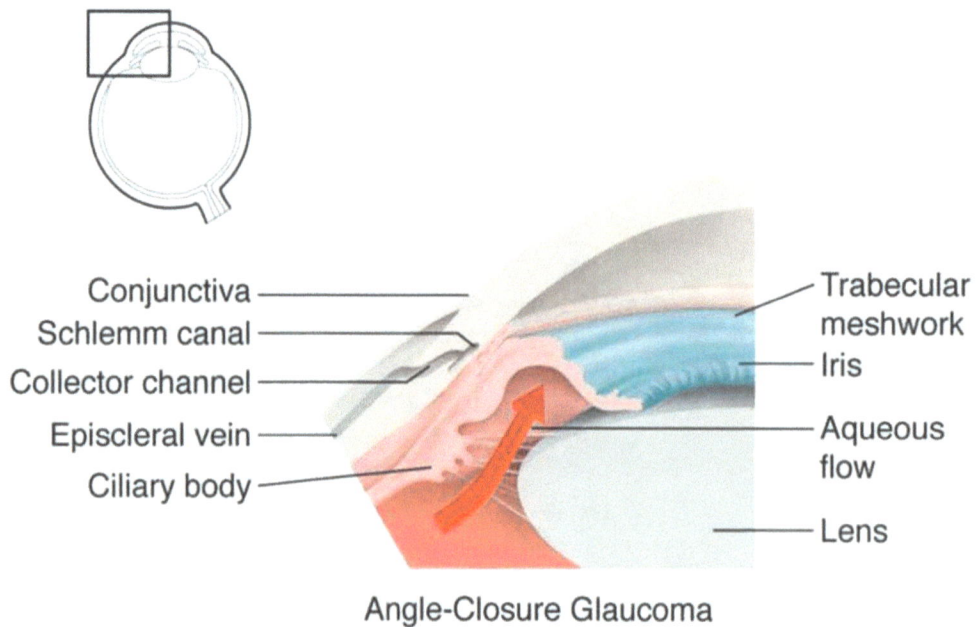

Angle-Closure Glaucoma

SACG: Examples include

-Phacomorphic, where the lens pushes into the iris and the TM. This is seen in Marfan's, homocystinuria, and exfoliation.

-Neovascular Glaucoma (NVG): Causes include vein occlusions (CRVO>HRVO>BRVO), PDR, CRAO and least commonly, carotid occlusive disease. Tiny neovascular strands reach from the iris base over to the TM forming a reddish membrane along the TM, which contracts and pulls the iris over to the TM, forming PAS and closing the TM off permanently. This reddish membrane along the TM is what you are looking for as the individual tentacles can be hard to see since they are so small. Remember, 10% of CRVO patients will have angle neovascularization (ANV) and not neovascularization at the pupil of the iris (NVI),[148] so do gonioscopy on all CRVO patients Q month for the first 6 months. Urgent PRP needs to be done on all CRVO patients the first time you see any ANV or NVI as it will eventually go all the way around the TM causing complete angle closure with eventual blindness. Here is a photo showing NVI at the pupil border (© AAO 2014):

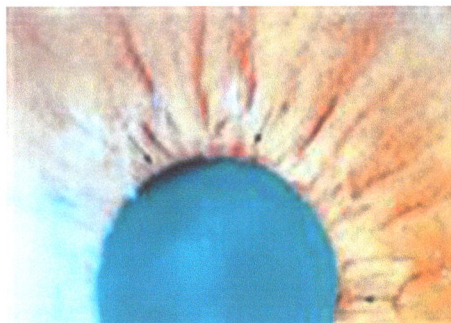

-Plateau iris is present when the iris is still blocking the TM after an LPI has been performed; this is due to an anterior insertion of the iris into the corneal-iris angle. The eye has a normal anterior chamber depth centrally and mid-peripherally but in the far periphery, the iris is up against the TM despite an LPI. An Iridoplasty is necessary and I have performed this in about 8 eyes over the last 12 years. This is one reason why gonio should be done after an LPI is done. This is a diagram showing iris plateau (© AAO 2014). You can see that the narrow angle would probably be missed unless gonioscopy is done.

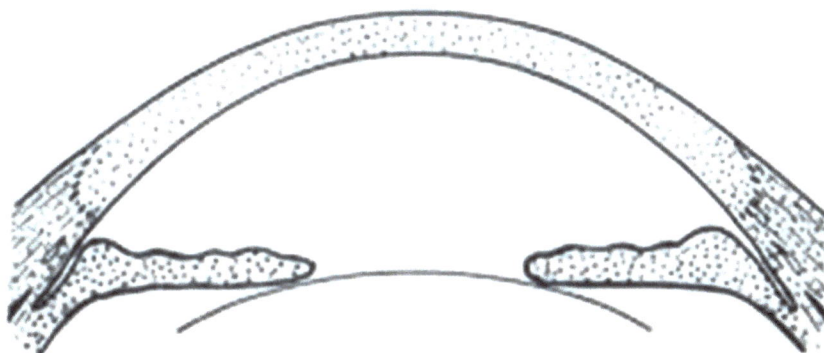

-ICE syndrome (Iris-nevus or Cogan-Reese, Chandler and Essential iris atrophy): An abnormal corneal endothelial layer grows from the TM to the iris and contracts, pulling the iris over to the TM, causing PAS and ACG. This is similar to NVG. The glaucoma is almost always unilateral, has an onset at age 20-50, has an irregular pupil,[49] and has PAS above Schwalbe's line. About 50% of ICE patients get SACG. ICE is usually not familial. It is rare; I have never seen a case. You treat it like you do POAG. Here are two photos of ICE showing the iris atrophy and the pulling of the iris over the TM: © AAO 2015.

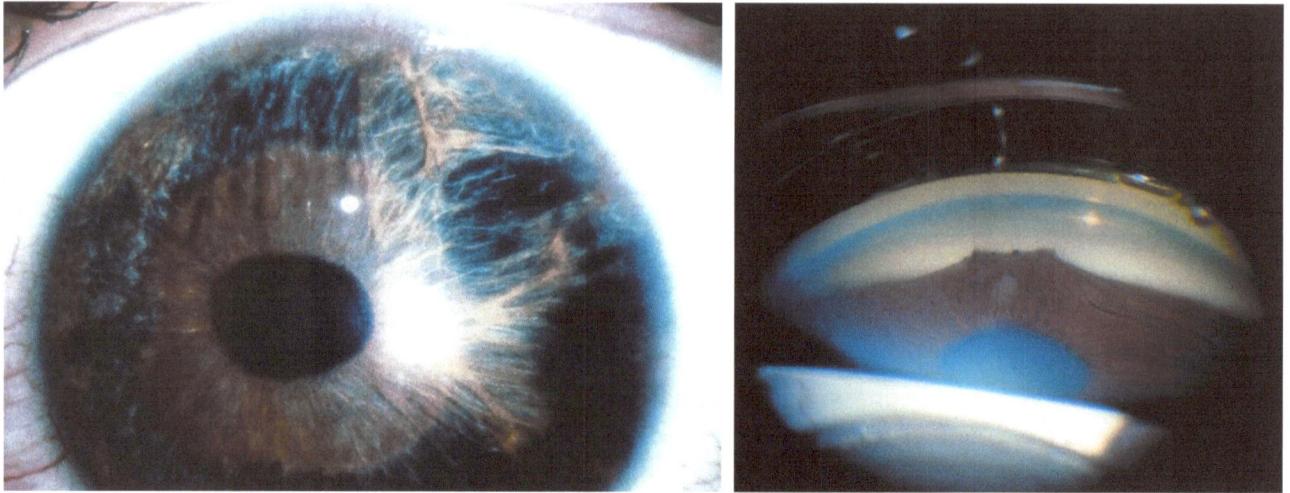

-Ciliary block (malignant) glaucoma: Rare, results from the ciliary body rotating forward and the aqueous shooting into the vitreous, pushing the vitreous/lens/iris forward into the AC.[50] This typically presents after intra-ocular surgery as a flat anterior chamber both centrally and peripherally with high IOP and without iris bombe. It is difficult to distinguish ciliary block glaucoma from a choroidal effusion or supra choroidal hemorrhage (which are both diagnosed on a B-scan), as they all present with a flat anterior chamber. A trab or CE postoperatively can also have a flat AC but the excessive aqueous leakage seen on a seidel test and the low IOP help differentiate this from the above 3 entities. If the IOP is high, place the patient on Xalatan and refer them back to the surgeon for an LPI or a YAG disruption of the vitreal face. The fact that ciliary block occurs is one reason why I don't recommend co-management for cataract extraction or trabeculectomy. Here is a photo of a patient with malignant glaucoma. Notice the anterior chamber is closed peripherally and centrally (© AAO 2014):

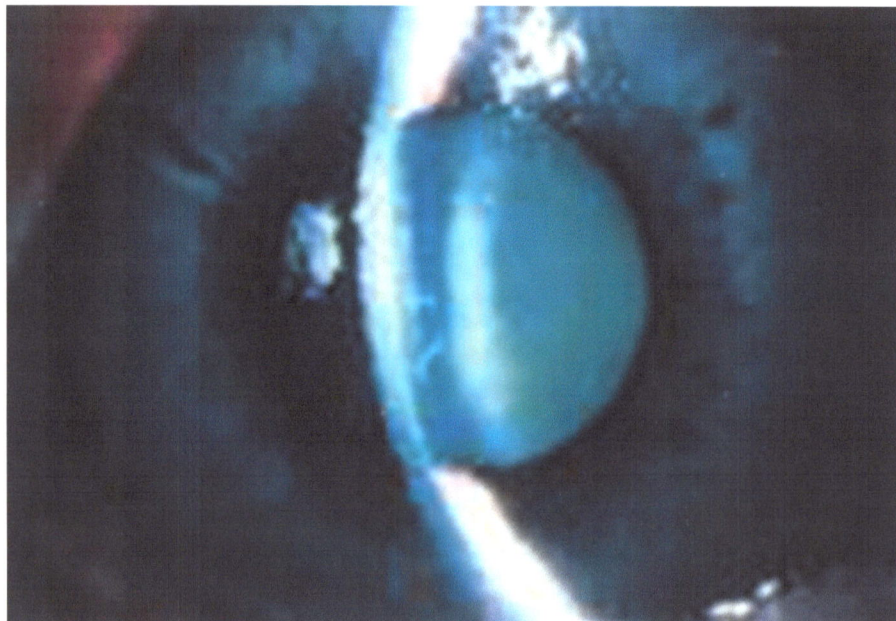

-Rare etiologies: Masses pushing the lens forward can also cause ACG, such as in a non-rhegmatogenous retinal detachment (RD), bleeding in the retina, bleeding in the choroid, the UGH triad from an anterior chamber IOL, epithelial ingrowth through a surgical incision (I saw one case years ago and you won't likely see one at all), scleral buckle with anterior rotation of the lens and air/silicone injections. All of these are reasons not to co-manage surgical patients as they typically occur within the global period, the first 90 days after surgery.

-PO Topiramate (Topamax) is used in seizures, headache, and depression treatment. It can cause a severe choroidal effusion with an anterior rotation of the lens leading to bilateral ACG and myopia, usually within the first 2 weeks of starting the medication.[51] You treat this by stopping the Topiramate and starting Xalatan/Cosopt and Atropine. The ACG usually resolves within 2 days, the myopia within 2 weeks. An LPI is not indicated since there is no iris bombe. Reversible VF defects have been recently reported. A 24-2 HVF Q 6 months should probably be done while on Topamax.

Patients that have both conditions, angle closure with a patent LPI, and persistent high IOP (in the absence of plateau iris) have *combined mechanism* glaucoma. You treat it as you do POAG.

WHAT ARE THE RISK FACTORS FOR ACG?

The major risk factors for **P**ACG include:

-Hyperopes above the age of 50, with an LPI or CE indicated if PAS is present on gonio.

The major risk factors for **S**ACG include:

-**Central Retinal Vein Occlusion (CRVO),** both ischemic and non-ischemic can lead to ACG: **Both eyes after a CRVO need to be treated for glaucoma**. A retrospective study examined 492,488 patients for the onset of CRVO and BRVO, (mean follow up was about 2 years).[184,185] On a multivariate analysis, the **risk factors for CRVO** were, in descending order:

> hypercoagulable state
> hypertension
> POAG
> ARMD
> stroke but *not diabetes* [184,185,85]

In the same patient population, the risk factors for BRVO were in descending order:

> hypertension
> stroke, but *not POAG or diabetes*

A CRVO has the classic "blood and thunder" appearance with the hemorrhages going to the periphery of the retina. This is in contrast to diabetic retinopathy which typically does not have hemorrhages to the periphery. (OIS also has hemorrhages in the periphery but has mild bleeding and non-tortuous veins and needs a carotid duplex). CRVOs come in two kinds, really just two kinds of severity.

-Ischemic, the worst ones with BCVA < 20/200, have an APD and are at high risk for NVG. These produce the 90 day glaucoma situations where NVG develops 90 days after the CRVO event. Treat all CRVOs with IOP reducing drops for life.

-non-ischemic, mild CRVOs that have tortuous veins, mild hemorrhages to the periphery, and can look like DR. The asymmetry in bleeding between the two eyes is also what signals a non-ischemic CRVO rather than DR. Here is a diabetic patient of mine with a non-ischemic CRVO OD on top of BDR OU (BCVA OD 20/80, OS 20/25). The OD is on the left, the OS is on the right as usual. She shows all of the classic signs of a non-ischemic CRVO OD. The tortuous veins, much worse bleeding in the CRVO eye, asymmetry in bleeding between the two eyes, and bleeding to the retinal periphery. Contrast this to the OS with its very mild bleeding only in the macula and straight veins: Consideration should be given to placing patients like this on glaucoma drops to reduce the chances of an ischemic CRVO in the future.

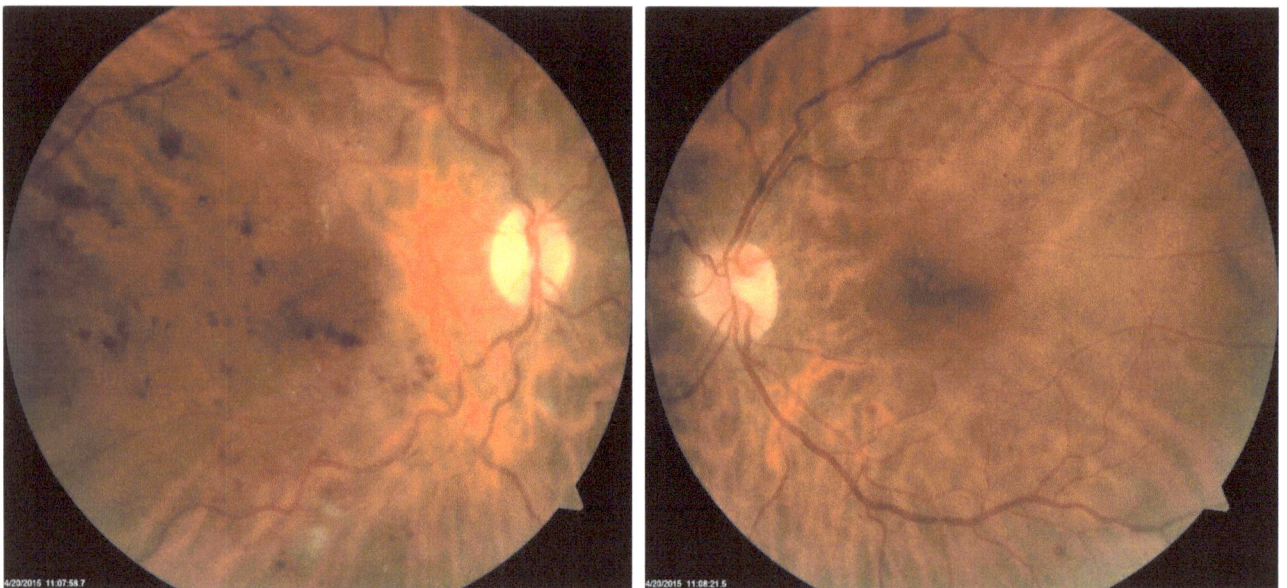

Here are more photos of the OD with the non-ischemic CRVO:

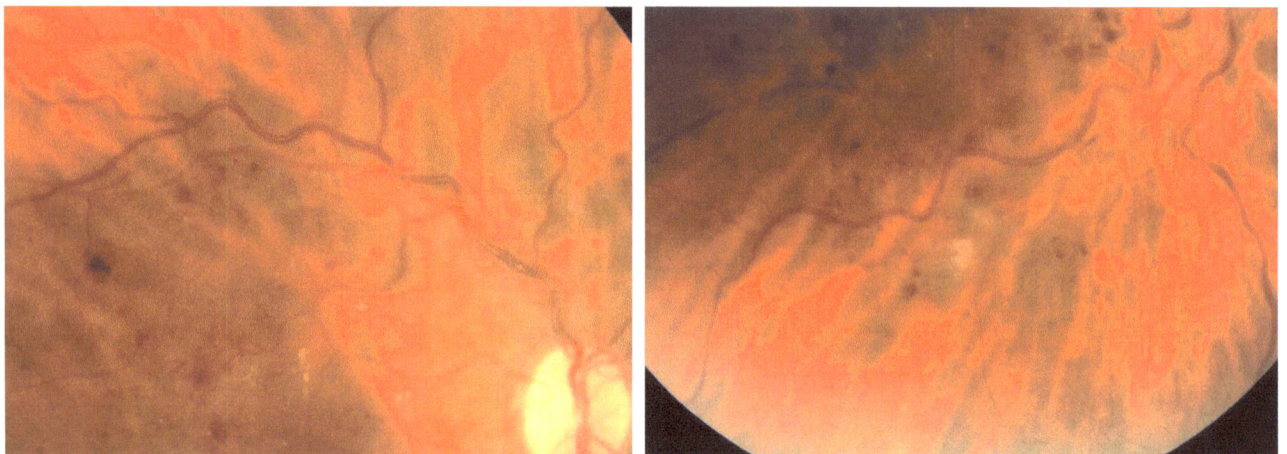

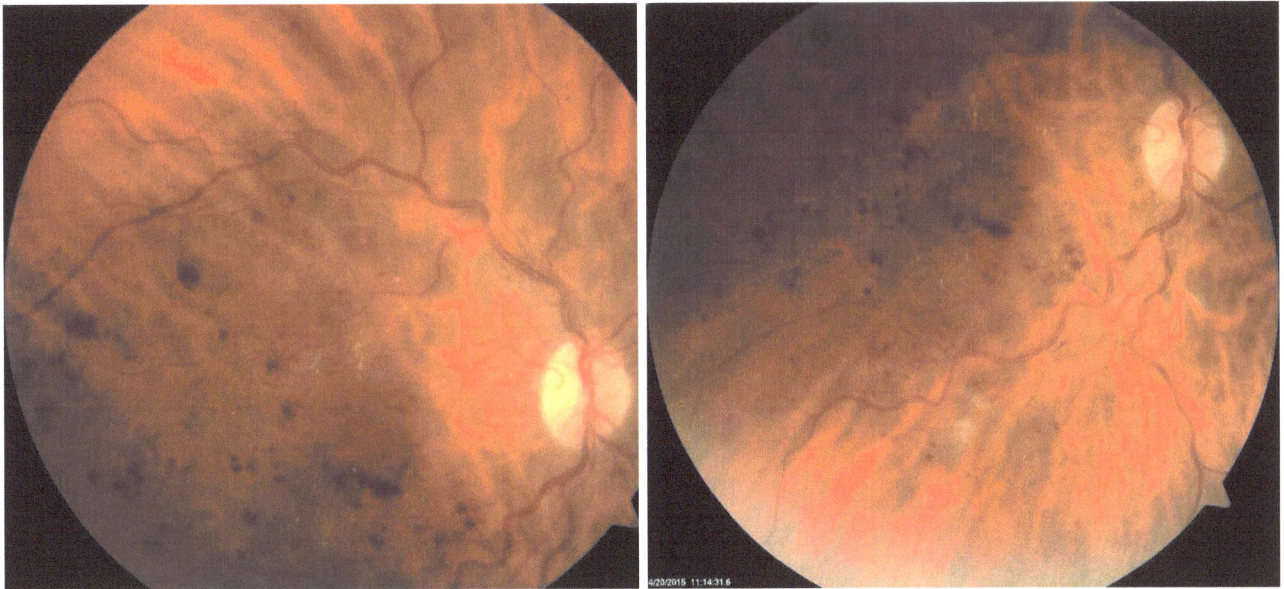

Ischemia from a retinal vein occlusion can cause neovascularization of the retina and of the iris. Neovascularization of the iris in the angle (ANV) and at the pupil edge (NVI) will lead to neovascular glaucoma (NVG). NVG is more common in CRVO > HRVO > BRVO. Retinal neovascularization is also more common in CRVO than BRVO, with BRVO essentially having only retinal neovascularization. Macular edema from a BRVO is watched for 4 months and then treated if reduced BCVA (< 20/40) has not returned to baseline. PRP is performed at the first sign of neovascularization either of the retina (NVD or NVE) or the iris (ANV or NVI). Aside from the treatment of POAG after a vein occlusion occurs, there are two main ocular concerns once a CRVO occurs, with its sudden onset and severe and usually permanent reduction in BCVA. These are:

1. NVG developing 3-5 months after the symptoms began[86] (and good luck determining exactly when the CRVO occurred in some patients). This is the so-called 90 day glaucoma.
2. An ischemic CRVO developing in the other eye.

"Thus, consideration should be given to treating elevated IOP in patients with a history of CRVO, the aim being to reduce the risk of a CRVO in the fellow eye."[87] Given wild card #1 below, treat both eyes in every patient for POAG who develops a CRVO, ischemic or non-ischemic. You don't want to be defending yourself in a court as to why a now bilaterally blind patient wasn't started on something as simple glaucoma drops to either stop a non-ischemic CRVO or HRVO from developing a sequential ischemic CRVO in the same eye or in the other normal eye. (Remember: the onset of NVG mandates PRP). Only 10% of ischemic CRVO Patients have VA better than 20/400[86] so pull out all of the stops and protect the initial eye with PRP and the second eye by treating it for glaucoma.

-**Hyper coagulation states** are defined as a tendency of the blood to form clots more readily. Protein C and S, and Anti-thrombin 3 are natural occurring proteins involved in the lysis of clots. A reduction in the levels of these molecules results in an increase in clot formation. Cancer, pregnancy, obesity, BCPs, prolonged inactivity (long plane rides), low platelet numbers, syphilis, high levels of homocysteine in the blood all increase the risk of clot formation primarily in the veins but sometimes in the arteries as well. We don't typically order a hyper-coagulation work up with arterial occlusions. We order for CRVO's (Wills eye pg. 305 and 303):

CBC
PT/PTT
RPR, FTA-abs
Protein C and S (genetic)
antithrombin 3 levels (genetic)
homocysteine levels (acquired and genetic)
Presence of **Antiphospholipid antibodies (anticardiolipin antibodies and lupus anticoagulants)** an acquired condition

Here are two photos of an IVFA on a BRVO/HRVO patient. The first is the initial IVFA and the second photo is 4 months later, showing the permanent nature of the damage from vein occlusions (both © AAO 2014):

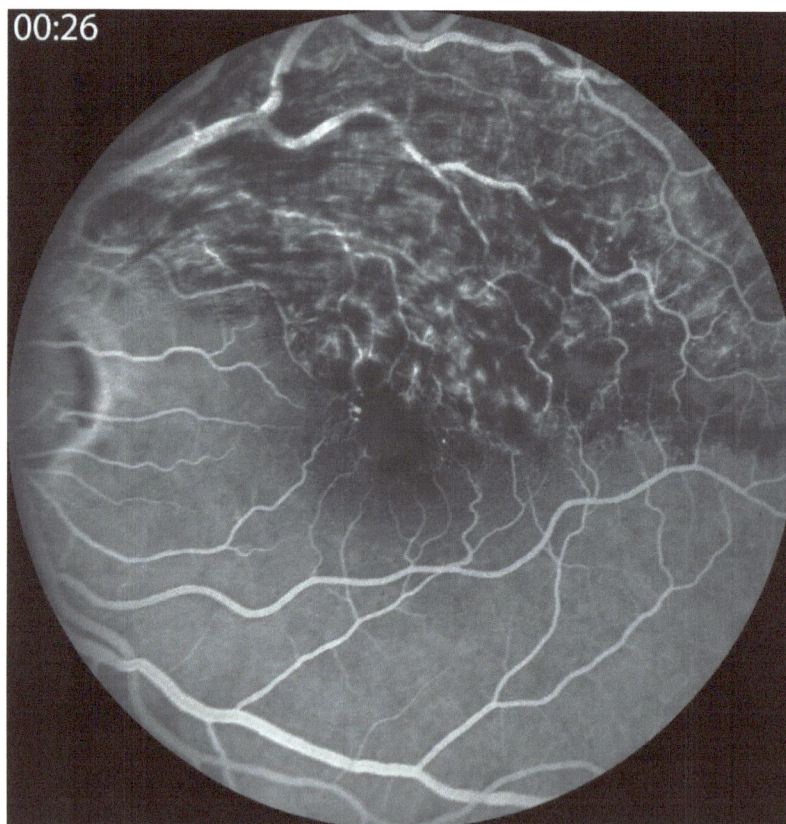

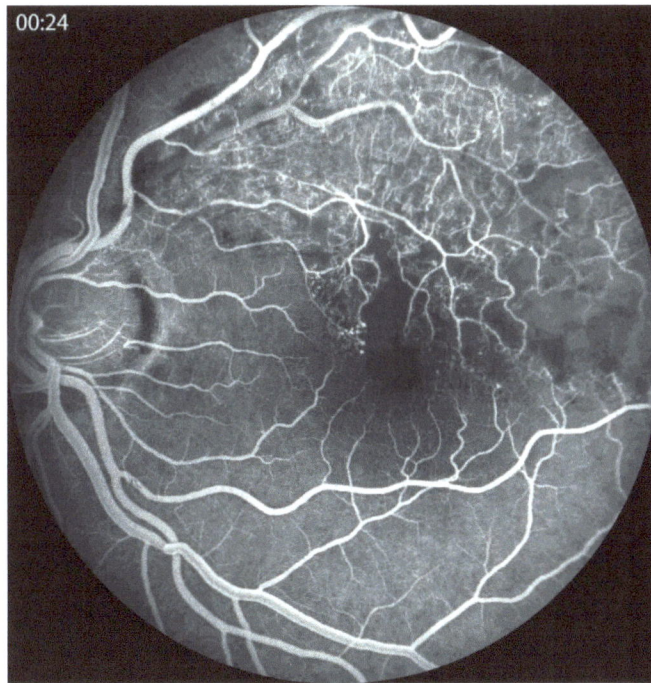

Here are photos of a CRVO patient. Photo A and B (IVFA) are initial tests and D and E are some months later. Notice the slight progressive nature of this CRVO on the IVFA (© AAO 2014):

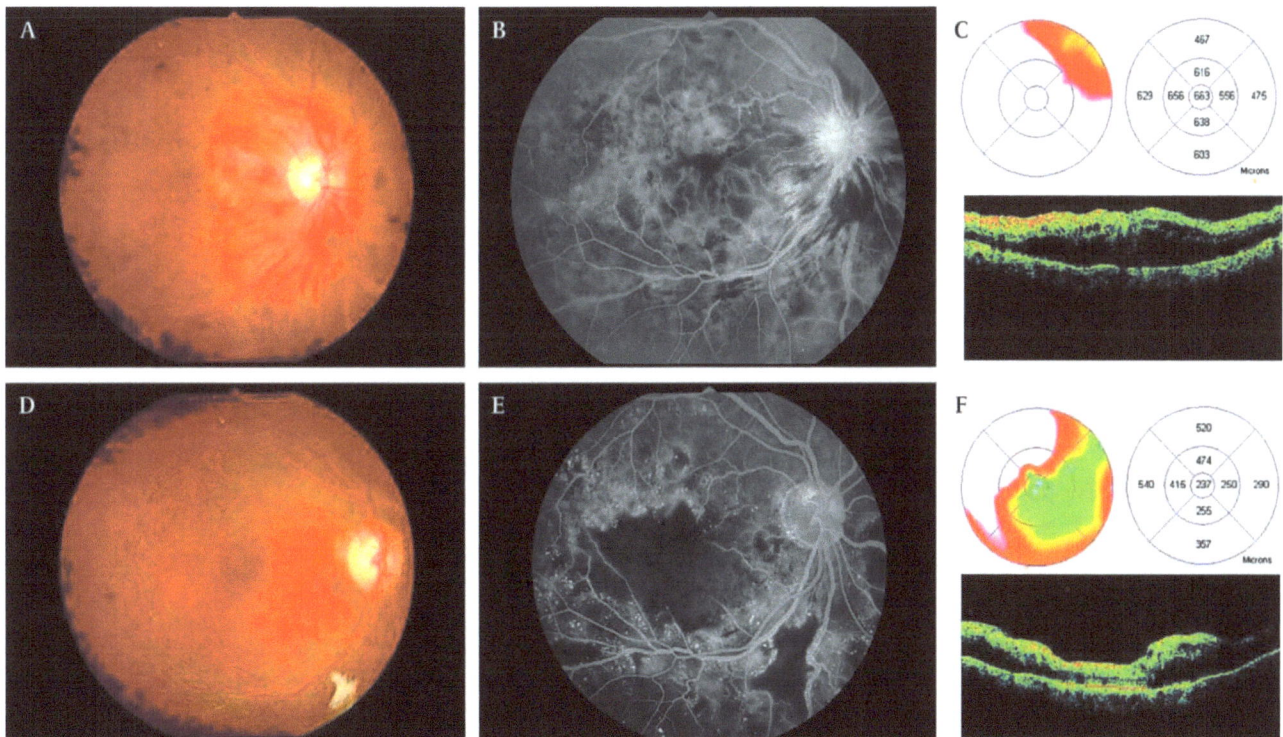

Here is a BRVO patient of mine that also has BDR. The difference is that the BRVO hemorrhages are in the sector of the BRVO and go into the retinal periphery of that sector. BDR is mainly in the macula and certainly isn't asymmetric in its distribution when it does extend outside of the macula.

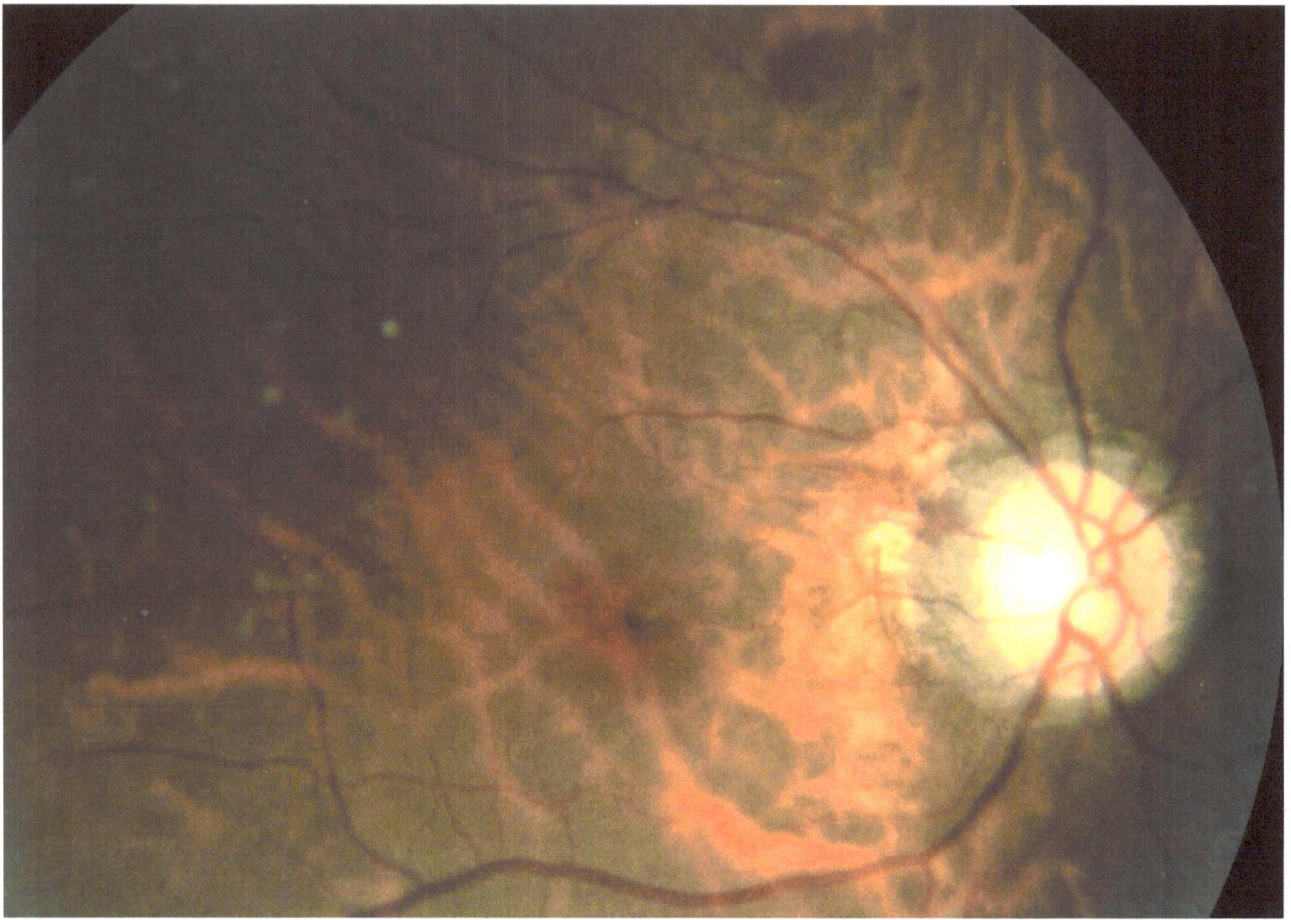

Unilateral high IOP (>40): think Posner –Schlossman, NVG from CRVO, steroid-induced from unilateral drops or intra vitreal steroids, Sturge –Weber (stains visible on face exam), PDG, Ghost Cell glaucoma, ACG.

Wild cards in glaucoma diagnosis and treatment

1) **Diurnal variation**: IOP can vary up to 6mm normally with the high IOP in most patients occurring in the early morning hours, i.e. 4 AM. In untreated glaucoma patients, it can vary greater than 10mm.[88]

The following graph shows that patients with OAG and a high IOP fluctuation have a higher risk of developing worsening glaucoma. This graph, from the prospective, randomized, multicenter Advanced Glaucoma Intervention Study (**AGIS**), plots the risk of worsening OAG over time between two groups of patients with uncontrolled OAG (N = 501 eyes of 401 patients total for both groups).[90] One group received treatment in the order of ALT-trab-trab and the other group received treatment in the order of trab-ALT-trab. For both groups, those patients with ≥ 3mmHg showed a significant worsening of HVF (as noted as a higher AGIS score) over a mean follow up of about 7.5 years (p = 0.0006). It shows that IOP variation in OAG patients is statistically significantly correlated with worsening of OAG, despite treatment.

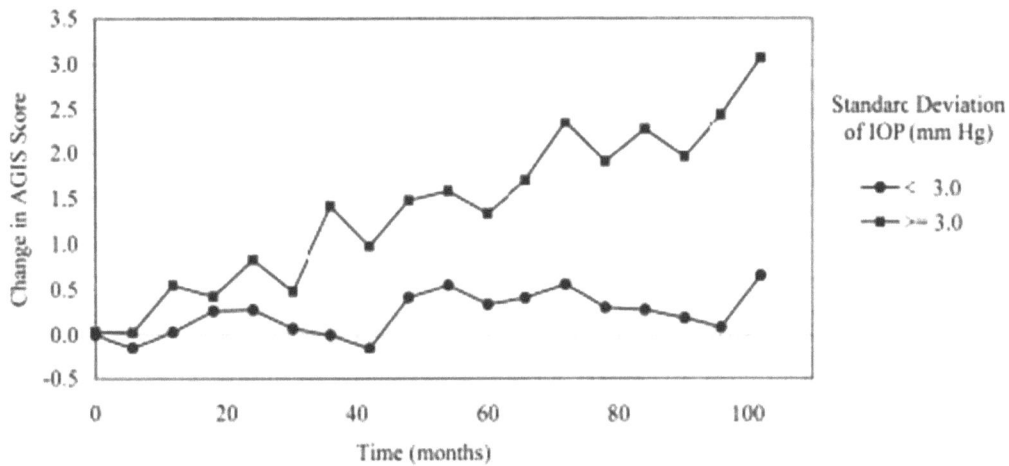

FA Medeiros, M.D. (figure below, 252 eyes of 126 patients)[91] found a similar trend in ocular hypertensive glaucoma suspect patients measuring conversion to OAG but did *not* find a statistical significance (P = 0.620) between patients with ≤ 3 mmhg of fluctuation and > 3 mmhg over a follow-up period of about 120 months. There was a trend that may have become statistically significant with more patients or a longer follow-up.

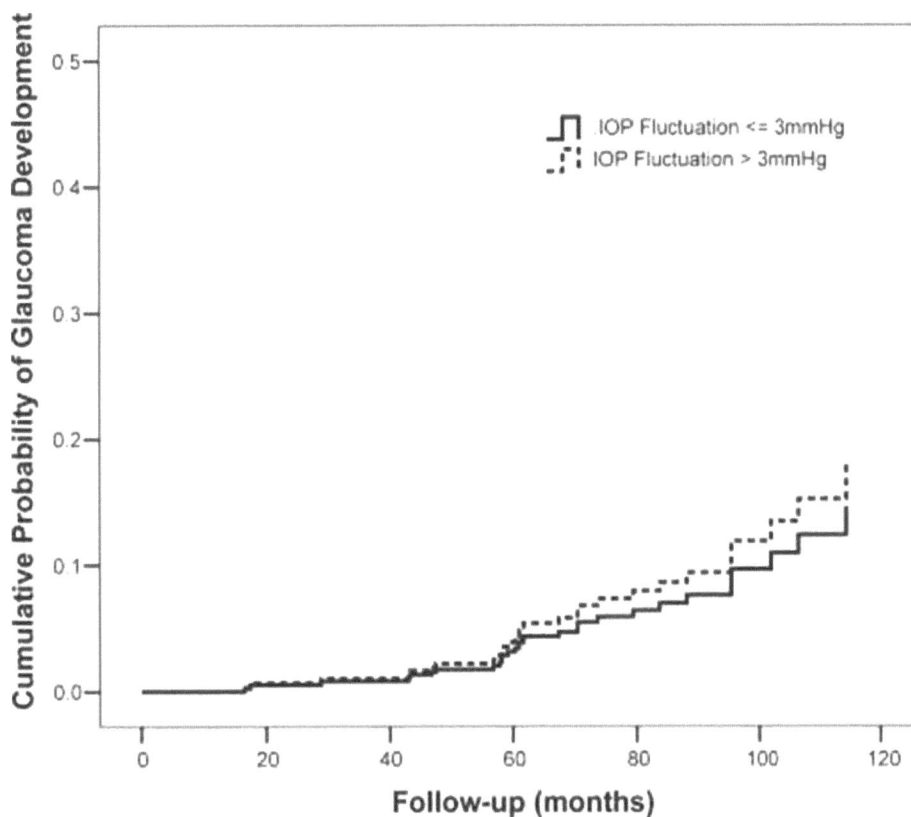

To me, diurnal IOP fluctuations represent the biggest unknown risk to a glaucoma suspect patient as well as to a glaucoma patient with progressing visual loss despite "good" IOP control.

Recent research by JHK Liu, M.D. at the University of California at San Diego (UCSD) has shown that this diurnal variation is probably due to patients being in the supine (lying down on your back) position rather than due to the time of day.[89] Recent studies looking at monocular trials for glaucoma treatment have inadvertently shown interesting results for diurnal curve IOP measurements.

The following graph shows IOP measured every 2 hours over a 24 hour period in healthy individuals in a sleep lab. The 24 hour test period was broken down into a 16 hour daytime ("wake") period with measurements in both the supine and sitting up positions and an eight hour supine only period. All positions required the patient to be in that position for at least 5 minutes.[89] In order to see if age had an effect on these diurnal measurements, two separate groups were studied, a younger age group (<38 years old, n = 38) and an older group (40 to 74 year old, n = 53). The graph below is for the **younger** age group:[89]

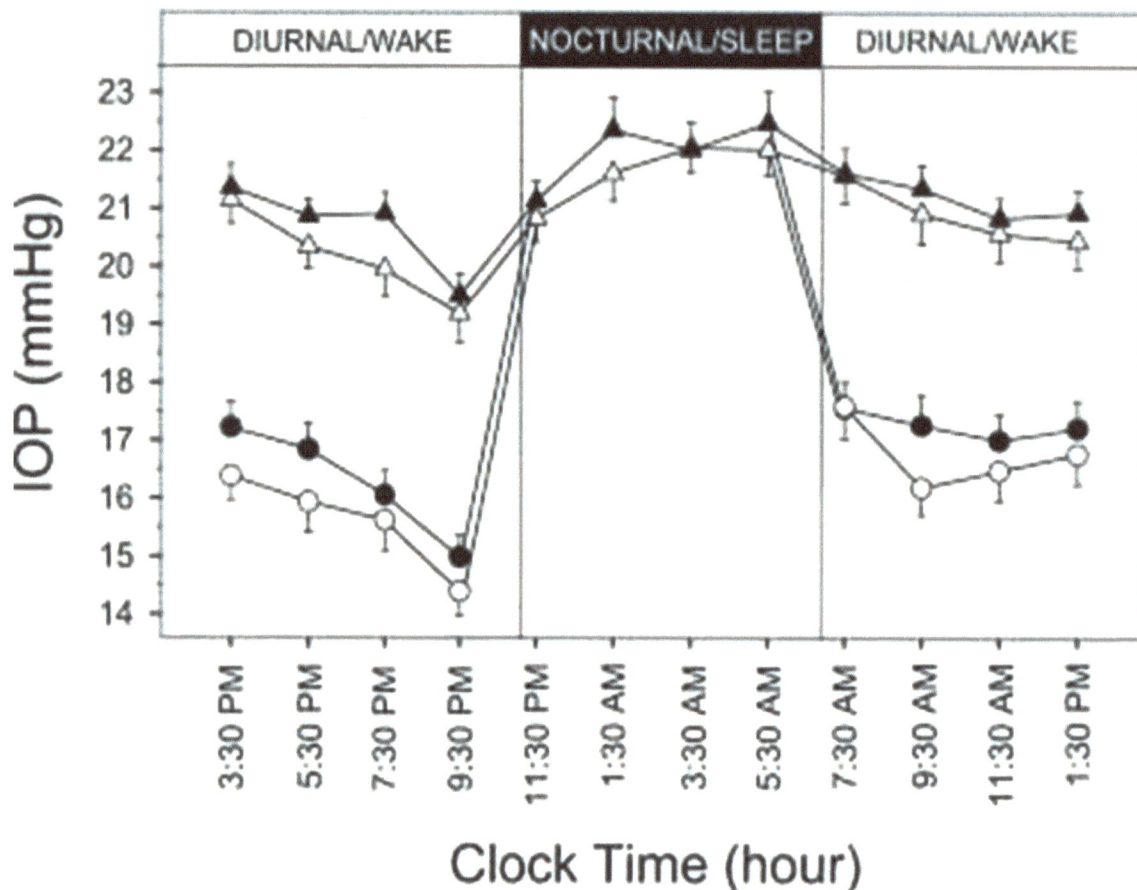

The triangles are IOP measurements in the supine position and the circles are IOP measurements in the erect position (erect IOP wasn't measured in the nighttime). Solid symbols represent IOP in the right eye, open symbols represent IOP in the left eye. (That differentiation isn't germane to the current discussion so we will ignore it). You can see that supine IOP measurements in healthy patients are much higher than erect IOP measurements no matter what time of the day it is. Looking at the supine only data, there was no statistically significant difference between awake supine and sleeping supine IOP measurements.[89]

Here is the data for the **older** group showing the same trends as the younger group:

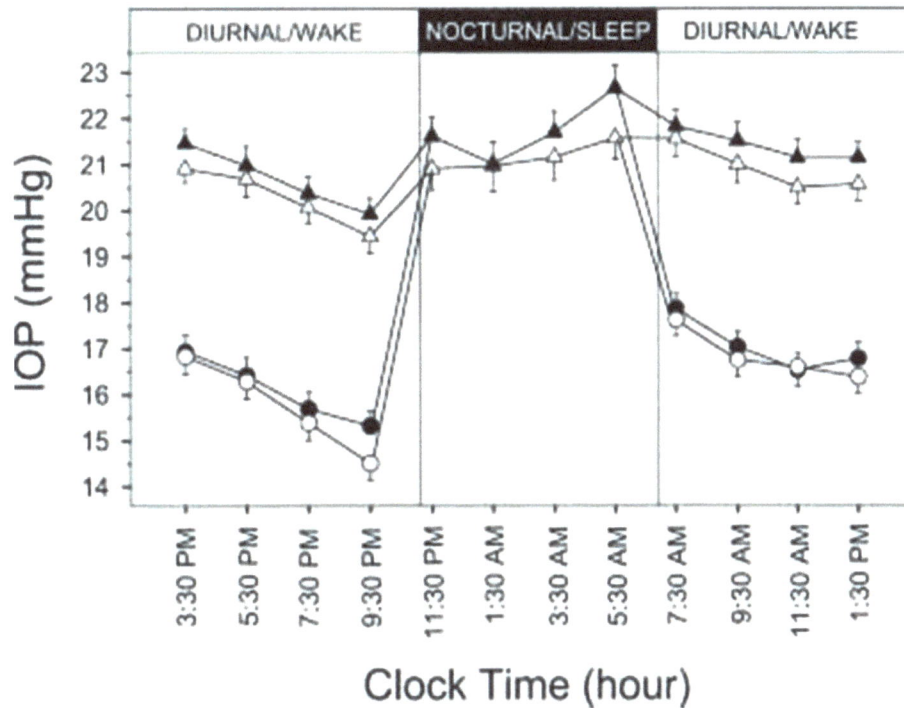

So, the diurnal curve is related to body position and not hormonal differences, and age does not affect the curve.

_____

Is diurnal variation exaggerated for patients with *untreated glaucoma* as so many reports have suggested over the years? This graph is from patients with untreated OAG using the same techniques as above (n = 41):[158]

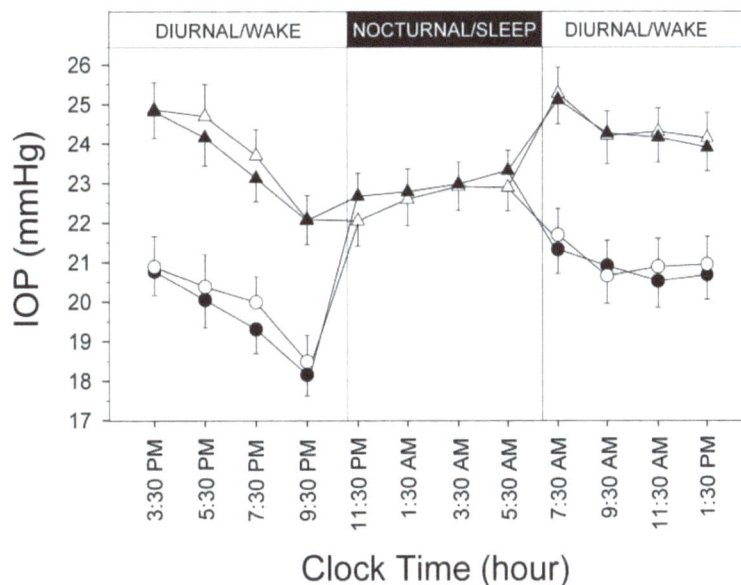

The graph above shows the same erect/supine trends as in the healthy patients. Again, the triangles are supine IOP measurements and the circles are erect IOP measurements. (Remember, we are ignoring the right eye/left eye data in our current discussion). The 24 hour supine data shows a surprisingly *higher* IOP during day time hours than night time hours in patients with OAG that have not received treatment. Unfortunately, this study did not analyze the erect/supine data statistically, but the erect/supine diurnal trend is still clearly present.

How about the diurnal curve in patients with *treated OAG*? The following graph shows erect and night time *supine only* data from patients (n = 28 eyes) being treated for OAG having recently undergone SLT and kept on the same drops after SLT as before SLT (mean number of drops before SLT = 2.6)[159] The circles represent pre-SLT data and the triangles are post-SLT data for the same patients. It shows the same strong trend between erect and supine IOP measurements. So, treatment for glaucoma does not eliminate the diurnal curve.[159]

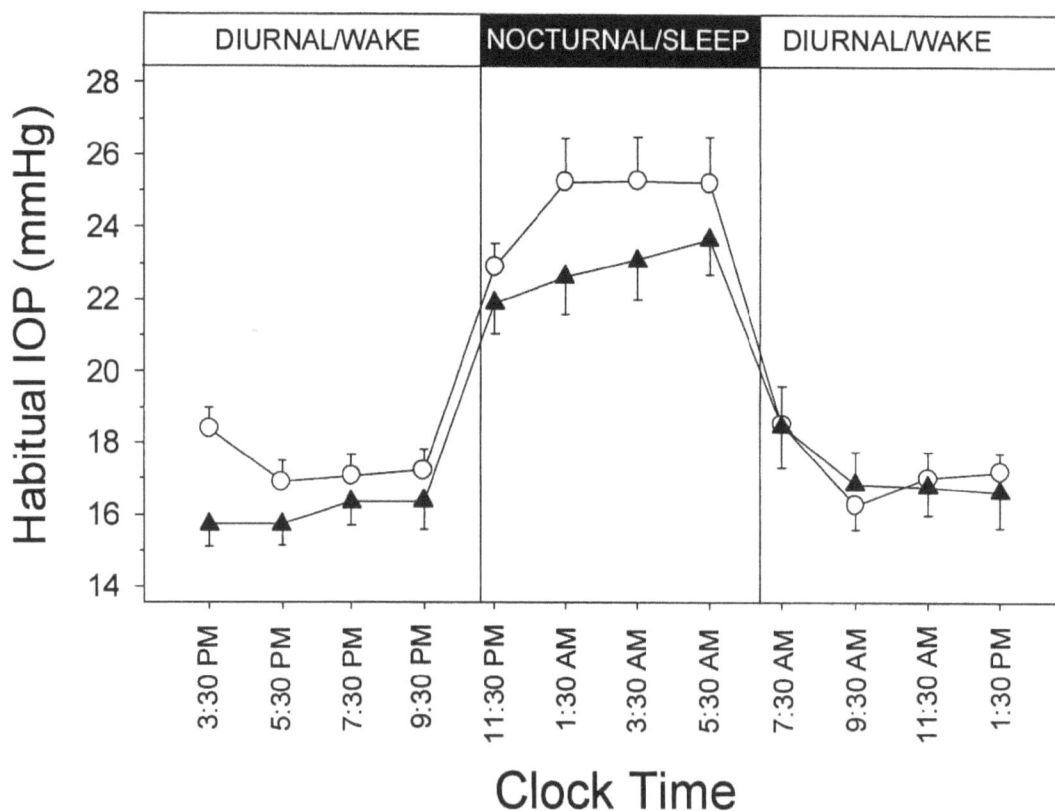

How about comparing the day time supine IOP to the night time supine IOP in the same cohort of treated glaucoma patients? The following graph shows the IOP for the supine position for all 24 hours (n = 28). Again, the circles represent pre-SLT data and the triangles are post-SLT data. SLT statistically reduced the night time IOP with the medically only treated patients having a significantly higher IOP in the night time supine position.[159]

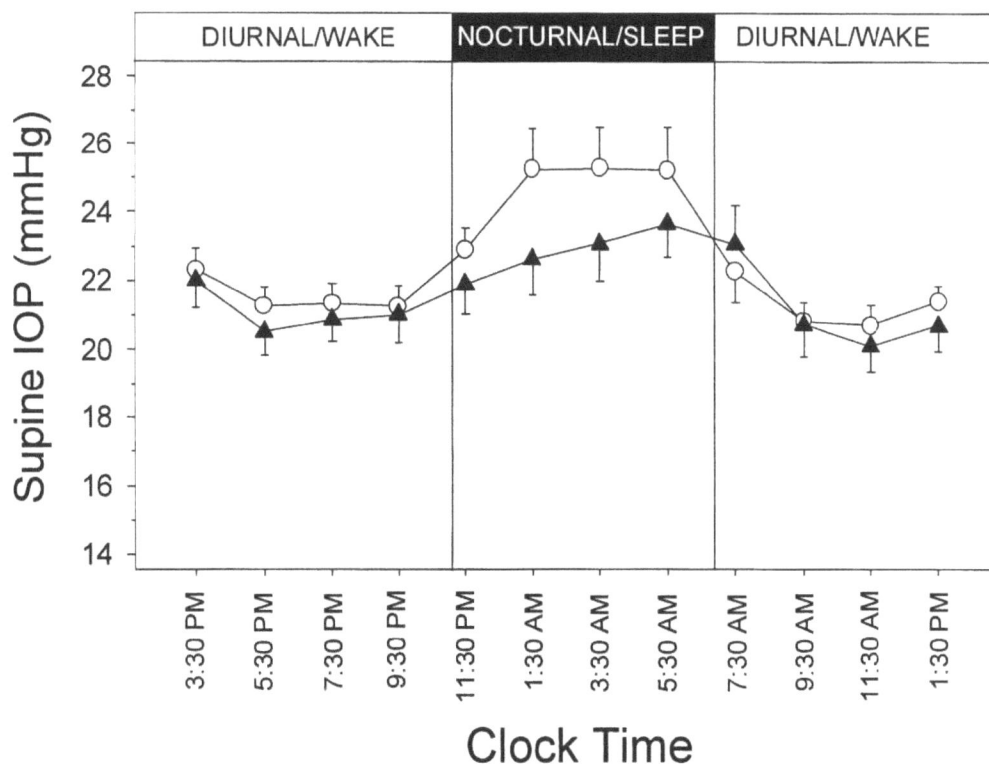

All of this diurnal data confirms that IOP is higher in the supine position for healthy, untreated glaucoma and treated glaucoma patients with SLT having a significant reduction in the night time supine IOP in treated glaucoma patients. To me, the increase in IOP in the supine position (the diurnal curve) represents the most significant unknown factor in the treatment of glaucoma.

2) That **Optic Nerve damage** is almost always **permanent** has been the mainstream opinion for decades. Recent research shows that a small percentage of patients treated for newly diagnosed glaucoma can actually improve their VF tests.

The Collaborative Initial Glaucoma Treatment Study (n = 604 who finished the study, mean follow up was 7.2 years)[182] was designed to compare two different treatments for newly diagnosed OAG. Which two treatments are irrelevant for our discussion here. The study showed that just as many patients improved their VF result as those that worsened over a 5 year follow up period (p > .20). In those patients that had Improvements in their VF, improvements stopped after 5 years of treatment while those with worsening VF results continued to worsen (p < 0.0053). Here is a scatter plot of all 604 patients broken out by the VF mean deviation (MD) at the baseline exam versus the 5 year VF MD change over the 5 year period showing that no matter how significant the damage was at baseline, before diagnosis and treatment (as evidenced by a more negative MD value on the plot), there were patients above and below the substantial gain and loss lines. I.e., the effect of VF improvement and VF worsening existed for all ranges of baseline MD values over the 5 year period. [182] (There are asterisks in the substantial gain and loss area for all measurements of baseline dB). I don't have a p value comparing worse baseline MD with better baseline MD over the 5 years of follow up but the results are obvious on the graph. Up to a 3 db of change on the VF MD was not considered significant and only those VF results with higher and lower than a 3 db change over a 5 year period were considered "substantial". You can see that the bulk of patients neither improved nor worsened but rather stayed steady over the 5 year follow up period of this graph.

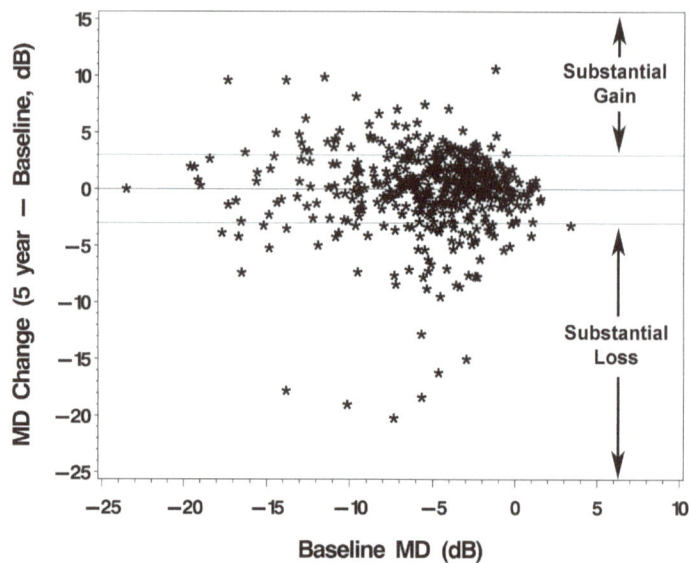

3) **Visual Field loss lags** significantly **behind loss of ganglion cells**. Harry Quigley, M.D. found in autopsy studies that 20% of the ganglion cells are destroyed before a 5db VF loss occurs on automated perimetry and 40% of the ganglion cells are gone when a 10 db VF loss occurs.[92] Earlier studies using Goldmann kinetic perimetry showed a 50% loss of ganglion cells before VF loss occurred.[93] So, permanent loss of the ganglion cells occurs before the first reliable VF loss occurs and HVF is better than Goldmann kinetic VF testing at detecting early loss.

Also, as the damage from glaucoma progresses, ganglion cells get destroyed faster. The graph below shows the loss of rim tissue against VF loss in dB and you can see that the rate of VF loss increases rapidly as the rim tissue goes from 50% loss to 90% loss.[94] This is why a lower IOP is needed when the glaucoma loss is greater, such as when the C/D ratio enlarges to 0.9 or if there are significant VF losses already in the patient:[94] I think that this graph is one of the most important graphs in this entire manual on glaucoma.

- In early disease, dB function changes relatively slowly compared to structure

- In later disease, dB function changes relatively faster compared to structure

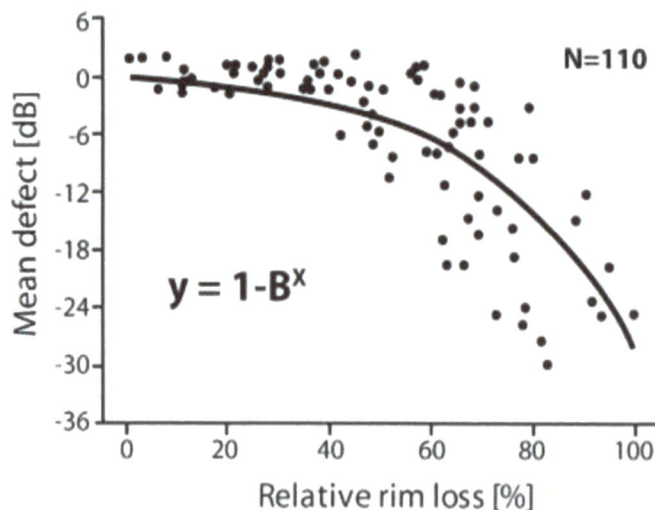

$$y = 1 - B^x$$

Here is another diagram showing the relationship in 1,245 eyes of normal, glaucoma suspect and glaucoma patients, of their VFI and the number of ganglion cells remaining. It also shows a rapidly descending number of ganglion cells as glaucoma damage increases (Copyright as shown).

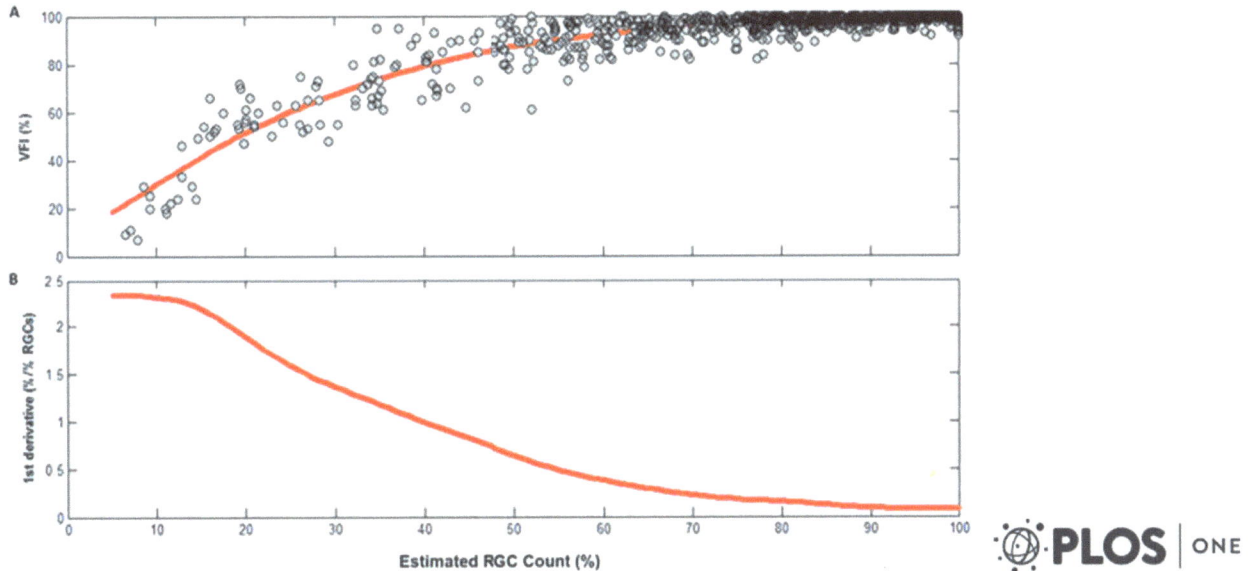

The Olmstead study showed that the most important risk factor for the development of blindness was the presence of VF loss of optic nerve damage at the time of initial diagnosis of glaucoma[95] (60,666 patient's charts reviewed, n = 295 patients with OAG, mean follow-up of 15 years). This is probably related to patients being on the rapidly descending slope at the time of diagnosis (my words).

4) **Low Tension Glaucoma** is treated with pressure lowering drops just like other forms of Glaucoma.[59] The prospective, randomized, multicentered collaborative Normal Tension Glaucoma Study (NTGS) found reducing IOP by 30% with drops reduced the 5 year chances of developing glaucoma from 35% to 12% (P < 0.0001, n = 140 eyes, follow up was 7-8 years, half received treatment, half did not). [96] So, lowering IOP works in normal tension glaucoma just as it does in high IOP glaucoma.

"In any individual patient, there is no characteristic abnormality of the optic disk or VF that distinguishes NTG from POAG with higher IOPs" .[54]

When a VF loss does not match the appearance of the optic nerve, the corrected IOP is normal or any other aspect of the patient's tests just don't match up with each other, you have to think about other conditions that may be present instead of glaucoma. Conditions that can mimic glaucoma and are typically non-progressive include:[97]

Compressive lesions of the ON or chiasm
Prior high IOP episodes such as steroid induced/Possner-Schlossman/uveitic/PDS/ACG
AION or prior *low* BP episodes such as trauma/heart surgery/MI found on PMHx
ON drusen seen as pseudo papilledema and rarely progressive found on exam/ B-scan
Retinal conditions seen on exam
Optic atrophy with pale rim tissue, not a pale cup seen on exam
Optic Nerve pits/myopic disks/coloboma seen on exam

You can see that most of these conditions are found on either the exam or in the medical history. The main condition not seen on exam are compressive lesions of the ON or chiasm or prior transient glaucoma episodes. So, if you get a little nervous about a patient that seems to have "glaucomatous damage" but has something that doesn't add up, get an MRI of the orbit and chiasm or refer it to a general/glaucoma ophthalmologist. Call first to ensure that they know that you want a second opinion/consult on a rather unusual patient, not a complete transfer of all eye care. Follow any recommendations or work-ups that the Ophthalmologist suggests. This is why you are getting their opinion in the first place. Here are MRIs of two patients with optic nerve meningioma(both © AAO 2014). The first shows the "train track" appearance of the right optic nerve indicative of an optic nerve meningioma. The second photo shows another meningioma with an MRI cut above the right optic nerve. MRI and CT images are presented as if the patient is lying on their back and you are at the patient's feet looking up towards the rest of the body. So, the patient's right eye is shown on your left in this MRI image.

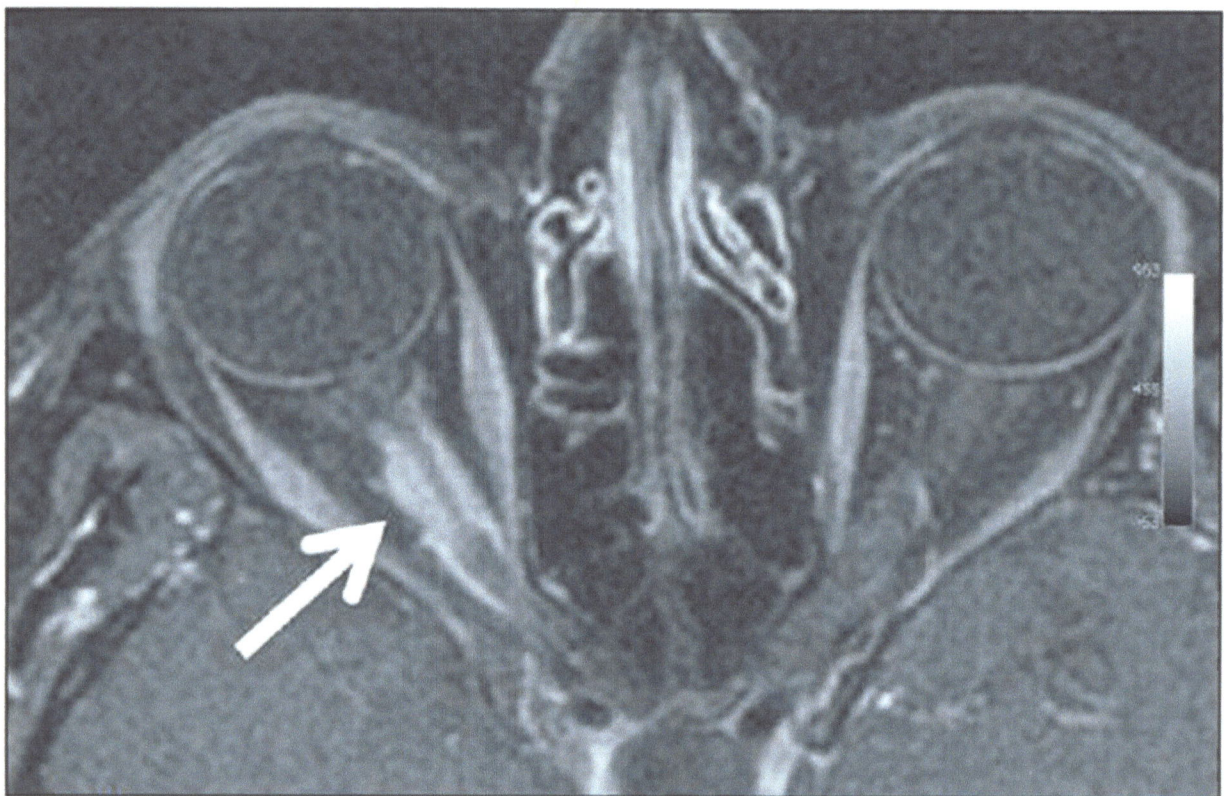

Figure 12. Axial, post-contrast T1-weighted image with fat saturation of the orbits of a female, 53-year-old patient with progressive loss of visual acuity of the right eye. Observe the optic nerve involved and compressed by a contrast-enhanced circumferential mass surrounding the optic nerve sheath (arrow).

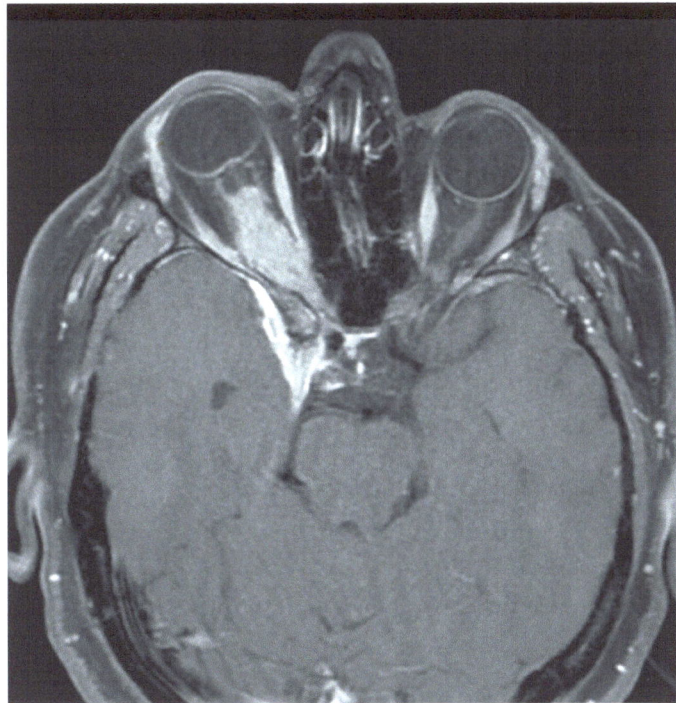

5) **Nearly half of POAG patients** will have a **normal IOP** measurement on at least one test.[79,97] Again, this is based on pre CCT data.

6) **Variability** in HVF/HRT/IOP test results are common, to be *expected* and will make your job more difficult. **86% (604/703) of the abnormalities on the initial and deemed *reliable* HVF tests in the OHTS were not confirmed abnormal upon retesting**.[98] Several studies have shown a 2.8 db improvement during the first two VF tests but little significant change after that. [183]

What constitutes a glaucoma work up?

1) **Tonometry:** Always do Goldmann tonometry for all glaucoma visits, not NCT. Your endpoint in patients with a pulse movement of the mires is to have the mires as much on one side of the inner mires as the other. This means that the correct end point is really the average of the pulsation (Zeiss.com). (© AAO 2014).

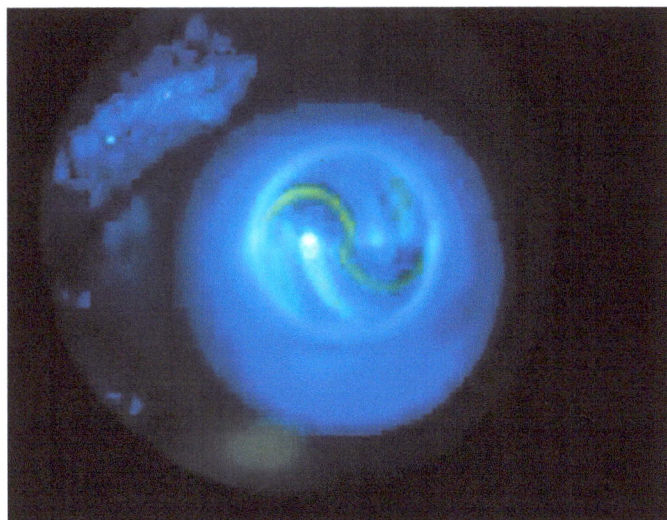

2) **Corneal pachymetry** (CCT) is done on the first visit so you can accurately measure the true IOP. I cannot find a longitudinal study documenting the stability of a single patient's corneal thickness over many years, but if there was a trend towards corneas getting thicker or thinner as they age and if you had enough CCT readings from patients of varying ages, it might show up on a graph. The European Glaucoma Prevention Study[99] shows a scatter plot of CCT for 854 patients versus the age of the patients. It shows a very gradual thinning of the cornea as the age of the patient's increases but not an obvious trend. This is not the best design for a study to answer this question, but it's the best I've got.

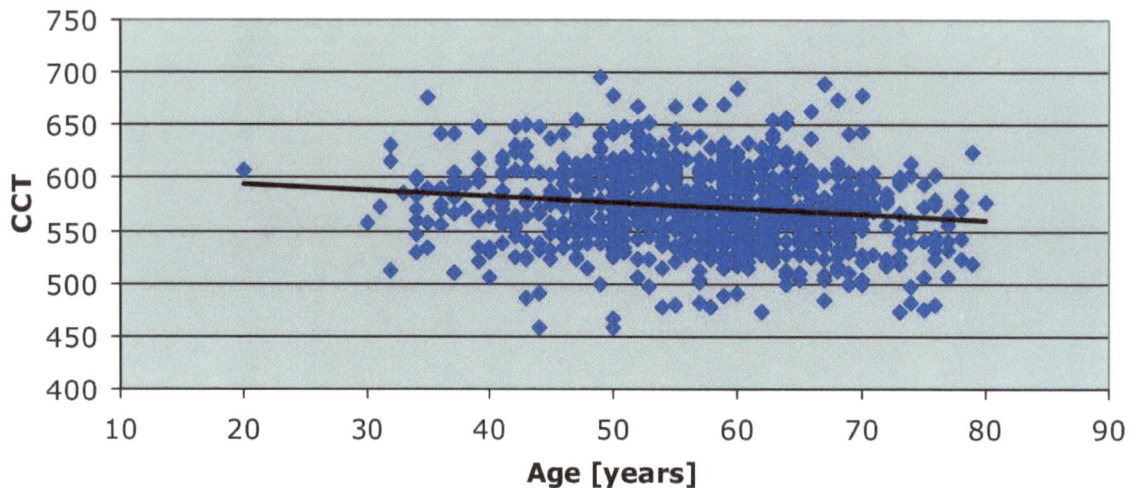

"Because CCT is relatively stable over the lifetime of an adult, a single measurement of CCT is adequate in most patients".[100] We generally do pachymetry once with the assumption that the CCT for a patient won't change significantly over time. Some insurance companies pay less than $7 for you to do corneal pachymetry but you have to do it anyway.

3) **Disk Photos: Disk photos are done yearly with the most recent photo compared to the initial photo in front of the patient during your exam**. We look for the items mentioned above under risk factors for OAG.

4) **Gonioscopy:** (All 3 diagrams © AAO 2014).

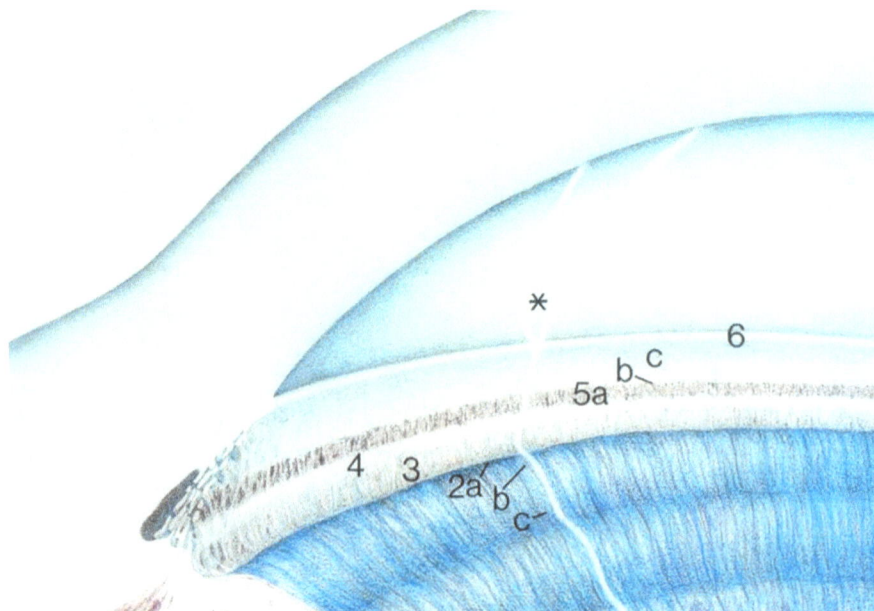

Heavy pigment

Light pigment

No pigment

Schwalbe's line

Trabecular meshwork

Trabecular pigment band

Schlemm's canal

Scleral spur

Longitudinal muscle

Iris processes

Ciliary body

Circular muscle

20° angle (Grade II)

45° angle (Grade IV)

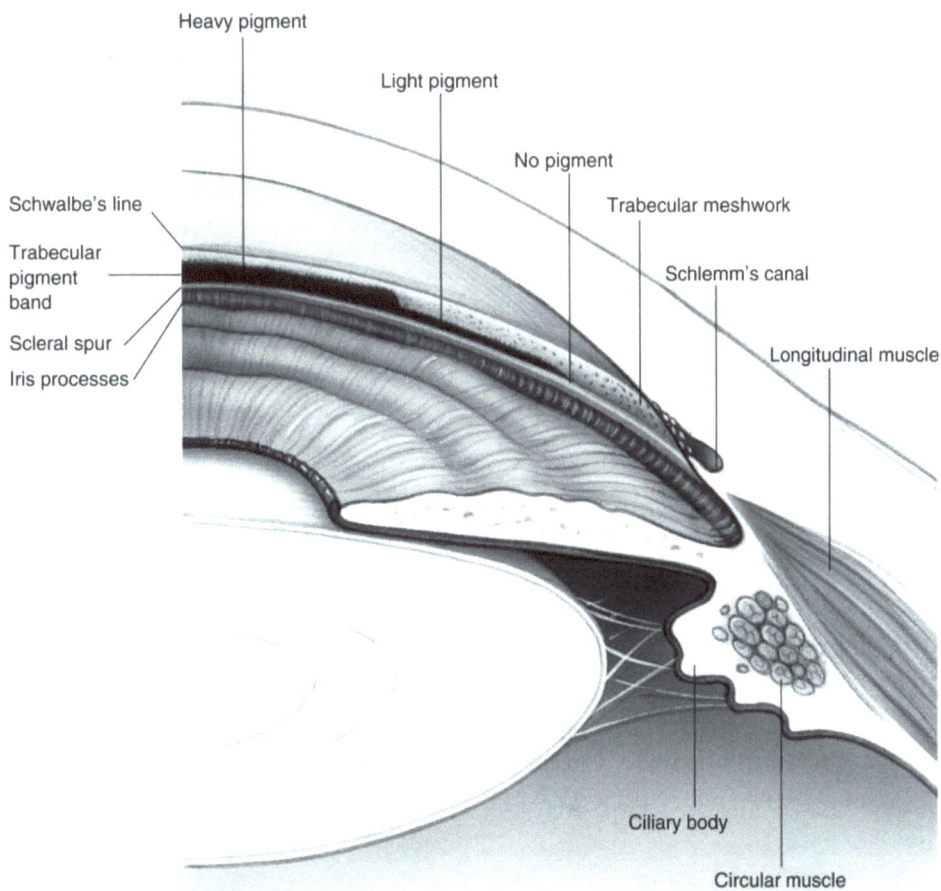

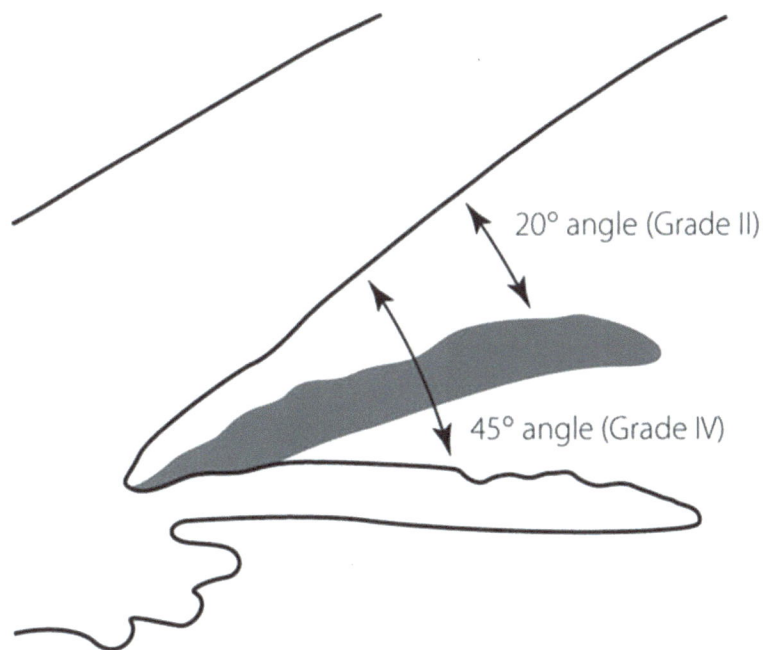

Use the Sussman 4 mirror as the patient's own tear layer is all that is needed to view the TM. Pushing too hard with the Sussman artificially opens the angle up, which is what dynamic gonioscopy is. Dynamic gonioscopy is sometimes done on purpose to determine if PAS are oppositional or permanent and to sometimes open an angle closure attack. You can guess whether the angle is grossly narrow or not by using the slit lamp and Von Herrick's method, but formal gonioscopy *has* to be done in order to determine how open the angle really is as well as to determine the following:

-Pigmentation level, + to ++++. With 3+ and higher being pathologic[101] and raising the concern of PDS (continuous pigmentation) and PXE (pigmentation worse inferiorly).[102]

-Whether ANV is present or not as in DM/CRVO/HRVO patients. Again, ANV are fine mesh-like BVs crossing from the iris to the TM and fan out along the TM giving a reddish band look to the TM. Eventually, the ANV contract and pull the iris to the TM closing it off permanently. These PAS do not open with dynamic gonioscopy. ANV usually develops 3-5 months after a CRVO occurs. Most neovascularization in a CRVO is in the angle as opposed to the retina. HRVOs have distribution of neovascularization that is split 50/50 between ANV/NVI and the retina. BRVO neovascularization is mostly in the retina. PRP of the retina is the definitive treatment/cure for ANV! DM patients tend to get NVI not ANV, if they get iris neovascularization at all.

-Whether iris bombe is present with or without PAS. Dynamic gonioscopy opens reversible PAS but not permanent PAS. PAS in the setting of a narrow angle requires an LPI. Refer narrow angles without PAS out for a potential LPI.

-Angle recession. Be suspicious of angle recession in unilateral glaucoma. Angle recession increases the risk for SOAG 10-20 years after the injury ripped open the CBB in the angle.[44] The more angle recession there is, the higher the risk of developing glaucoma. 180 degrees of recession has been traditionally used as the minimum needed to develop glaucoma.

Sometimes it is difficult to know where the TM is in a lightly pigmented eye. There is a technique to determine where the TM meets Schwalbe's line. Thin the light down to the thinnest slit possible with the light on its brightest setting. Place the light source off to one side. You will see two reflections off of the cornea, one off of the front surface of the cornea and one off of the back surface. These two lines meet at the junction of the anterior TM and Schwalbe's line.[104] Do this technique for the first 500 gonioscopies that you do in order to not miss an angle closure mistaking Sampaolesi's line for the TM. This is the most common error in gonioscopy. If you do not see the two lines meeting, then you have PAS and the iris is closing the TM off completely in that area. At that point, a referral for an LPI is mandatory.

Gonioscopy is usually done every 5-7 years, more often if you are suspicious of changes. I draw a big X for each eye and document the most posterior structure I see in all 4 quadrants with the elements noted above if present.

5) **Humphrey Visual Field (HVF):** This has been our gold standard for detecting glaucoma for 100 years. When glaucoma starts to damage the optic nerve, it will not affect your straight ahead vision nor your side vision laterally. It will initially decrease your peripheral vision nasally, just above and below your straight ahead gaze. You can describe this early loss to patients like this:

"Imagine that you are looking at the center of a door. Initially, glaucoma will decrease your ability to see the top and bottom of the door. At first, the decrease will be very subtle and you will not notice it. As glaucoma progresses, the peripheral field loss can become more noticeable".

The HVF machine uses the eye's natural blind spot to detect when you are not looking straight ahead during the test. During the test, the machine places a light into the blind spot and if you see it, the machine knows that you are not looking straight ahead. Tell the patient not to worry if they don't see a light for a while, they aren't supposed to see about 50% of the lights anyway. The read-outs for VF tests are described above. I believe in using the best visual field machine available, the Humphrey by Zeiss. Visual fields are typically done twice a year with the same 24-2 test pattern, taking advantage of the new VFI capability.

Make sure that other conditions aren't simulating a glaucoma VF defect. Here are two examples of this. The first is a patient that had an RRD with PPVx and SB and is, coincidently, a POAG suspect. The second is another POAG suspect that developed a choroidal melanoma after the first two HVFs. She got treated with brachytherapy.

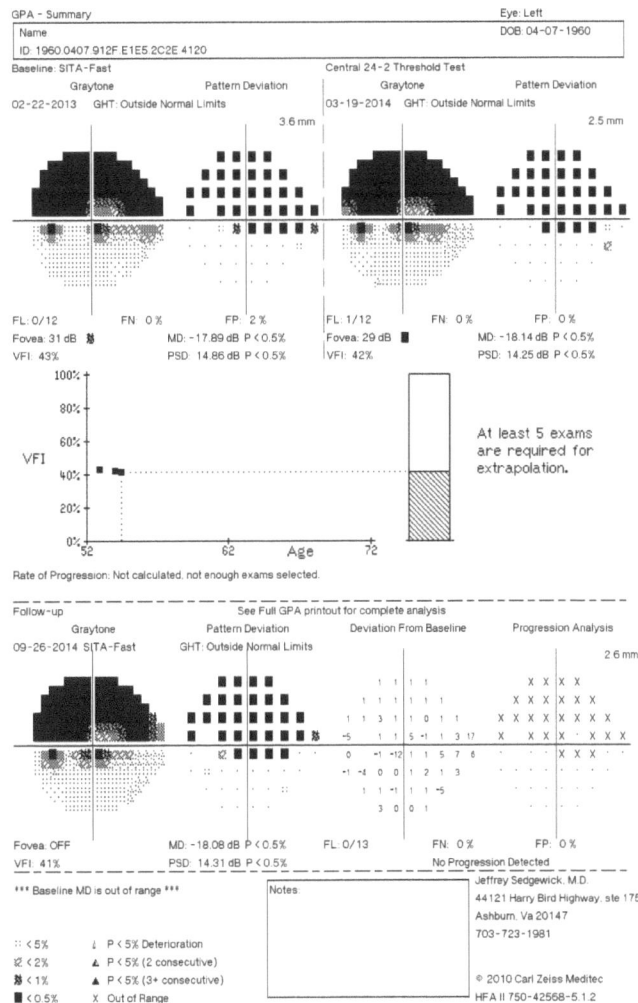

GPA - Summary   Eye: Left
Name   DOB: 04-07-1960
ID: 1960.0407.912F.E1E5.2C2E 4120

Baseline: SITA-Fast   Central 24-2 Threshold Test

Graytone   Pattern Deviation   Graytone   Pattern Deviation

02-22-2013   GHT: Outside Normal Limits   03-19-2014   GHT: Outside Normal Limits

3.6 mm   2.5 mm

FL: 0/12   FN: 0 %   FP: 2 %
Fovea: 31 dB   MD: -17.89 dB P < 0.5%
VFI: 43%   PSD: 14.86 dB P < 0.5%

FL: 1/12   FN: 0 %   FP: 0 %
Fovea: 29 dB   MD: -18.14 dB P < 0.5%
VFI: 42%   PSD: 14.25 dB P < 0.5%

VFI chart (100%, 80%, 60%, 40%, 20%, 0%) vs Age (52, 62, 72)

At least 5 exams are required for extrapolation.

Rate of Progression: Not calculated, not enough exams selected.

Follow-up   See Full GPA printout for complete analysis

Graytone   Pattern Deviation   Deviation From Baseline   Progression Analysis

09-26-2014 SITA-Fast   GHT: Outside Normal Limits   2.6 mm

Fovea: OFF   MD: -18.08 dB P < 0.5%   FL: 0/13   FN: 0 %   FP: 0 %
VFI: 41%   PSD: 14.31 dB P < 0.5%   No Progression Detected

*** Baseline MD is out of range ***

Notes:

:: < 5%
≥ < 2%
♦ < 1%
■ < 0.5%

↓ P < 5% Deterioration
▲ P < 5% (2 consecutive)
▲ P < 5% (3+ consecutive)
X Out of Range

Jeffrey Sedgewick. M.D.
44121 Harry Bird Highway. ste 175
Ashburn. Va 20147
703-723-1981

© 2010 Carl Zeiss Meditec
HFA II 750-42558-5.1.2

GPA - Summary   Eye: Left

Name.   DOB: 11-23-1959

ID: 1959.1123.3913.119F.F723.5BD1

Baseline. SITA-Fast   Central 24-2 Threshold Test

Graytone   Pattern Deviation   Graytone   Pattern Deviation

02-28-2007 GHT: Within Normal Limits   03-24-2008 GHT: Outside Normal Limits

                            6.5 mm   6.0 mm

FL: 0/11   FN: 0 %   FP: 6 %   FL: 0/10   FN: 1 %   FP: 0 %

Fovea: 37 dB   MD: -1.52 dB P < 10%   Fovea: 37 dB   MD: -2.02 dB P < 5%

VFI: 99%   PSD: 1.39 dB   VFI: 98%   PSD: 2.07 dB P < 5%

VFI chart — values: 100%, 80%, 60%, 40%, 20%, 0%; Age axis: 46, 56, 66; 5 years

Rate of Progression: -1.7 ± 1.5 %/year (95% confidence)

Slope significant at P < 5%

Follow-up   See Full GPA printout for complete analysis

Graytone   Pattern Deviation   Deviation From Baseline   Progression Analysis

02-24-2015 SITA-Fast   GHT: Outside Normal Limits

                                                                        5.1 mm

Fovea: OFF   MD: -10.69 dB P < 0.5%   FL: 0/12   FN: 0 %   FP: 0 %

VFI: 82%   PSD: 11.34 dB P < 0.5%   Likely Progression

Previous Follow-up Exams: Notes: Jeffrey Sedgewick, M.D.
      02-25-2014   08-25-2014   44121 Harry Bird Highway, ste 175
:: < 5%   ⩗ P < 5% Deterioration   Ashburn, Va 20147
⩗ < 2%   ▲ P < 5% (2 consecutive)   703-723-1981
⫸ < 1%   ▲ P < 5% (3+ consecutive)
■ < 0.5%   X Out of Range   ® 2010 Carl Zeiss Meditec
                                                              HFA II 750-42568-5.1.2

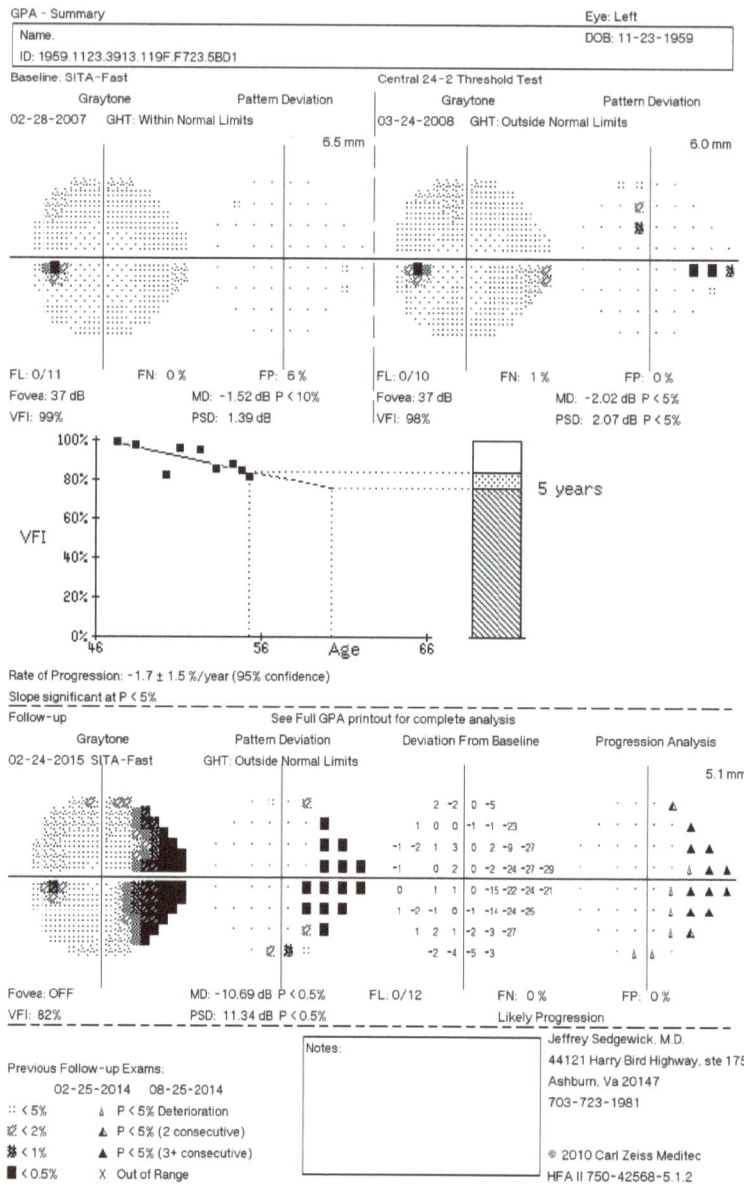

6) Heidelberg Retina Tomography (HRT): Reliably detecting *early* ON damage has been an elusive goal for many years. HRT (or OCT) testing now offers this ability. I prefer the HRT over the OCT, even though the HRT, OCT, and GDX machines have all shown equivalent reliability in detecting early glaucoma.[105]

"There are three types of computer-based imaging devices currently available for glaucoma: confocal scanning laser ophthalmoscopy [HRT], optical coherence tomography [OCT], and scanning laser polarimetry [GDX]. In a systematic review, the versions of these devices that were studied were similar in their ability to distinguish glaucoma patients from controls."[105]

The OCT has the same CPT code as the HRT and retina ophthalmologists use the OCT to analyze retinal diseases, something that you are going to refer out. Insurance companies are refusing to pay for the same tests done by different doctors when performed too close together. At least you can make the case in an audit that the HRT is designed to detect different diseases than the OCT.

The optic nerve will suffer permanent damage before the first loss is detected on a visual field test. We know this from the autopsy studies done by Harry Quigley, M.D. noted before. The HRT uses a laser and computer software to measure the thickness of the optic nerve at and around the optic disk. Being able to compare later, subsequent readings to the patient's baseline reading is what gave the HRT the edge over the OCT, in my opinion. Newer OCT machines can now also do comparisons over time but this is an add-on feature to the OCT, not the reason for their development. The HRT was developed specifically for glaucoma, comparing sequential test results. Single measurements, at one point in time, have limited value in detecting early glaucoma from normal thicknesses due to variations in the normal nerve fiber layer thickness. Trending a patient's thickness over time has much more value in detecting early glaucoma since glaucoma is a progressive disease. Repeated measurements will tell us if the thickness is staying the same, which is normal, or getting thinner, which is potentially an early sign of glaucoma. I recommend that the HRT be done every 6 months for 3 or 4 visits in order to see if there is an early downward trend or not. After this, I reduce the visits to once a year in low risk patients but in those patients with a high risk of developing glaucoma or for patients diagnosed with glaucoma, staying on a 6 month regimen is more appropriate. As in HVF testing, variability is common and you are looking for the overall trend. Insurance will not pay for a HRT, an OCT or a fundus photo performed on the same day so you have to do them on different days.

The HRT estimates the ganglion cell thickness close to the optic nerve rim tissue, or the peripapillary area. Its main value comes in looking for a trend over time in the thickness since glaucoma is a progressive disease. I don't look at the baseline reading too closely for the reasons mentioned above. The photo below is a POAG suspect's first HRT. The cup, which is very large, is coded red. The red/green bars with the multiple lines through them represent the probability that the thickness is normal at each of the points. I don't use this data too much. I concentrate on the displays over time, which is shown on the next page.

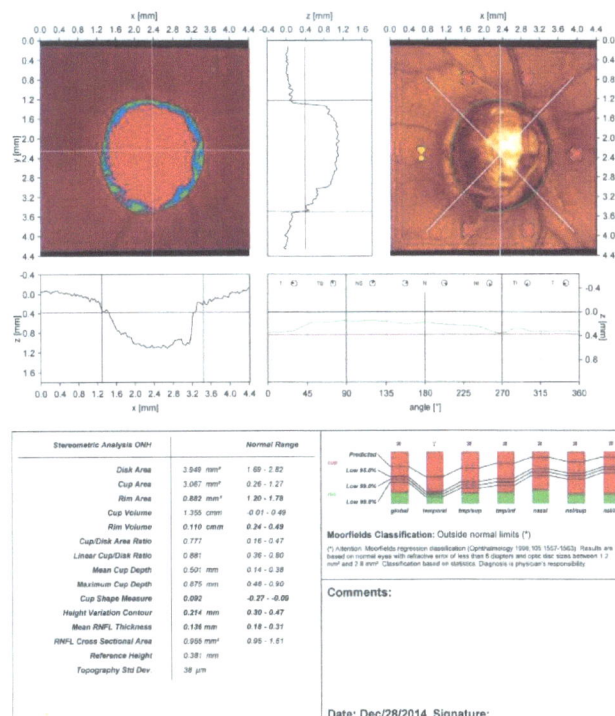

The value of the HRT is to look for thinning trends over time. Time is on the X axis. The zero point on the Y axis is the baseline reading and you want subsequent readings to be at or above the zero point on the Y axis. A downward slope over time is suggestive of thinning and the presence of POAG. Variation in values from test to test can be large and are normal in a lot of patients. Here is an example of a stable HRT OD (OD is on the top of the pair) in a POAG suspect showing a normal amount of variation. Due to this variation, you need 3-4 readings, each at least 6 months apart to see if early damage is occurring. The OS reading (bottom one) in this patient suggests possible thinning over time:

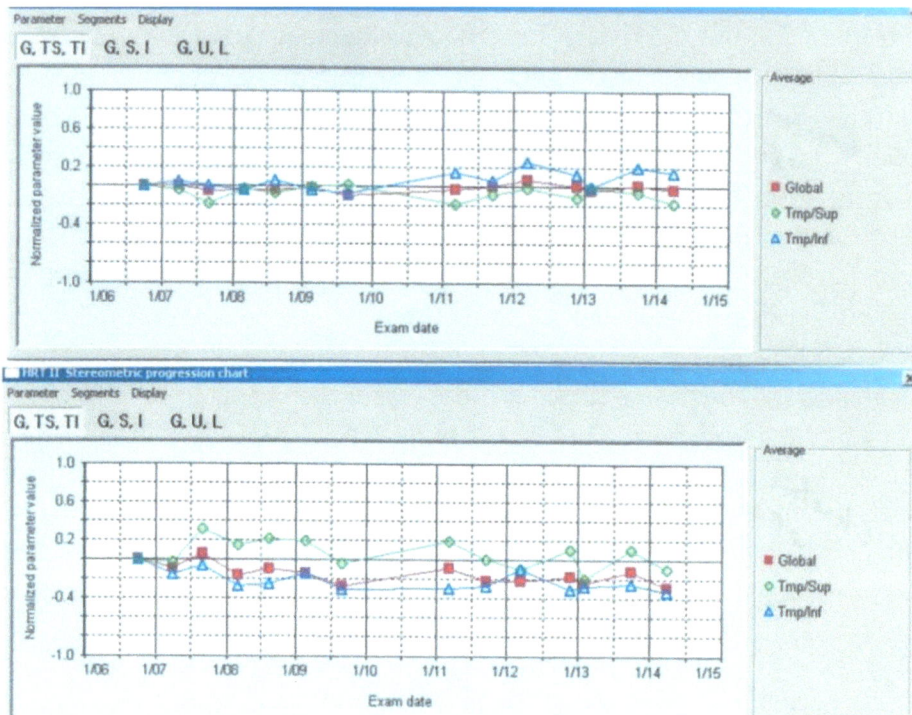

Here are more examples of normal variations in HRT readings. When the variation is large between the temporal superior (blue) and temporal inferior (green) areas, concentrate on the red square or the global value. A cataract can cause this variation as well as anything else that may scatter the red laser light:

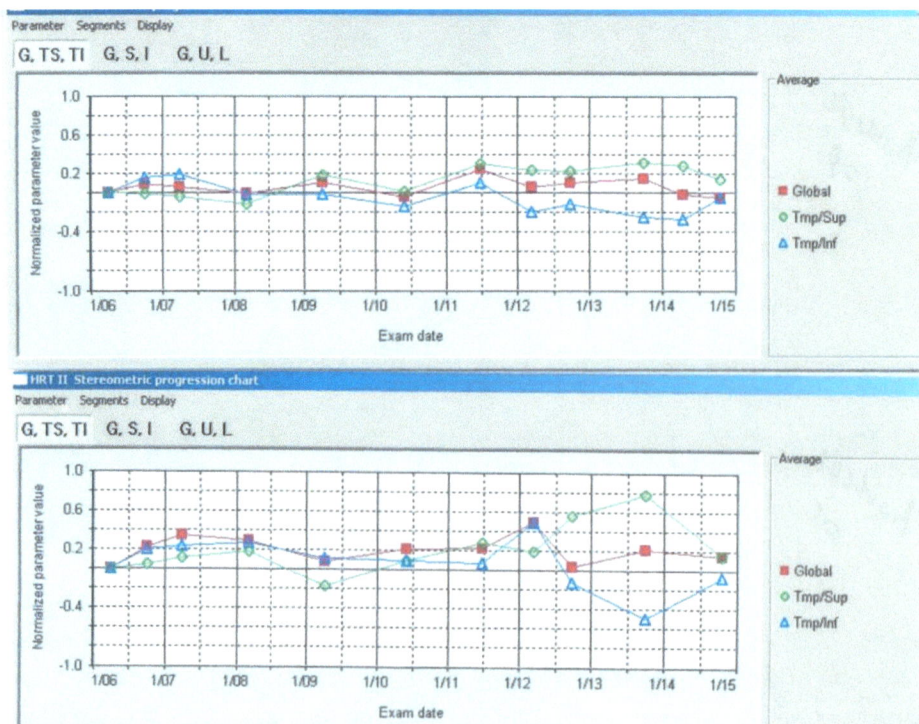

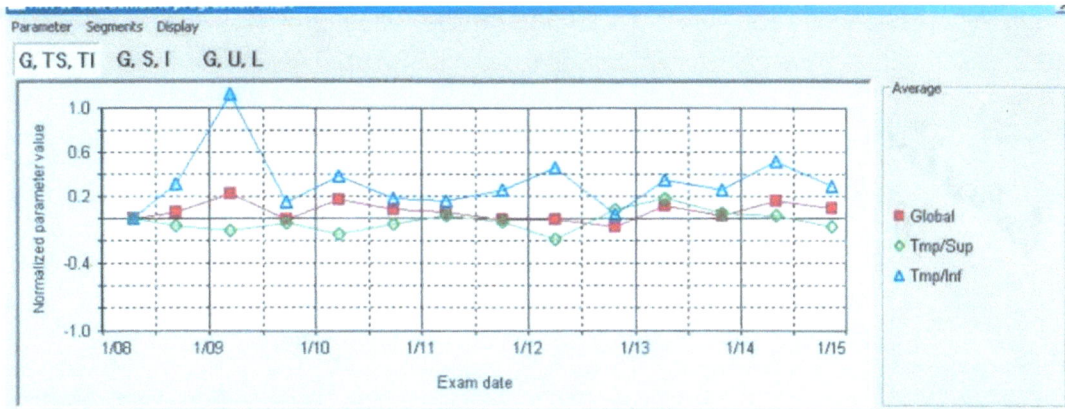

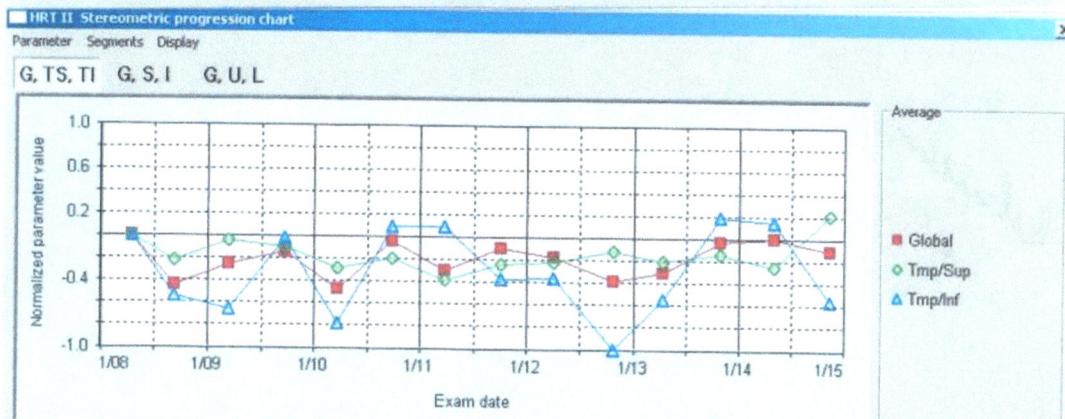

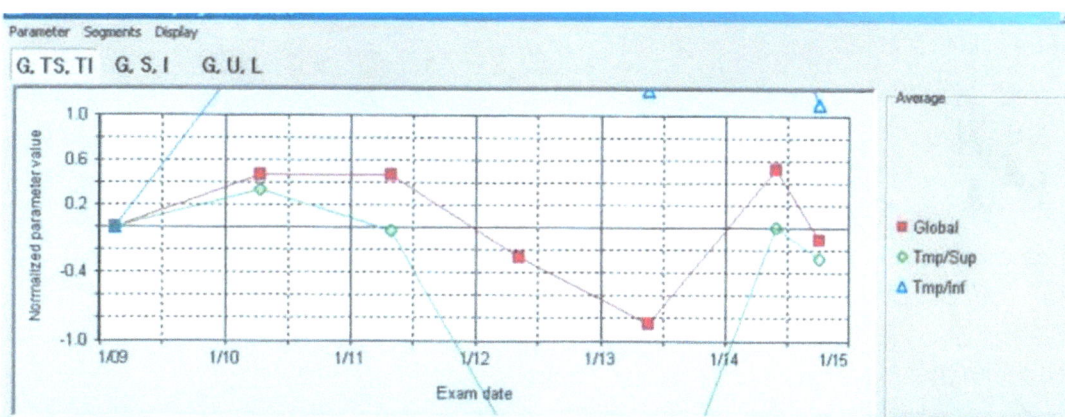

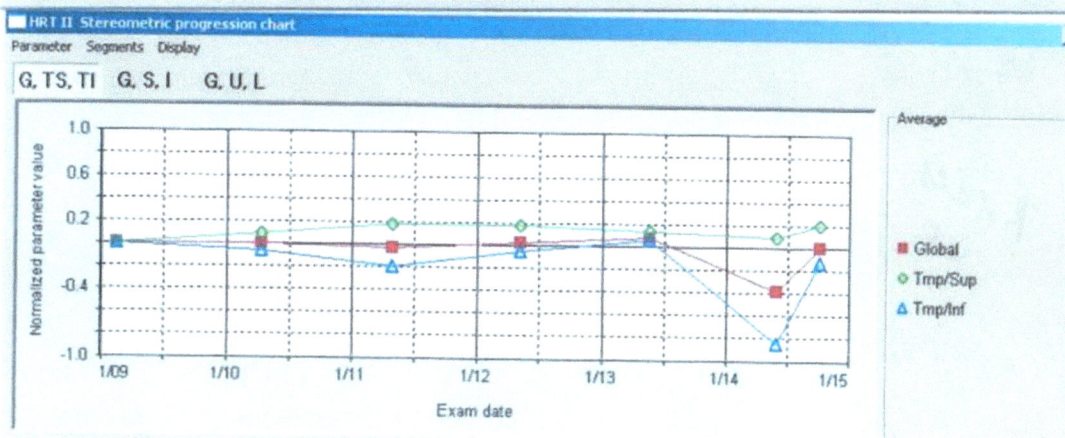

The next patient probably has a significant thinning on the readings in the right eye (top graph) and therefore represents progressive glaucoma damage (assuming no other disease is thinning the axons):

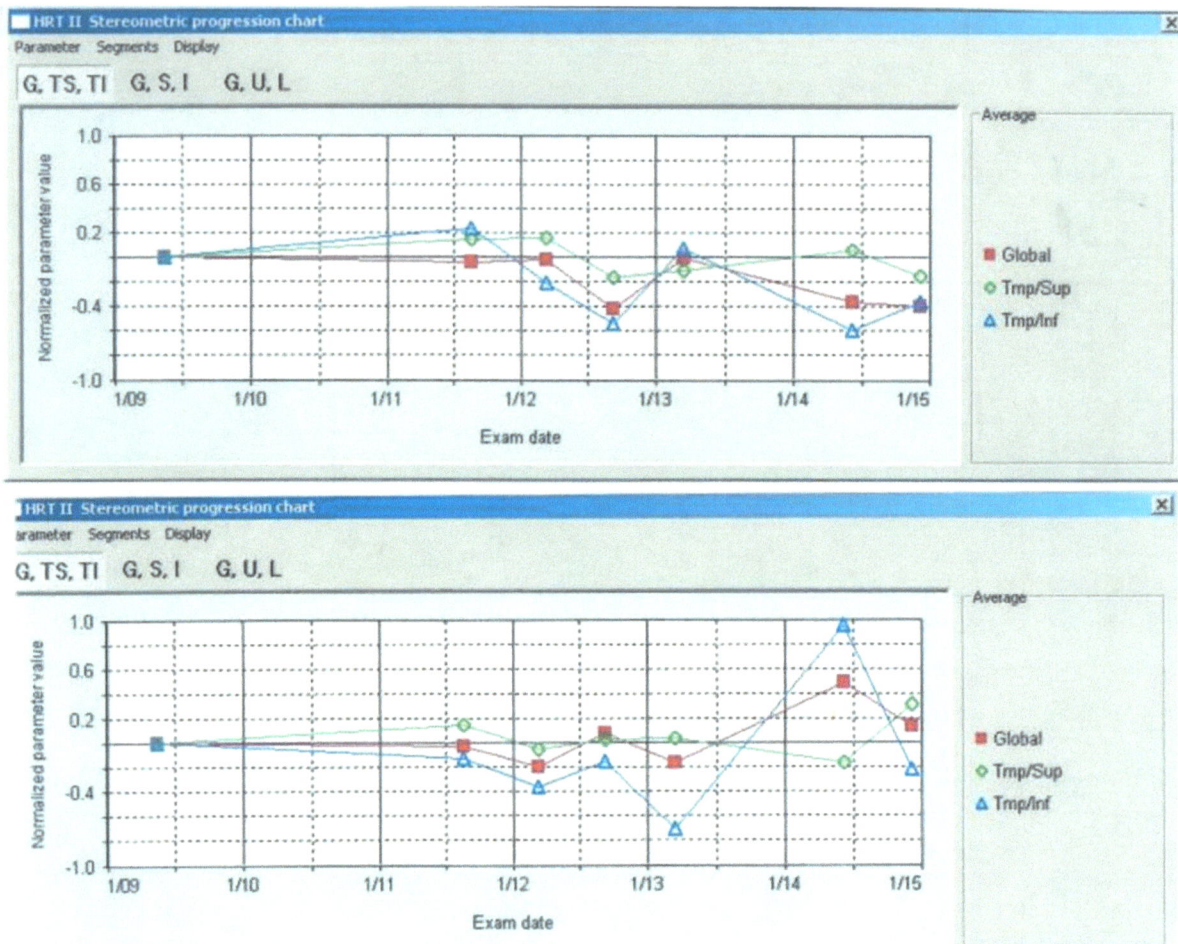

Recent research is looking at measuring ganglion axon damage away from the optic nerve. In a recent article, 104 normal, non-glaucoma suspects patients had their peri-foveal and peripapillary areas measured with OCT.[196] 40% had abnormal readings on the peri-foveal area and 31% had abnormalities on the peri-papillary area. [196] Multivariate analysis revealed that axial myopia was significantly associated with abnormal peri-foveal and peri-papillary OCT measurments.[196] The diagram below shows 3 examples of normal patient OCTs in A,B and C. The red free photo, the HVF, the peri-foveal and peri-papillary OCTs are shown for all 4 patients. D is a glaucoma proven OCT pattern showing how easy it is to falsely identify a patient as having glaucoma when they don't. © AAO 2015.

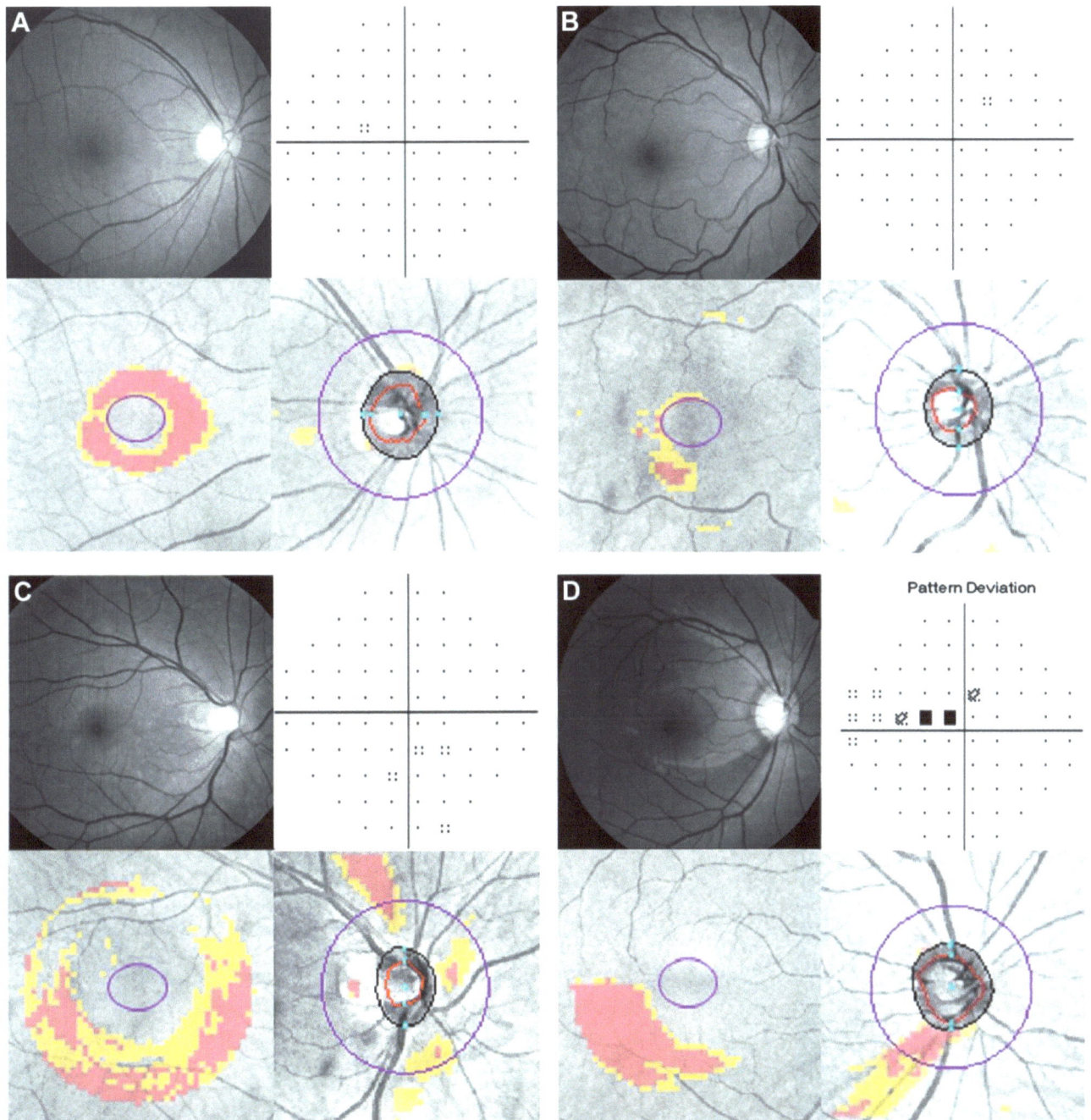

WHY AND HOW DO WE TREAT GLAUCOMA?

A number of studies have shown that reducing the IOP reduces the risk of developing glaucomatous damage. An early large, prospective, randomized, multicenter study that looked at lowering IOP to see if there was a benefit to treatment was the Early Manifest Glaucoma Trial (**EMGT**). 44,243 patients were screened for glaucoma, 255 had newly diagnosed OAG via loss on HVF. Half were randomized to "treatment" with ALT and Betaxolol or to "no treatment." The results showed a 50% reduction in progression of OAG with treatment, progression was reduced from 76% to 59% (P < 0.0001. n=227 completed the median follow up of 8 years).[106, 197] Notice that the treated group continued to have VF loss despite treatment and that this study did not measure or randomize CCT initially.

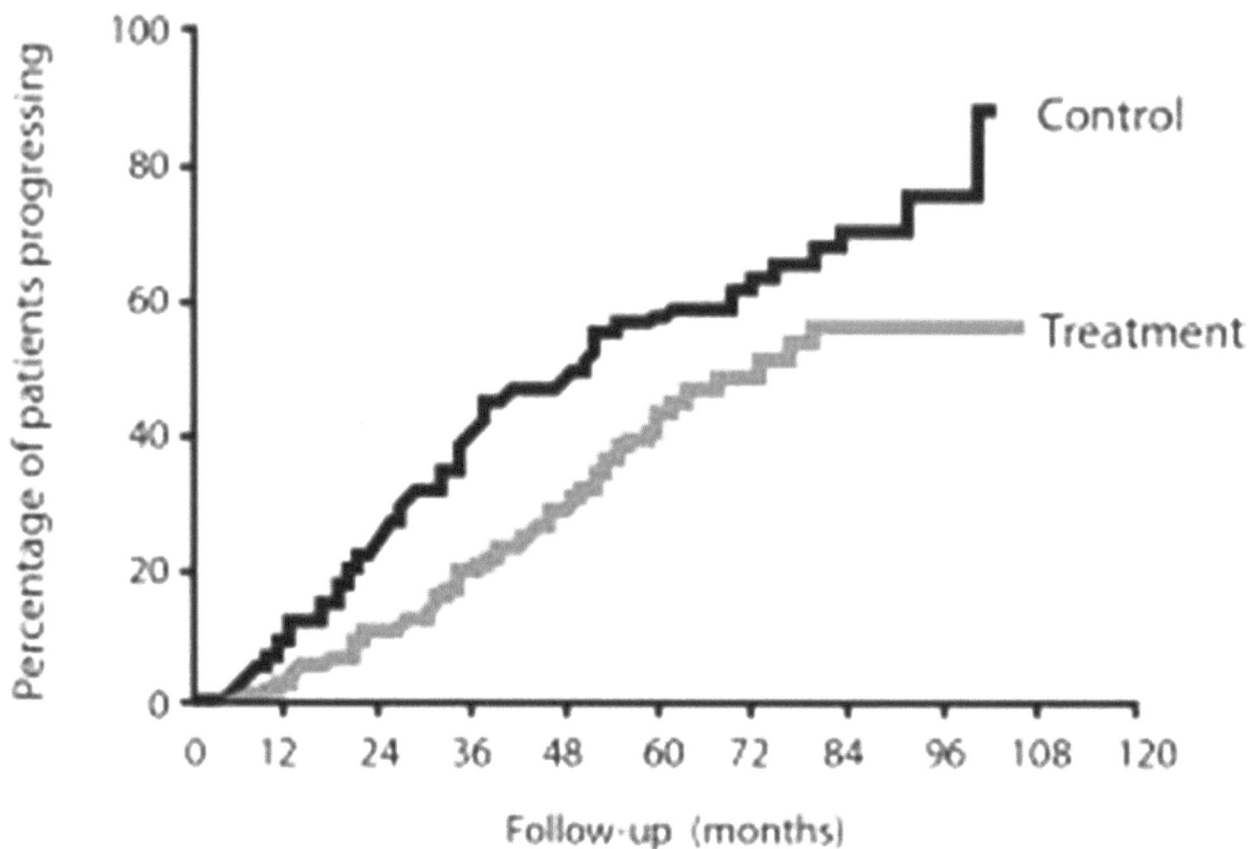

What about treatment in uncorrected-for-CCT ocular hypertensive patients?

The Ocular Hypertensive Treatment Study (OHTS), a prospective, multicenter, randomized study that randomized a total of 1,408 patients with ocular hypertension to two groups, half received OAG treatment and half did not. Patients completed 60 months of follow up. Treatment reduced the risk of developing OAG by over 50% (9.5% vs 4.4%, P < 0.0001).[65] This completed phase 1 of the OHTS. After receiving 60 months of follow up, both groups were given treatment for OAG. They were followed for an additional time, for a total of 13 years, to determine if the non-treatment group had a penalty for delaying treatment. They did not (P < 0.77). [107] This was phase 2 of the OHTS. In the graph below, you can see that the two groups, represented by the two lines, diverge before the gray area, which represents phase 1. The time between phase 1 and phase 2 is represented by the grey column. Phase two, the area after the grey line, also shows that the two groups had their risk of developing OAG parallel, meaning that delaying treatment in the non-treatment group did not induce a penalty on them.[107]

The 13 year reduction in onset of glaucoma in the prospective, randomized OHTS study. Phase 1: n= 1,408, follow up= 60 months.[65] Phase 2: n = 1,159, follow up = 13 years for both phase 1 and 2 combined. [107]

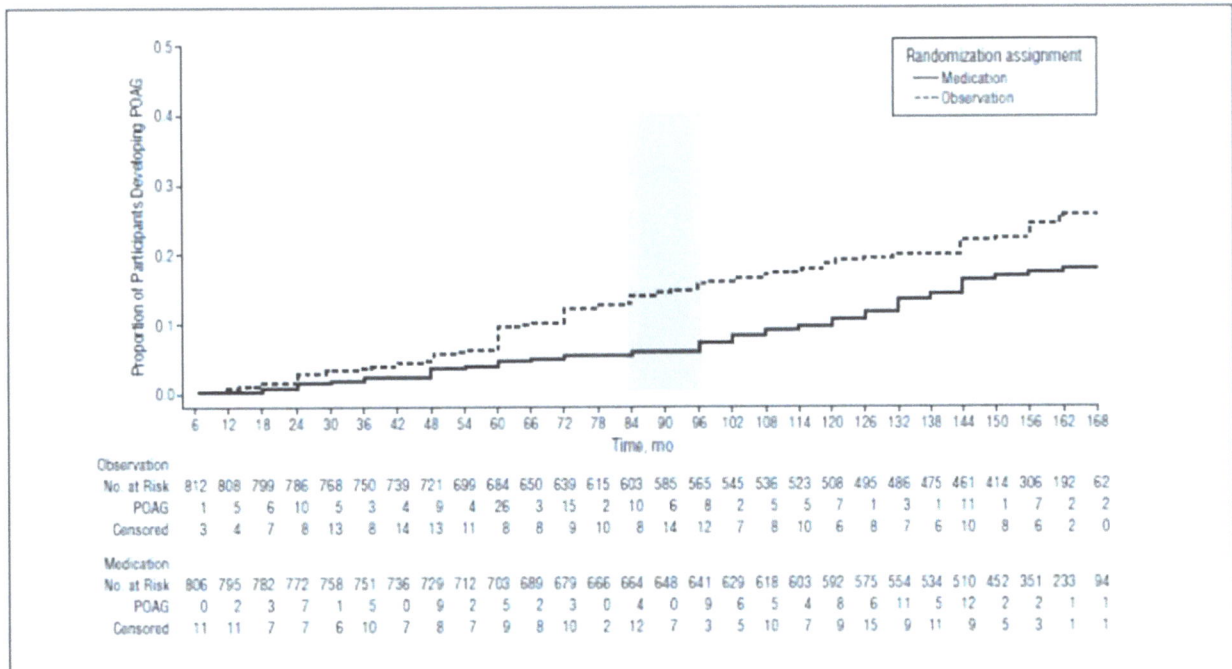

Figure 4. Survival plot of the cumulative probability of developing primary open-angle glaucoma (POAG) over the entire course of the study (February 1994 to March 2009) by randomization group. The number of participants at risk are those who have not developed POAG at the beginning of each 6-month period. Participants who did not develop POAG and withdrew before the end of the study are censored from their last completed visit. Participants who did not develop POAG and died are censored at their date of death. The shaded column indicates initiation of medication in the original observation group.

So, lowering IOP works in patients with early OAG and in CCT uncorrected, ocular hypertensive patients. What about CCT-uncorrected normal tension glaucoma patients? The graph below is from the Normal Tension Glaucoma Study (NTGS), a prospective, randomized, multicenter study that 140 eyes randomized to treatment or no treatment and followed patients for an average of 5.6 years. The "proportion surviving" is defined as patients that maintained no further damage (i.e. did not progress in glaucoma damage) after a number of covariates were factored out (P < 0.0001).

Prospective, randomized, multicenter study, the Normal Tension Glaucoma Study (NTGS) showing a reduced risk of further damage from NTG in patients already diagnosed with NTG (n=140 eyes)[108]

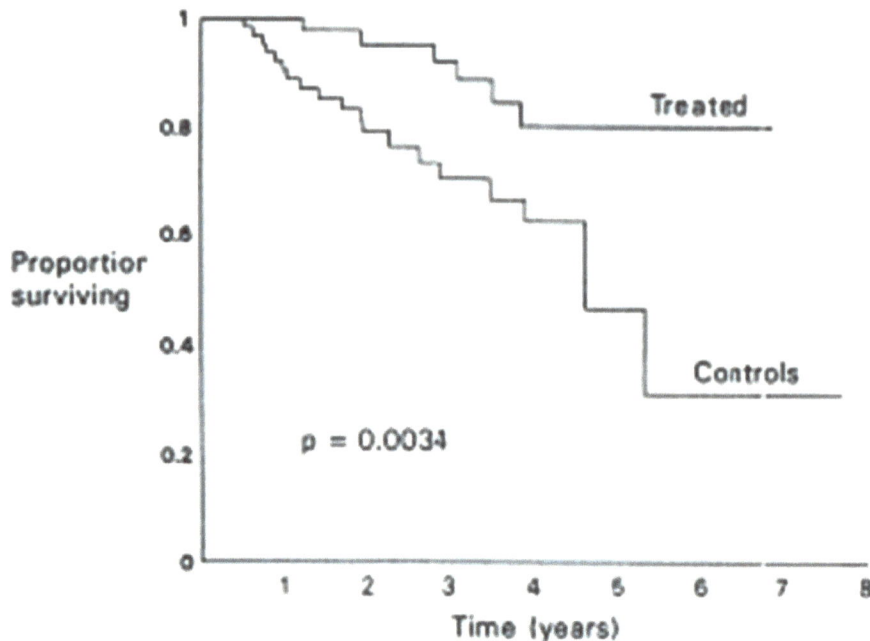

So, lowering IOP works in low tension glaucoma patients as well. Note that this study did not use corrected tonometer readings and we don't know if the true IOP was much higher than measured. We can also look at treatment in different types of OAG as was done in the early manifest glaucoma trial (EMGT).

Time to progression of glaucoma in patients with existing glaucoma for 3 different types of OAG (n = 118) in the "no treatment" of the EMGT cohort (3 graphs above); high tension glaucoma (HTG), normal tension (NTG) and pseudo exfoliation (PEXG) in the EMGT[109]

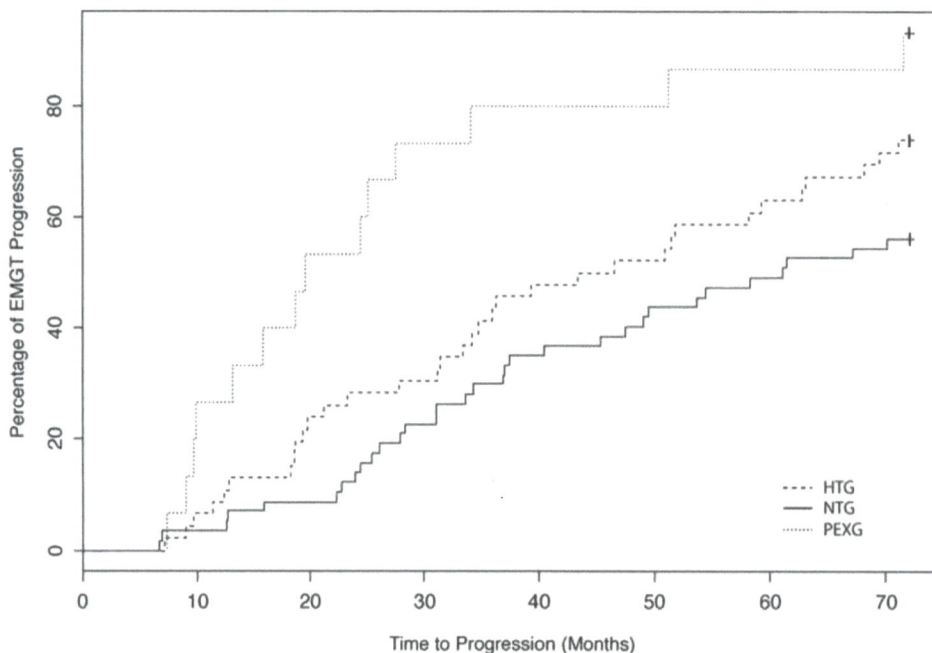

So, treating all forms of OAG; normal tension, early stages and ocular hypertensives, are benefited by the lowering of IOP. You will notice the higher risk for PEX in the study above. This is probably where PEX gets the reputation of having a high risk for the development of glaucoma.

GOAL: The goal of treatment for Open Angle Glaucoma is to lower the IOP enough to reduce or eliminate the chances for further optic nerve damage. The more advanced the glaucoma is (see the rim loss study above), the younger the patient is, or the closer VF defects are to the fovea, the lower the IOP needs to be. The younger the patient, the lower the IOP has to be since they have a lot more years left to develop further optic nerve damage. The more aggressive the glaucoma, the lower the IOP needs be. Various books state that a 25% or 30% reduction in IOP is advisable. Another guideline is to try to reduce the IOP to at least 18mmHg "with an upper limit that is unlikely to lead to further optic nerve damage."[101] You try to go lower than 18mmHg if LTG or advanced glaucoma is present.[110] It is not clear if this is the corrected or uncorrected IOP but I always use the corrected IOP and watch closely for progressive damage with the HVF/HRT/ERG/VEP tests. I feel that with today's tests, we have a decent chance of detecting further optic nerve damage at an early point. **Signs of further optic nerve damage warrant further reduction in the IOP**. However, it is well known that the more drops a patient is supposed to take, the lower the compliance is with taking the drops. You do the best that you can. *Glaucoma surgery is reserved for when drops and SLT fail to reduce the IOP to a safe level.*

Treatment of OAG:

IOP lowering drops are one of two initial treatment options. Today, we have very effective drops, unlike 30 years ago. Compliance with taking the drops is directly related to how many times a day they need to be administered. The more drops you need to take every day, the less likely you are to take them. Fortunately, new drops have been developed that lower IOP significantly and have once a day dosing. Multiple types of drops may be needed in order to get the patients' IOP to a safe level. For all drops, significant side effects are rare but can be reduced further by having the patient close their eyes for a few minutes after administration or, better yet, by pressing on the puncta through closed eyes after administration.

-xalatan (prostaglandin analogs): Works by increasing uveoscleral outflow. Dosing is QHS. Xalatan can permanently darken green or light brown irises, but probably not blue or dark brown irises,[111] and can cause local lid and conjunctival redness that should slowly resolve. Xalatan can also induce eyelash growth and is marketed as Latisse for this effect. You need to document that you told the patient these side effects. Xalatan has also reported to cause CME and might exacerbate HSV keratitis. Xalatan is contraindicated in pregnancy so tell all your pre-menopausal female patients not to use it if they are pregnant or trying to get pregnant.

-beta Blockers: I use the XE gel since it is once a day dosing, TXE 0.5% QAM. Beta blockers reduce aqueous production. Beta blockers can worsen asthma, heart block, and depression. Ask the patient if they have heart block. If they do, they will typically know. Document any history of reactive airway disease and, if they have ever used inhalers, consider not using beta blockers.

-carbonic anhydrase inhibitor: Trusopt is used TID. It decreases aqueous production and is commonly used with timolol and marketed as Cosopt BID. Trusopt is a sulfa-containing drug so check for allergies to sulpha and don't use it if there is a history of a sulfa allergy. Do not use PO Diamox or the PO hyperosmotic agents as the side effects with both of these drugs can be significant.

-alpha adrenergic agonists: Alphagan P .15% BID or TID, decreases production and increases aqueous outflow. When Alphagan is used with the monoamine oxidase inhibitors anti-depressants, (nardil, marplan, emsam, parnate) it can lead to a hypotensive episode (fainting). Alphagan can also cause a significant local dermatitis, which can be chronic and also lead to iritis. Greater than 20% of patients become intolerant to chronic use of alphagan.[112] Iopidine, the other alpha agonist, have a higher allergic response than alphagan so it is only used in pre-treatment before laser surgery.

-Selective Laser Trabeculoplasty (SLT): This is an in-office laser treatment that opens up the drainage meshwork so that the aqueous drains out of the TM better. SLT has been shown in the Glaucoma Laser Treatment Trial (GLT)[113], a prospective, randomized, multicenter study (n=203, median follow up 7 years) to be even more effective than drops in the initial treatment of OAG. If it does not work, drops are as effective after failed SLT as before SLT, so we have little to lose by trying SLT first. The success of SLT seems to be related to how much pigment is in the drainage meshwork, with more pigment increasing the chance for a response. Overall, 70% of patients will respond to SLT. The effect of SLT lasts for an average of 3-5 years and slowly wears off. SLT does reduce the need for drops during those 3-5 years and for many patients this makes non-compliance less of an issue. I treat 360⁰ initially. I have had some patients with IOP's above 40 respond to SLT with IOP going down to 15 within 1 week after SLT. This is an exceptional response, however, and is not typical. As you can tell, I am a fan of SLT!

I typically start treating glaucoma with Xalatan QHS or SLT. Add TXE QAM or Cosopt BID, then Alphagan P 0.15% BID as needed. If the combinations of these drops and SLT don't control the patient's IOP adequately, I refer for glaucoma surgery. Treating glaucoma is usually as straight-forward as this!

Cyclodestructive procedures performed during cataract surgery are a convenient option. Unfortunately, Medicare pays for the Cyclodestructive procedure at half its normal rate when it is performed during CE so don't expect this to be become a widespread option. Recent techniques using micro pulse long wave laser treatment on top of the conjunctiva promises to be as important as SLT in our arsenal.

Surgery (trabeculectomy and tubes): trabeculectomy or tube surgery is the option of last resort in the US due to a significant failure rate and a high complication rate compared to drops and SLT. In the Tube vs. Trabeculectomy Study (n=157, half had trab, half had a tube),[114] a prospective, randomized, multicenter study, 30% of the tube patients and 47% of the trab patients had a failure to control IOP after 5 years of follow-up (P= 0.002):

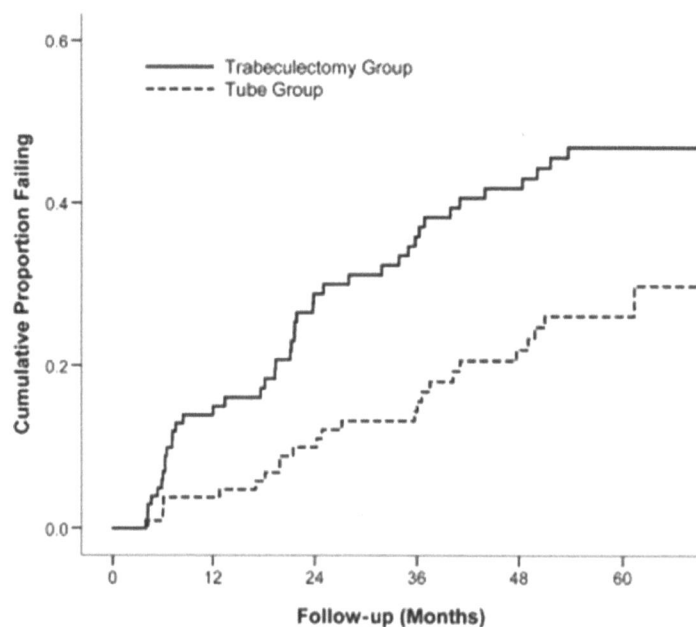

Further, 50% of both groups lost two or more lines of BCVA once "other" causes of reduced vision related to the surgery (such as endophthalmitis and corneal edema) are factored in. And, this doesn't take into account the significant lifelong risk of endophthalmitis that exists after both of these surgeries. This is why we reserve glaucoma surgery as a method of last resort to lower IOP in the US. In Britain, with its socialized medicine, it is cheaper to do trabs as a first line therapy so they do a lot more glaucoma surgery. Recent new technologies such as Minimally Invasive Glaucoma Surgery (MIGS) promise to reduce complication rates but so far have not lowered IOP as well as trabs and tubes do. Here is a photo of a trabeculectomy. Notice the peripheral iridectomy (© AAO 2014):

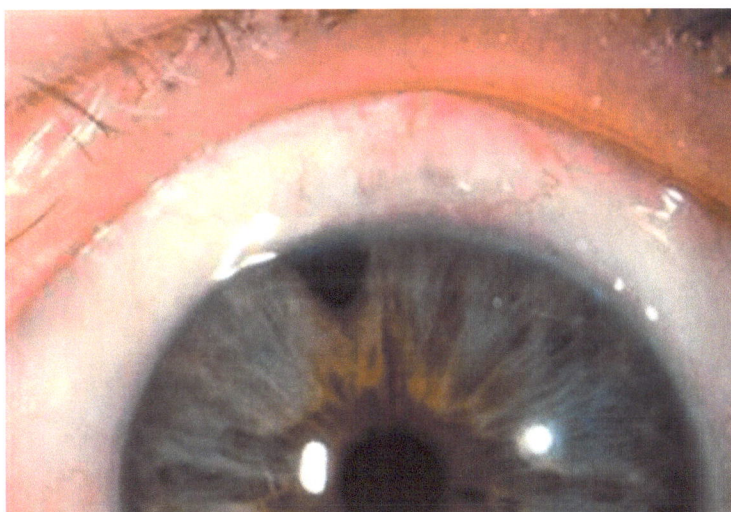

So, after all of this, how do I approach treatment initiation in a glaucoma patient?

My first priority is to classify the glaucoma as open angle or closed angle. You need to refer out for an LPI if only the TM is visible for most of the angle and PAS is present.

Second, I treat IOPs, uncorrected or corrected, if there are 2 or more IOPs at or above 30, regardless of what the other tests show.

Repeated VF losses consistent with glaucoma in the absence of other causes, warrants starting treatment.

Next, I look at the following features: the rim tissue, the C/D ratio along with the corrected IOP, the VF 24-2 results, the presence of a family history of glaucoma, DH, NFLD, if glaucoma is in the other eye, C/D and IOP asymmetry, disk photo changes, HRT trends and the age of the patient. I mentally come to a conclusion to start treatment now or not and to follow up in 6 – 12 months depending on the risk from all of these factors.

Starting medications is a lifelong commitment so make sure that there are *no other causes of the abnormal test results you are treating as glaucoma.*

Again, I start glaucoma treatment with either SLT or xalatan and then add TXE QAM or cosopt BID, and then Alphagan P as needed for IOP control. I aim for an IOP of 18 corrected for most of my glaucoma patients, keeping in mind that I may have to go lower depending on the VF and rim tissue appearance. If these 4 medications do not control the VF and HRT or if DH or NFLD occur while on treatment or the rim tissue is very thin, I consider referring out for MIGS or for a tube or a trab. That is it in a nutshell!

2. Choroidal Nevus

A choroidal nevus occurs in up to 10% of the population and has a pigmented appearance since it is buried beneath the retinal RPE.[115] Have you found that 10% of your patients have a choroidal nevus? Probably not. Imagine the number of nevi that you have missed over the years! The main concern with a nevus is its conversion to a melanoma, which *will* kill you.

Here are examples of choroidal nevi, some obvious, some subtle:

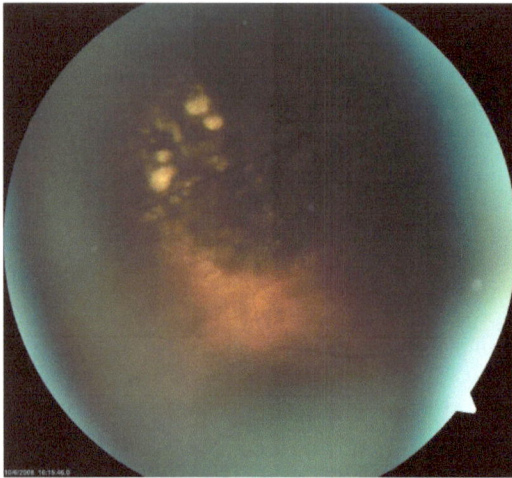

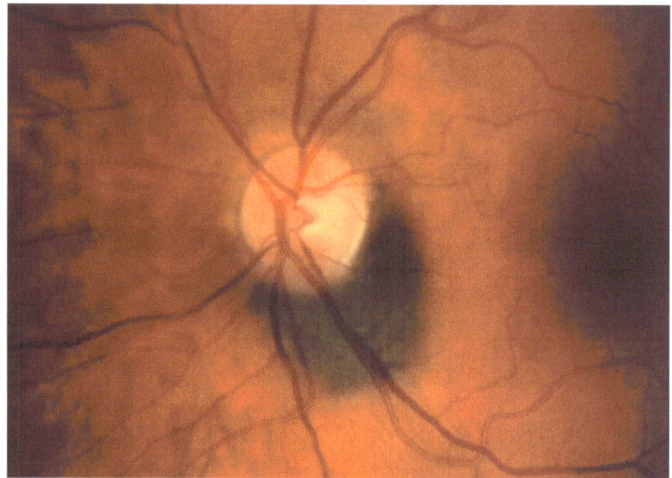

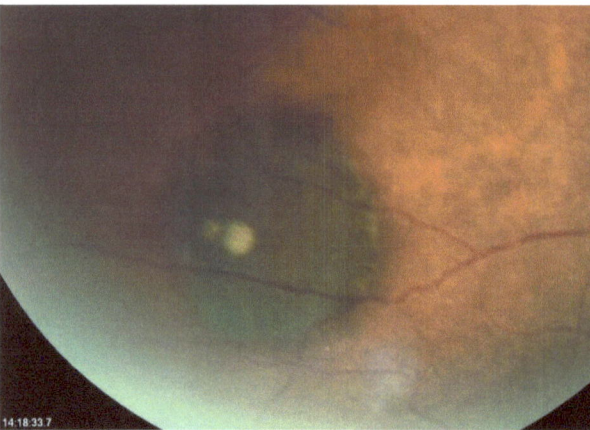

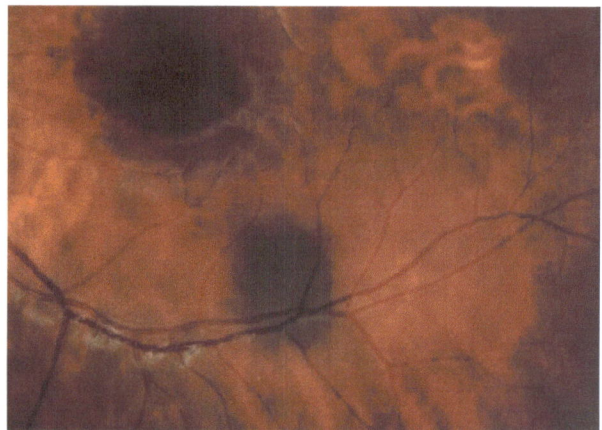

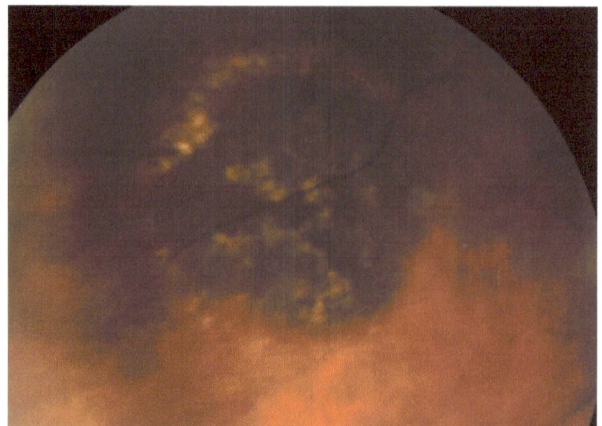

Here is a patient with a subtle, easy to miss sub foveal nevus OD that looks like normal pigment until you compare the patient's two fovea's to each other:

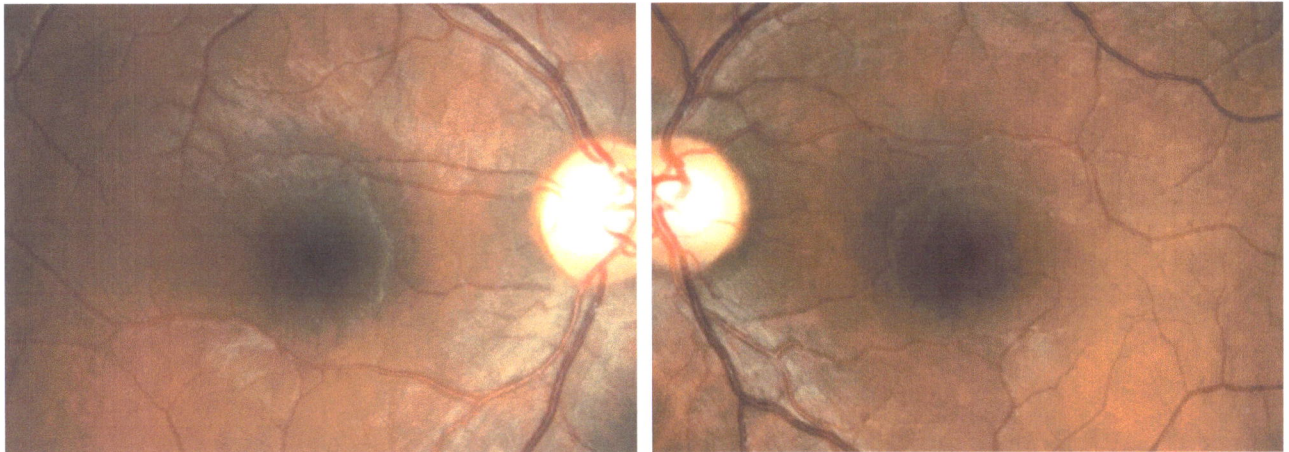

More nevi. The one on the left deserves a B scan for thickness:

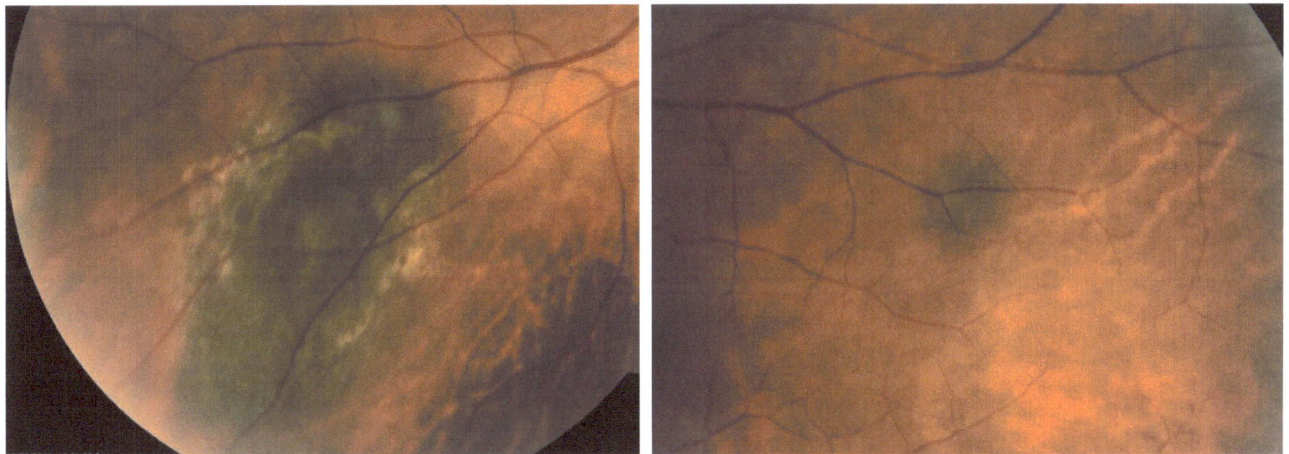

Another patient with a subtle choroidal nevi and a CWS that disappeared over 4 months:

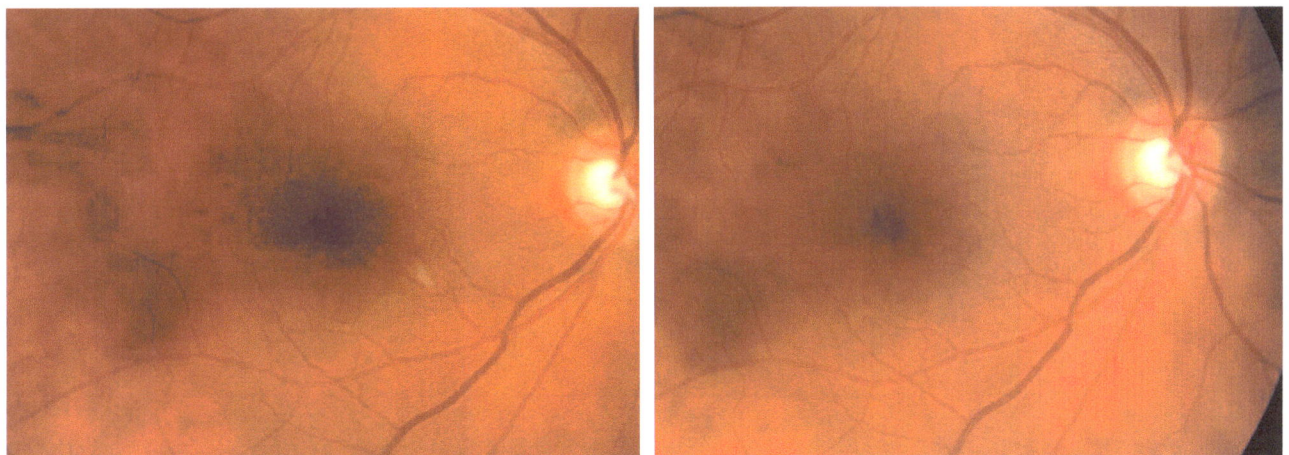

There are two concerns a patient has when the subject of an ocular nevus/melanoma arises. What are their chances of their nevus converting into a melanoma and once the melanoma is treated, what are the chances of metastasis.

First, what are the chances of conversion from a nevus into a melanoma? Large nevi are easy to spot. It is the small ones with a reduced chance of conversion but don't have zero risk. The Small Collaborative Ocular Melanoma Study (COMS), a retrospective study of 240 patients with lesions 1.0 mm to 3.5 mm thick followed for an average of 3.3 years, showed that the risk of a lesion <2 mm in thickness showing growth over a mean follow up of 3.3 years was 2% (not 0%), and the risk of a lesion ≥ 2 mm in thickness showing growth over a mean follow up of 3.3 years was 14%.[116] Documented growth is a real concern for conversion and deserves a referral to an oncology center.

The following graph shows the risk of growth versus the thickness (height) of the lesion being followed. It shows the higher risk for lesions ≥ 2.5 mm thick in the small COMS.[116]

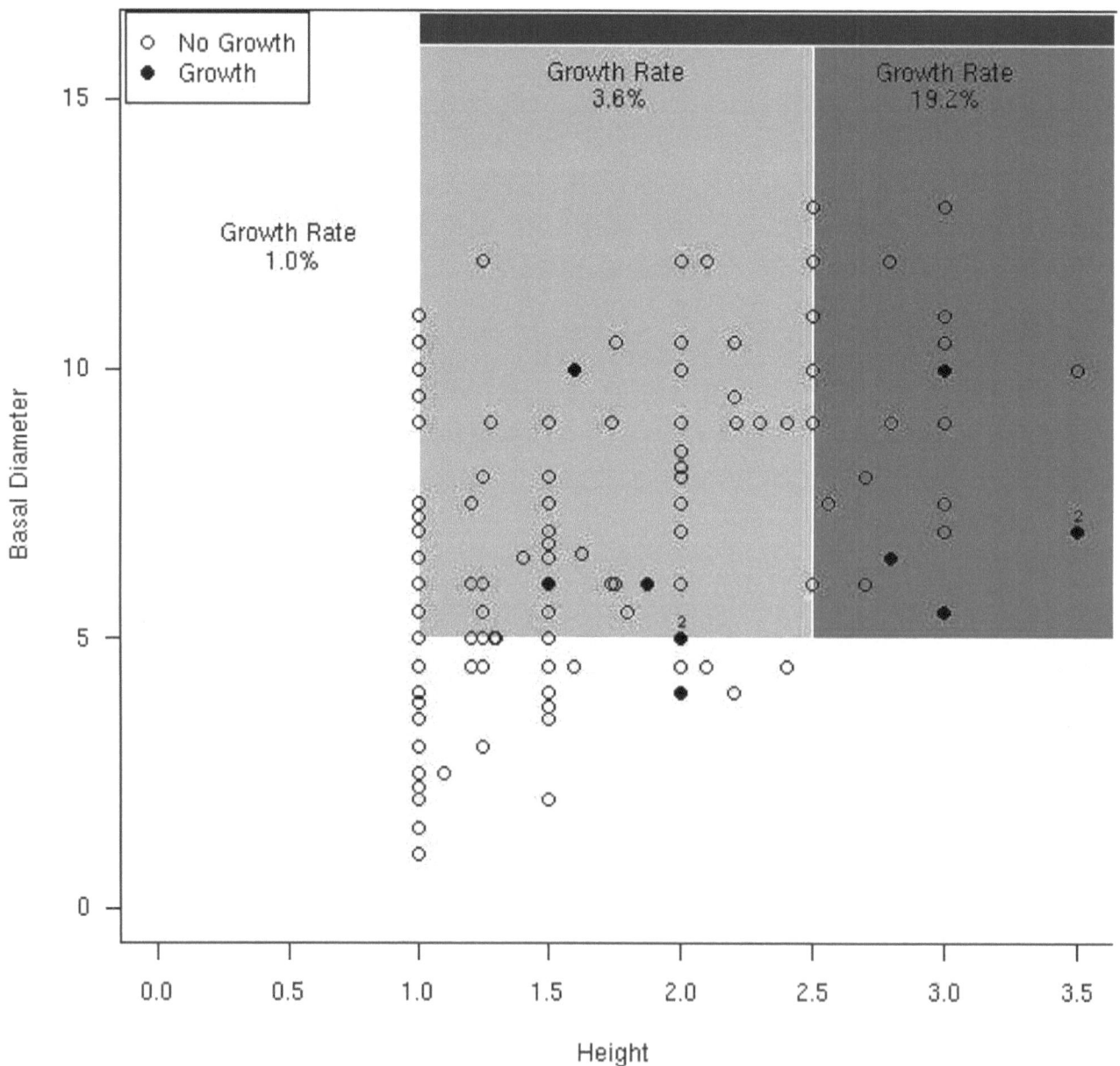

In the largest study to date of small nevus patients (1,287 patients retrospectively followed for an average of 62 months), Carole Shields, M.D. found that in a multivariate analysis,[117] the following factors were identified as increasing the risk of metastasis. The relative risk to lesions <1.0 mm thick presented as well:

Factor	(relative risk of metastasis)
- tumor thickness (1.1 to 3.5 mm in thickness compared to < 1 mm thick)	8.8 X
- documented growth	3.2 X
- margins of the lesion touching the disk	2.9 X
- symptoms of blurred vision	1.9 X

Eighteen percent of the small lesions in this study overall exhibited growth and 3% metastasized.

"Virtually all choroidal melanocytic lesions thicker than 3 mm are melanomas, and virtually all choroidal melanocytic lesions thinner than 1 mm are nevi. Many lesions 1-2 mm in thickness (apical height) may be benign, although the risk for malignancy increases with height" [176]

Ninety eight percent of uveal melanomas are found in Caucasions (n=4,070 in the NIH SEER data collection).[118] Eighty two percent of the uveal melanomas were choroidal with the other 18% arising from the ciliary body and from the iris.[118] The mnemonic "To Find Small Ocular Melanomas" (TFSOM) is used as a reminder of the risk factors for growth of a choroidal nevus. 3 or more of these findings have a > 50% chance to show slow growth and to likely be a small melanoma instead of a nevus.[119]

- **T**hickness: 2.0mm or more in thickness, basal diameter 8mm or more, as found on B-scan
- **F**luid: subretinal again found on B-scan
- **S**ymptoms: flashes or floaters
- **O**range pigment (lipofuscin): not yellow pigment
- **M**argin: less than 3mm from optic disk

Another way of looking at ocular melanomas is to ask, what are the chances of survival *after* treatment? In the largest study to date, Carol Shields, M.D. found that in 7,256 eyes with choroidal melanoma, metastasis at 10 years after treatment was strongly correlated to the thickness of the lesion. The following are the 10 year post treatment risk of metastasis: [191]

Thickness	risk of metastasis 10 years after treatment
0 – 1.0 mm	5%
1.1 – 2.0 mm	12%
2.1 – 3.0 mm	12%
3.1 – 4.0 mm	16%
4.1 – 5.0 mm	26%
5.1 – 6.0 mm	28%
6.1 – 7.0 mm	28%
7.1 – 8.0 mm	41%
8.1 – 9.0 mm	48%
9.1 – 10.0 mm	44%
> 10.0 mm	52%

In this study, 90.3% of the uveal melanomas were choroidal, 3.5% were iris, and 6.1% were ciliary body. New research is looking at predicting melanomas that are at higher risk to metastasize. Notice that the risk of metastasis is not zero for small lesions! Chromosomal monosomy 13 has been linked to a significantly higher risk of metastisis.[192] A biopsy of the lesion is required to diagnose a chromosomal abnormality, a technique that could increase metastasis. Patients electing to undergo biopsy at the same time as placement of the plaque or enucleation (still done for large tumors or ones encircling the optic nerve) can elect to receive chemotherapy if found to have monosomy 13. Suturing a radioactive plaque to the sclera has been shown to be as effective as enucleation for reducing metastasis.

Other lesions that can look like a nevus or melanoma include:

- melanocytoma, a darkly pigmented lesion at the optic disk
- RPE hypertrophy/bear tracks or CHRPE (not the same as in Gardner's Syndrome, which has variegated lesions with halos and tends to be multiple and bilateral. See below.)[120]
- metastatic carcinoma (has increased thickness on B scan, refer to ophthalmology if suspected)
- choroidal hemangioma with RPE hypertrophy
- choroidal osteoma

A nevus patient requires yearly photos _for life_. I document that I have told the patient that the nevus can convert to a melanoma and kill them so they realize the importance of yearly follow up. Treatment used to be enucleation of the eye but radioactive brachytherapy treatment is now the standard of care (for the COMS study see http://www.jhu.edu/wctb/coms/index.htm). Refer all large, suspicious nevi for a B-scan to determine the thickness and the basal diameter. Large-diameter nevus (greater than 2-3 optic disks) should also have a yearly B-scans done as well as photos. Make it clear to the retina doctor that they are not to do photos. Both of you will not be paid for taking photos.

A confusing topic has been the association of congenital hypertrophy of the retinal pigmented epithelium (CHRPE) and Gardner's syndrome. CHRPE, jet black in appearance as opposed to a choroidal nevus' grey appearance, has been described as occurring in 3 forms; _solid, grouped (bear tracks),_ or _multiple._[178]

"Bilateral or multiple CHRPE are disseminated over the entire fundus in both eyes and can be associated with familial adenomatous polyposis or Gardner's syndrome. However, solid and grouped CHRPE usually occur <u>unilaterally</u> and seem to represent distinct entities unrelated to familial adenomatous polyposis coli".[178]

Single CHRPE appear homogeneously black when seen in a younger patient and in older patients, develop lacunae (areas of depigmentation in the center of the lesion).[175] _Grouped_ CHRPE appear as bear tracks. In the original article, Gardner's syndrome was defined as the "presence of bilateral lesions, multiple lesions (more than 4), or both appear to be specific (specificity, 0.953) and sensitive (sensitivity, 0.780) clinical marker for Gardner's syndrome".[179] Further, Jerry Shields, M.D., states that "solitary CHRPE and congenital grouped pigmentation differ clinically from the multiple pigmented lesions seen with familial adenomatous polyposis and that patients with these [former] conditions, as well as their relatives, are not at a greater risk of developing intestinal cancer". [180] Bottom line, <u>bilateral</u> CHRPE patients are referred out for colonoscopy, not unilateral bear tracks or single, unilateral CHRPE. Virtually 100% of patients with Gardner's syndrome develop colon cancer so refer patients with _any_ suspicion of bilateral CHRPE whether it is solitary or grouped (bear tracks).

Here is a solitary CHRPE lesion in a younger patient with lacunae (© AAO 2014):

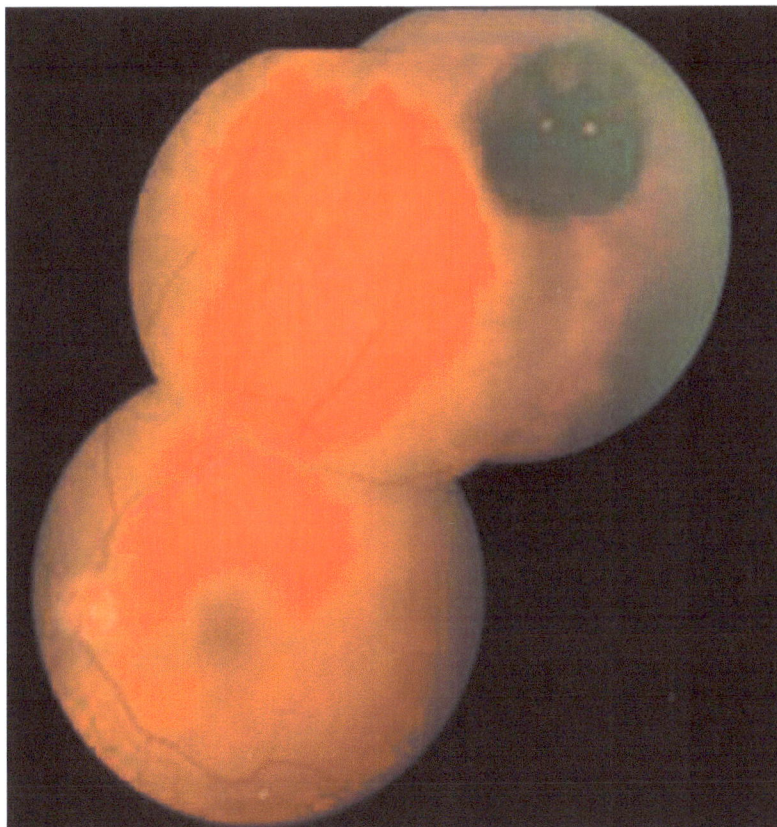

Here is a CHRPE lesion in a older patient with a lacunae (© AAO 2014).

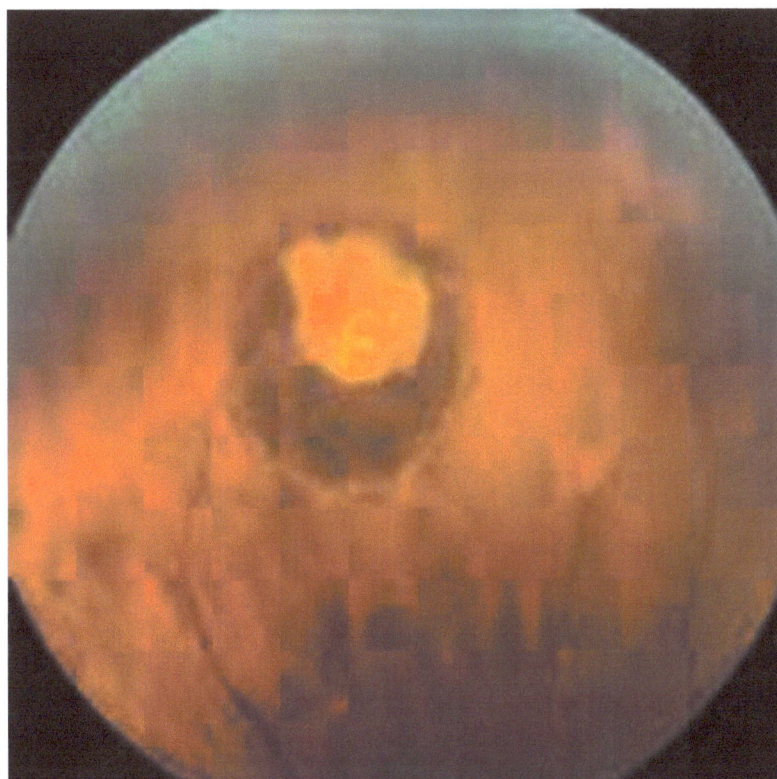

Here is a photo of bear tracks. This in both eyes can be Gardner's syndrome.

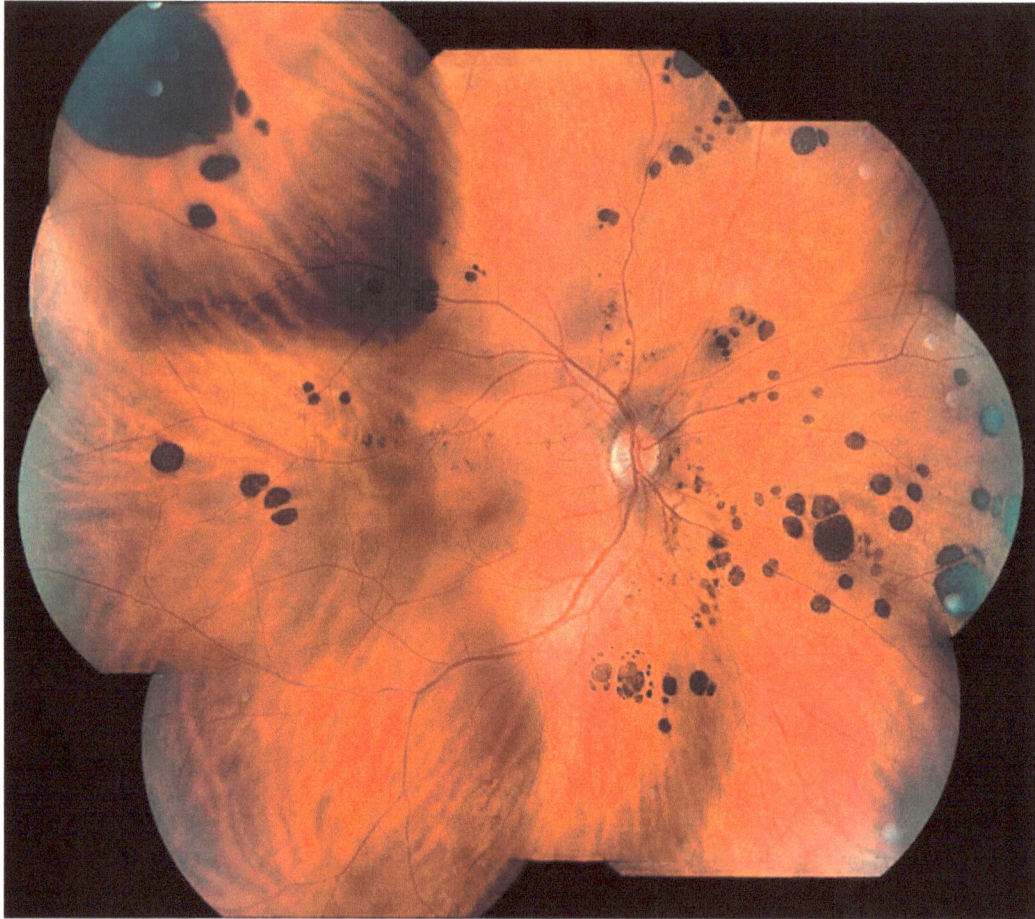

3. Meibomianitis/Blepharitis/Dry Eyes:

Meibomianitis (*posterior* blepharitis) is a very common condition. Fully 2/3 of my patients have this to some degree. Meibomianitis is an inflammatory plugging up of the meibomian gland openings that results in worsening of dry eyes, injected eyes, corneal infiltrates, or rarely, corneal ulcers. The corneal responses come from the toxins that the bacteria living inside the glands excrete. You diagnose meibomianitis by placing pressure on the lid margin by the eyelashes with a 6 inch Q-tip. Look for no discharge or worse, a whitish discharge coming from the meibomian gland openings. You treat it with warm compresses and lid scrubs (WC/LS) QD *for life*. Tell the patient that this is a condition they will have the rest of their life and that they need to do WC/LS *for the rest of their life*. You will be asked why they have it. We don't know. Some people just get it. 100 mg BID of PO doxycycline in non-pregnant adults over a 2-3 week period can help bring a bad case of meibomianitis under some control. However, do not rely on PO doxycycline for chronic control. People want a magic bullet to control meibomianitis forever. It doesn't exist. They have to do WC/LS QD for life. Do not rely on a chronic use of topical antibiotics to treat meibomianitis and use systemic antibiotics only for pre-septal cellulitis cases. Topical steroids are powerful and should only be used for stubborn corneal SEI cases that don't respond to WC/LS after 6-8 weeks. Do not use topical steroids for sterile ulcers from lid margin disease either!

Have the patient warm a wash cloth under hot water. Tell them not to burn themselves. Place the warm washcloth on their closed eyelids for about 3-5 seconds. The washcloth transfers its heat quickly. Re warm the washcloth and repeat the process for about 7-10 times. Really heat up the lids as this is probably where most of the relief comes from. Then, using Dove soap, wash the closed eyelids off, moving side to side so as to not get the soap

inside your eyes. This whole process is one warm compress and lid scrub (WC/LS). Significant cases of meibomianitis and blepharitis need to do this 2-3 times a day for 2-3 weeks to get the glands unplugged, and then once a day for life to keep them unplugged. Styes need to do WC/LS 3-5 times a day for 2-4 weeks to get the gland unplugged and then once a day for life. Topical or oral antibiotics aren't going to help. Tell the patients that they need to do WC/LS every day for life for control of these conditions. There are no magic bullets here.

Blepharitis (*anterior* blepharitis) is an inflammatory condition of the eyelash follicles and results in a dandruff-like material deposited at the base of the lashes. It can also give rise to red eyes, worsening dry eyes, corneal infiltrates or rarely corneal ulcers. It is also treated with WC/LS for life. Tell the patient that they will have the condition the rest of their life and that we don't know why some people get the condition. Here is a photo of blepharitis demonstrating scurfs (© AAO 2014):

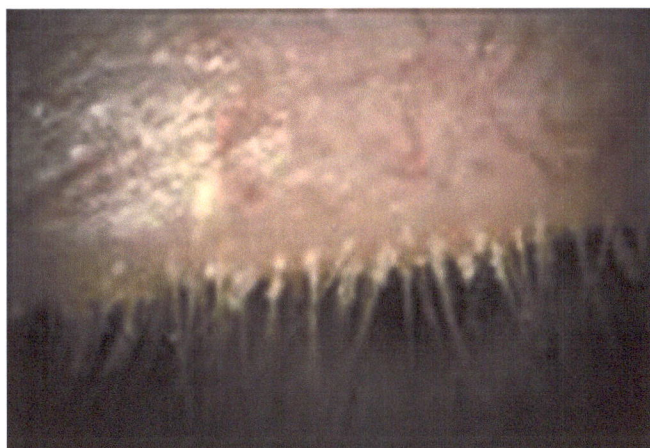

Here is a photo of a patient with angular blepharitis (© AAO 2014):

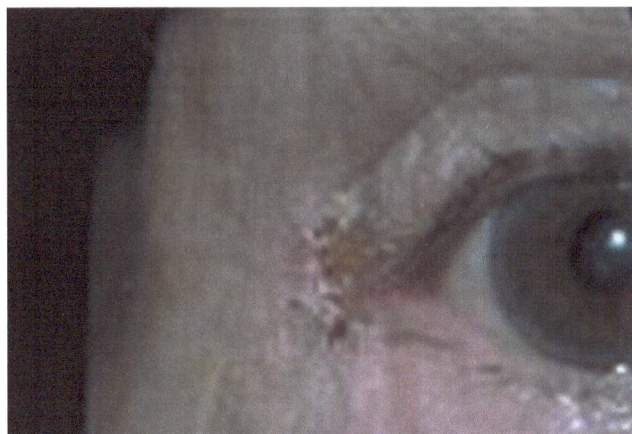

Here is a photo of meibomianitis (© AAO 2014):

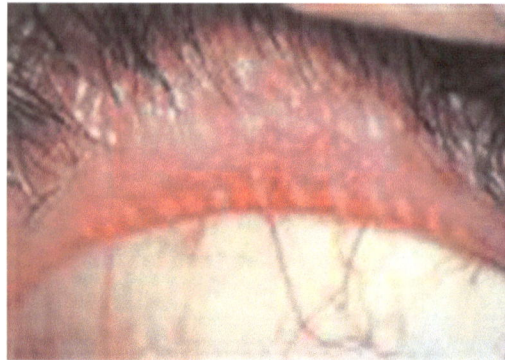

Ocular rosacea is another chronic inflammatory disease that plugs up the meibomian glands. The treatment is also WC/LS for life. Here is a photo of ocular rosacea (© AAO 2014):

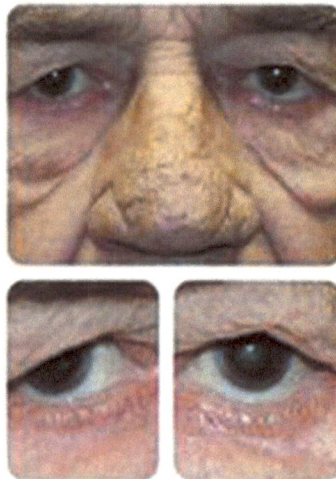

Most of the endophthalmitis from intra-ocular surgery comes from bacteria colonized in the eyelid glands and lashes. One of the most important things you can do for a patient about to undergo cataract surgery is to tell them to do WC/LS BID for 2 weeks before the intra-ocular surgery.

Recently, Dr. Scheffer Tseng, M.D. has popularized the idea that chronic blepharitis is caused by a Demodex mite infection, seen attached to eyelashes using a high powered microscope. He advocates tea tree oil lid washes and claims to have had good success. You can pursue this if you like and become the chronic blepharitis expert in your area. Here is a photo of a demodex infection (© AAO 2014):

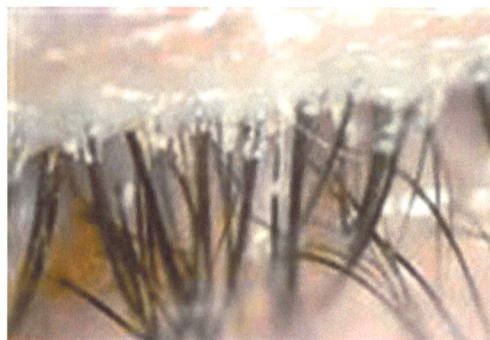

Dry Eyes (Keratoconjunctivitis Sicca):

Dry eyes is a very common condition; symptoms include:

- burning
- a sandy or foreign body sensation (FBS)
- watery eyes from reflex tearing
- symptoms worse in the PM after the eyes have been open for hours and the tears have evaporated away

Dry eye is due to an underproduction of tears made worse with age, meibomianitis, blepharitis, anti-histamines, and the inability to close the lids properly as seen in CPAP users and in Bell's palsy patients. Doing a lot of reading where you blink less, especially in dry air conditions can also exacerbate dry eyes as can SCL use.

There is a lot of buzz surrounding the treatment of dry eye lately but let's break it down simply and clinically. First you need to control any meibomianitis or blepharitis that can be present, with WC/LS QD for life. *This alone may be enough*. (I suspect that the anti-inflammatory effects of restasis may just be an indirect way of opening up the meibomian glands, something that can be accomplished with WC/LS without expensive medications. This is my opinion).

Next, artificial tears (AT) PRN are used. When the patient gets tired of using ATs and wants more permanent control, either Restasis BID *for life* or punctal plugs (PP) are the next option. Restasis takes 3 months to begin to have its effect, has to be taken BID for life, and is expensive for some. Assuming that you have any meibomianitis or blepharitis present under control, I prefer PP to restasis. I place a temporary set of 4 that lasts for about 1 day. The global period for PP is 10 days so if the patient likes the temporary PP, I bring them back for permanent PP 2 weeks after the temporary PP. I use to insert the smart plug but I had a problem with a chronic low-grade pussy discharge in the space above the smart plug and the punctal opening, requiring me to massage the smart plug out, so now I use the older style Oasis non-sterile PP. They can fall out if the patient rubs their eye too vigorously but I don't get a pussy discharge. Tell the patient not to rub their eye nasally as they may rub the PP out. They can wipe debris off of the inner canthus with the eyes closed if they must rub their eyes. PPs are easy to insert and only require topical anesthesia. You really should be inserting them. Again, make sure that you control any meibomianitis and blepharitis first before inserting PPs. Inserting PP in an active meibomianitis patient can trap the toxins that are produced by these two conditions and make the eye irritation worse. Do not use topical steroids or systemic meds to control dry eyes. Humidifiers, omega 3 and 6 fatty acid supplements, and hydration with water might help a little.

Diagnosis is by an anaesthetized (the only one to do) Shirmer's tear test, a reduced tear meniscus, the tear break up time (TBUT) test, or the Tear Lab test for osmolarity. Abnormal results include a shirmer's test < 15 mm, a TBUT test < 10 seconds (fluoress is good to use), an osmolarity greater than 308 or a difference between the eyes of 8 units or more, or a reduced tear meniscus. I usually do an anesthetized shirmer tear test or a Tear Lab test, document any symptom consistent with dry eye and document that ATs have failed to permanently control the symptoms in order to have the insurance pay for the PP. An operative note is required, as well as a consent form. I've been refused payment for not having a "separate and identifiable" operative note despite exam notes clearly stating that I did insert the plugs.

Severe cases of dry eye develop filaments or noninfectious ulcers. Refer these patients out. A tarsorraphy or bandage SCL, or even an amniotic membrane, may be needed.

You will have patients on Plaquenil. The risk of retinal toxicity rises sharply towards 1% after 5-7 years of use or with total doses of > 1,000 grams.[121] Patients have had toxicity on less than this as well.[122] To find the total dose, you multiply the daily dose in grams X (365 days/year) X (number of years on Plaquinil).

For example, a patient on 400 mg/day Plaquenil for the past 10 years would have: (.400 grams/day) (365 days/year) (10 years) =1,460 grams of total dose.

Current (2011) AAO recommendations are: these tests every 12 months after 5 years of use:
    a DFE (I do photos at baseline and yearly too)
    HVF 10-2 with white light
    And one or more of the following:
        - ERG macula centered, I use Diopsys
        - OCT
        - autofluorescence

I personally see patients every 6 months with a HVF 10-2 white light, a DFE, and a macular centered ERG (Diopsys). I'm going to add autofluorescence soon. Most patients want the tests every 6 months after starting Plaquenil and I oblige them. You may wonder why you are using so many resources following negative Plaquenil tests until you have your first patient lose vision or have continuous photopsias from plaquenil. These can even continuing to get worse even years after stopping the Plaquenil. It is used in the treatment of RA and Lupus and it does help a lot in those two conditions. The Amsler grid and color vision testing are no longer recommended. I use the Diopsys ERG in my clinic for Plaquenil testing. Here is a photo and IVFA of first chloroquine and then two plaquenil retinopathy patients (both © AAO 2014):

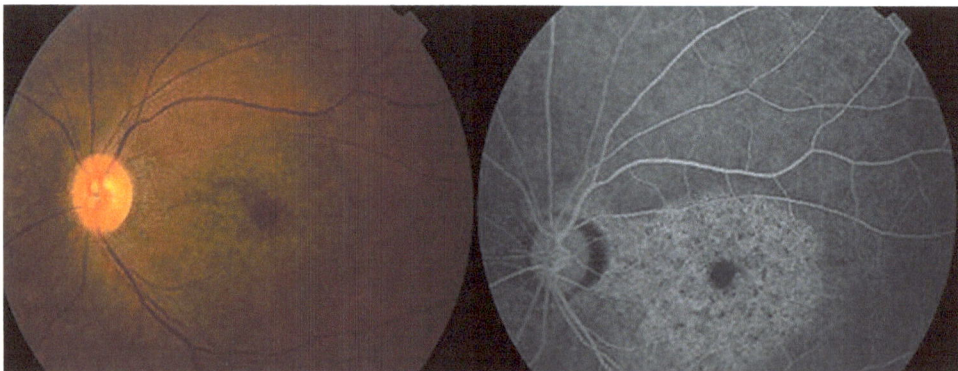

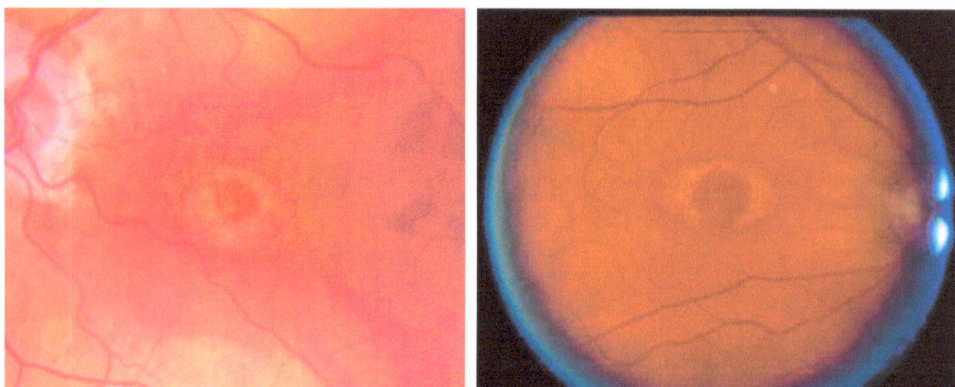

Let's look at a normal spectral domain OCT before we look at abnormal OCTs:

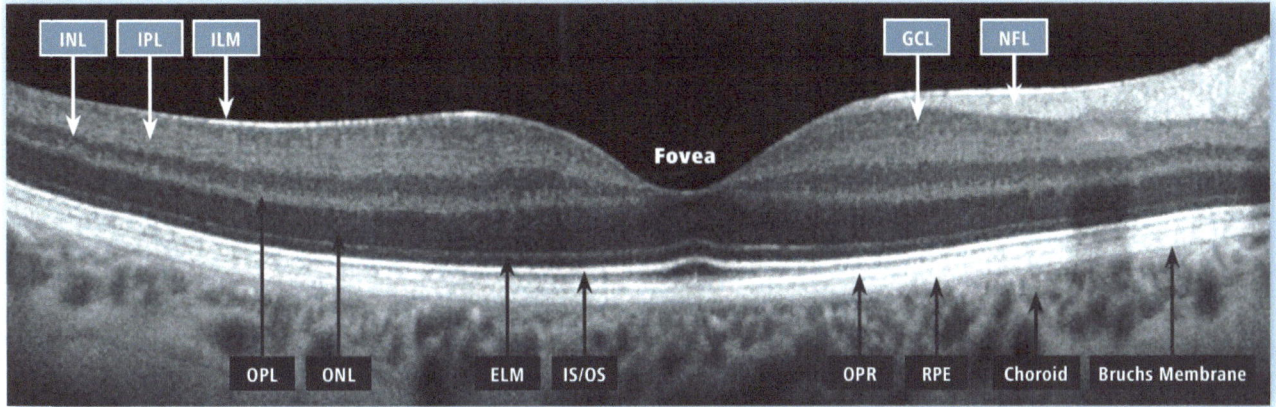

ILM: Inner limiting membrane
IPL: Inner plexiform layer
INL: Inner nuclear layer
OPL: Outer plexiform layer
ONL: Outer nuclear layer

ELM: External limiting membrane
IS/OS: Junction of inner and outer
photoreceptor segments
OPR: Outer segment PR/RPE complex

NFL: Nerve fiber layer
GCL: Ganglion cell layer
RPE: Retinal pigment epithelium
+ Bruch´s Membrane

Here are two sets of photos showing plaquenil toxicity on a 30-2 HVF exam (not a 10-2) with OCT and disk photos, followed by photos of a normal macula on autofluorescence (A) and then 3 different patterns of plaquenil toxicity (B,C and D). All © AAO 2015.

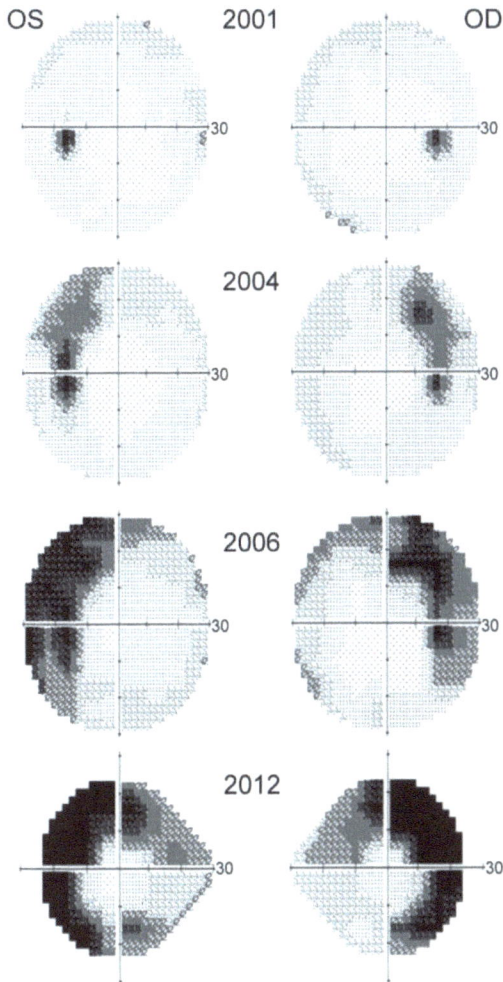

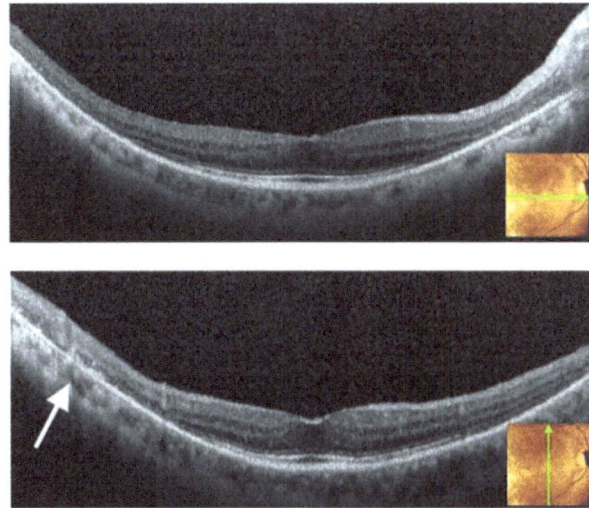

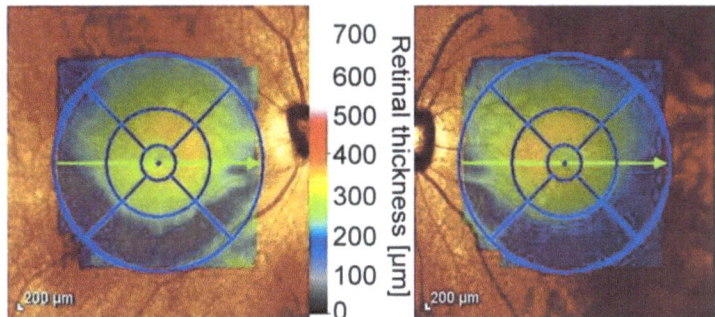

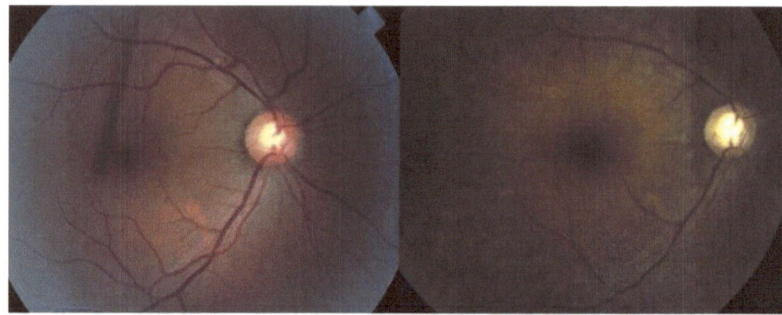

1993   2002

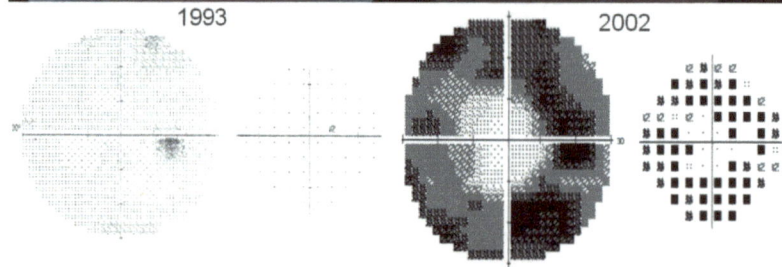

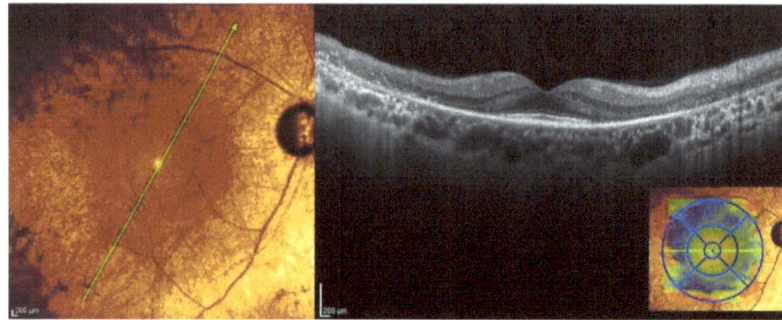

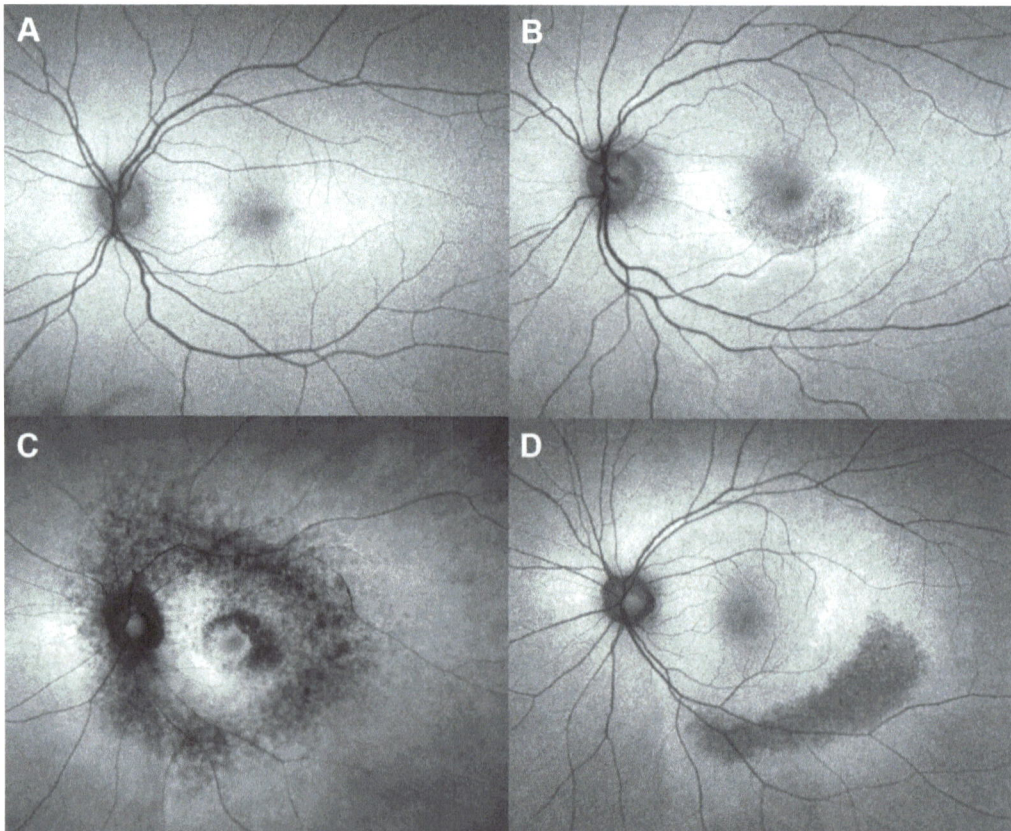

Plaquenil is a Chloroquine derivative with a lot less toxicity. Quinine, similar in chemical structure to chloroquine, was the first anti-malarial treatment. It was first used by new world natives for the treatment of non-malarial shivering (there wasn't malaria in the new world until Africans arrived) and then was used by Jesuit priests in Peru in 1630 that brought it to the old world for the treatment of shivering from malaria. Quinine is from the bark of the cinchona tree and it became the best treatment for malaria for the next 300 years. Quinine was placed in tonic water in 1771 by Joseph Schweppe and was mixed with gin to make the drink more palatable. Gin and tonic is still a popular mixed drink today, even when there aren't any mosquitoes to be found. Here is a photo of Tamoxifen and then mellaril retinal toxicity (both © AAO 2014):

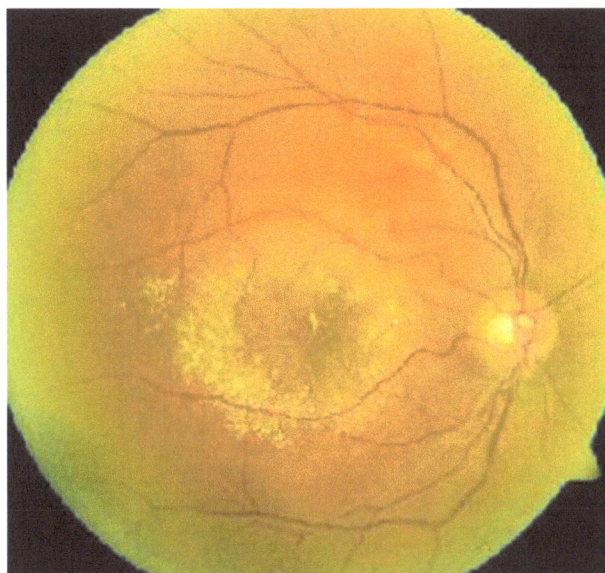

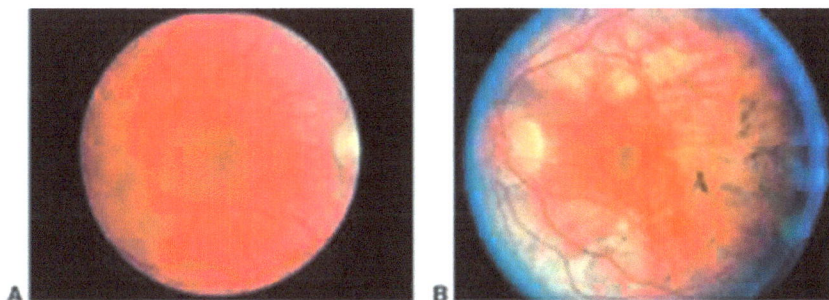

5. Cataracts:

It is rare to have a cataract removed for medical reasons, with phacomorphic and phacolytic glaucoma being the exception. Insurance will pay for a cataract extraction (CE) when:

- the patient is having symptoms caused by the cataract and *desires* CE
- there is a reasonable expectation of visual improvement
- the retina cannot be visualized by the eye doctor

If the patient does not desire CE, there are no symptoms caused by the cataract, the risk of a heart attack or stroke during surgery is too high for that patient, or appropriate post-op care cannot be arranged, then do not take the cataract out. (AAO, Preferred Practice Patterns [PPP] cataract guidelines, 2013).

Some patients are afraid of complications from cataract surgery. Your job is to recommend the best surgeon you know when a CE is indicated. Don't place your economic interests ahead of the patient's best interests. When the cataract is taken out, a small plastic IOL is placed inside the capsule where the cataract was. Choosing the correct IOL power by the surgeon results in elimination (or close to it) of the distance refractive error, except for corneal astigmatism. (I'm not a fan of toric IOLs. I like limbal relaxing incisions better). Elimination of the refractive error is not a reason to do CE but it is a very nice outcome as it is the same result, optically even better since it is closer to the nodal points of the eye, than LASIK. To determine the correct power of the IOL, you measure the length of the eye with an A-scan, factor in the keratometer readings and then use various formulas to determine the power you want for the desired post-operative refractive error, if any. Here is a print out of an A-scan. C is the cornea, L1 and L2 are the lens reflections and R is the anterior surface of the retina. The length in this patient was 24.08 mm (© AAO 2014):

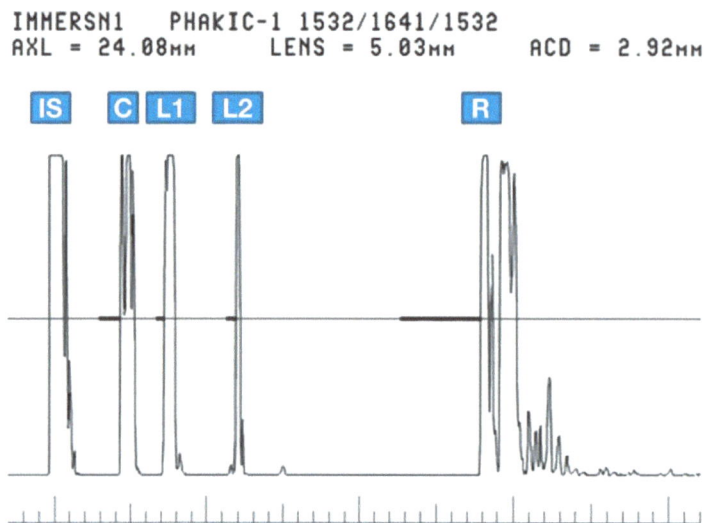

Complications of CE include endophthalmitis, which usually occurs within the first 4-7 days of surgery and CME starting 3-6 months after surgery. CME is difficult to treat and consideration should be given to keeping patients at risk for post CE CME (diabetics and patients who got CME after the first CE) on either brand name PF or NSAIDS for 6 months after a CE. Infectious endophthalmitis, defined as an infection in the vitreous, is an emergency and is treated with fortified antibiotics injected directly into the vitreous. Here is a photo of an injection into the vitreous (© AAO 2014):

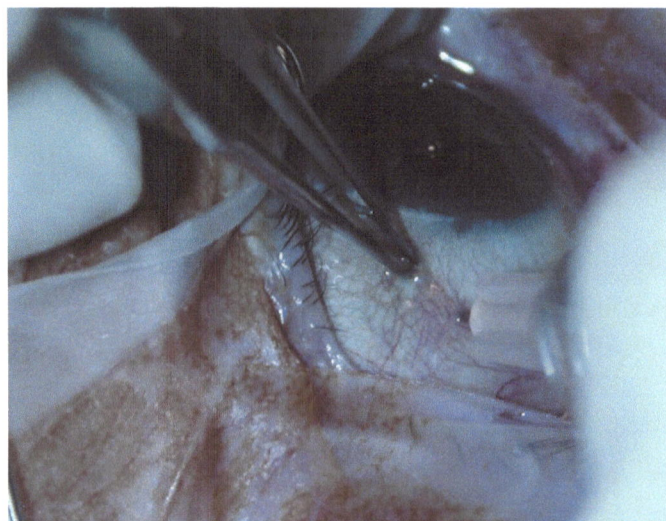

Non-infectious endophthalmitis, (also called toxic anterior segment syndrome or TASS) is thought to be related to persistent material on sterilized equipment from the prior surgery being exposed to the inner globe during surgery or from tainted solutions being injected into the eye during surgery inducing an inflammatory reaction in the eye. The Wilmer Institute made changes to their sterilization protocols and eliminated TASS.[123] TASS presents within 12-24 hours after surgery,[124] as opposed to endophthalmitis which presents 4-7 days after surgery.

Here is a photo of cystoid macular edema (CME) on an IVFA that can occur 4-6 months after a CE. You should suspect this, or a secondary cataract if the patient's vision unexpectedly gets worse 4-6 months after a CE (© AAO 2014):

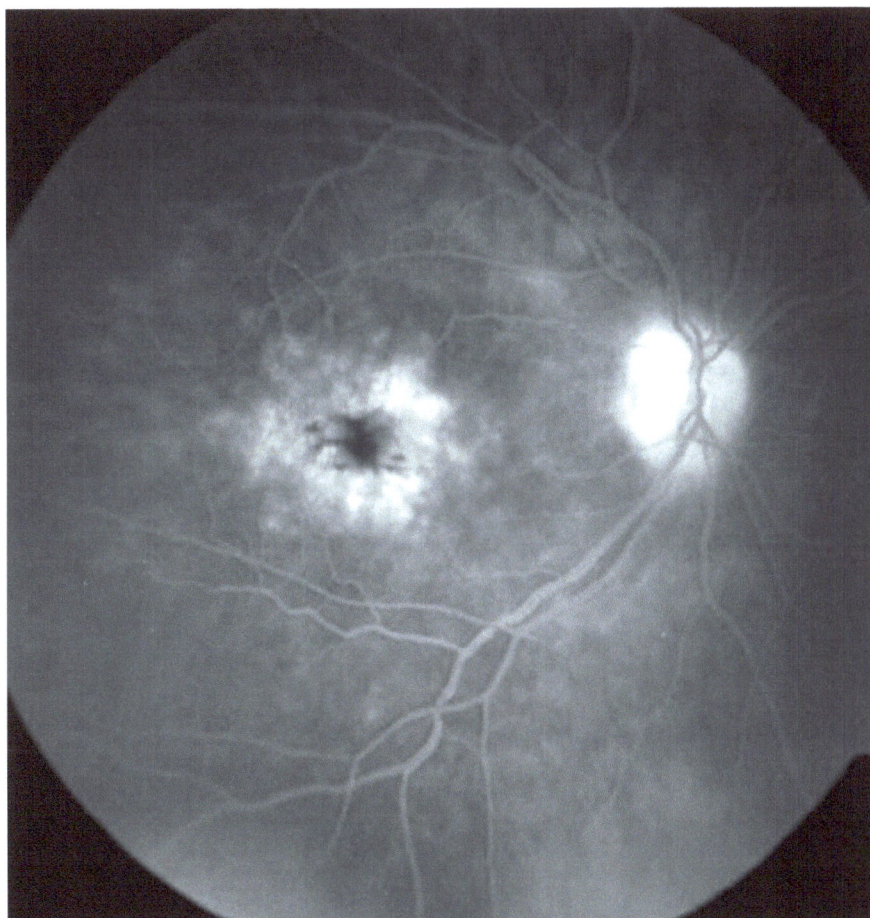

I don't recommend co-management of surgery patients with 20% of the fee coming to you (20% for the full 3 months care is the percentage mandated by Medicare, a higher percentage is illegal). Delaying treatment for an infectious endophthalmitis is an automatic lawsuit and it just isn't worth the exposure. This is why any co-managing doctor must be available 24/7 to examine any red eye or reduced vision complaints in CE patients. Co-management of LASIK is a different story and does present benefits for the co-managing doctor.

An after-cataract YAG capsulotomy has the same conditions for insurance coverage as a CE does. It is only justified if the patient has symptoms and wants the procedure. The low reimbursement paid today for a YAG capsulotomy does not justify the cost of the laser for all but the largest practices. The economics of cataract surgery has changed a lot in the past 30 years!

This is not an ophthalmoplegic third nerve migraine. Ophthalmic migraines present as a mirage-like distortion in the peripheral vision with oscillating kaleidoscope-like images, starting smaller, getting bigger and then going away. They typically have zig-zag edges, last from 3 minutes to an hour or longer and are not associated with a headache afterwards. Here is a diagram showing what one patient experienced with their ophthalmic migraines (© AAO 2014).

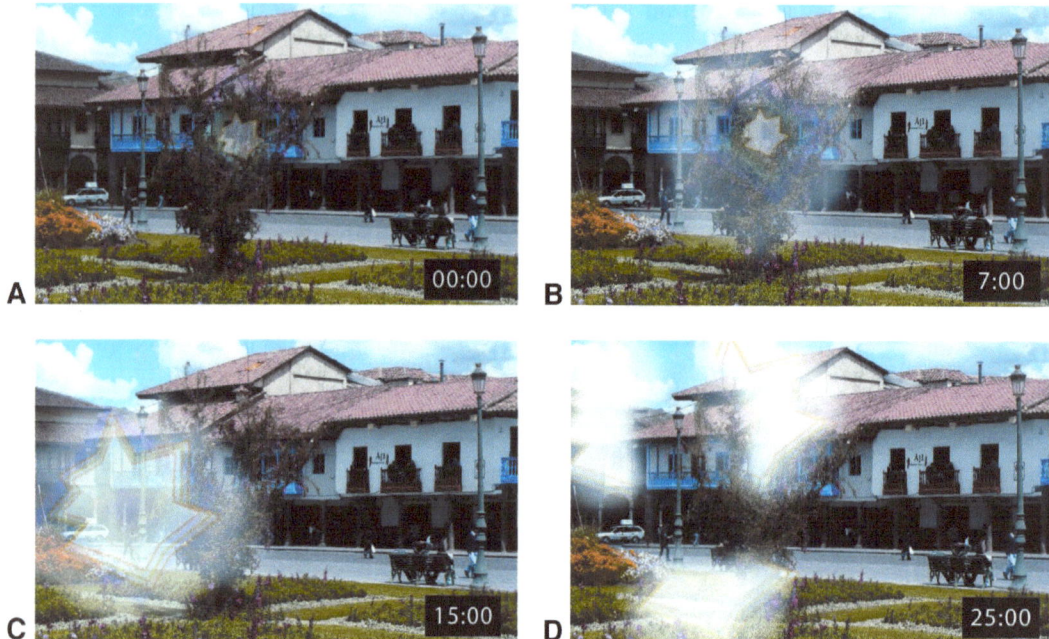

Here is how one of my patients drew his ophthalmic migraine as it progressed over time. It starts at the lower right and progresses to the up and left:

Ask the patient if they had any loss of muscle or sensory ability in their speech, face, arms or legs during the episode. If they did, a brain MRI is necessary with a referral to a neurologist. See Will's eye page 290. Otherwise, reassure the patient that this may occur again and have them pull off of the road if they get an episode while driving. Tell them that, in a way, they are lucky because they are experiencing a migraine vasospastic process without the headache.

You probably will not be involved in migraine headache work ups but it is important to determine if the aura (the physical symptom in a classic migraine) precedes the headache, which is the normal sequence in migraine headaches. An aura after the headache is abnormal and requires an MRI of the brain to rule out tumors and a referral to a neurologist. Also, ask if the headache occurs only on one side of the head, which is also abnormal. Even if only one headache was on the other side of the head, that is enough to reduce your concern about a referral to a neurologist and an MRI of the brain. *Strictly* unilateral migraine headaches are abnormal and require an MRI of the brain to R/O other possibilities such as a tumor. An MRI of the brain is also needed if the patient has had a permanent deficit associated with the migraine.

7. PVD:

Here are two diagrams of the vitreous and a PVD. (Both © AAO 2014)

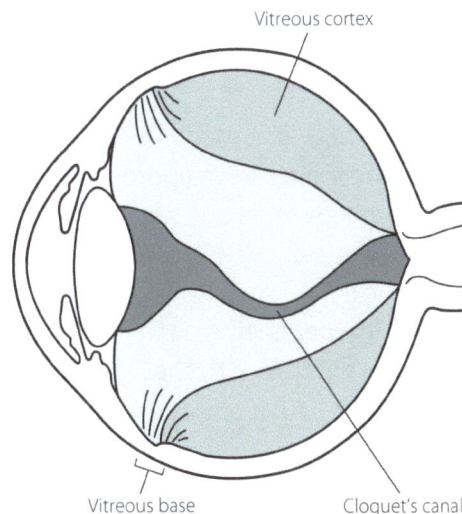

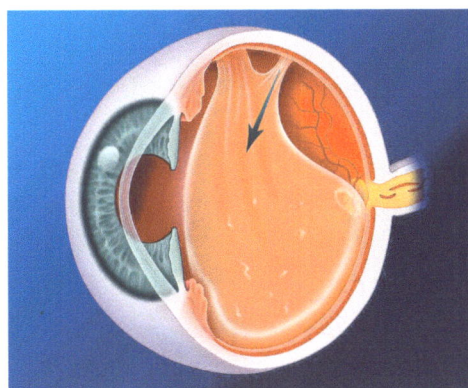

A *posterior* vitreal detachment (PVD) is a separation of the vitreous from the optic nerve and retina. An *anterior* vitreal detachment is a separation of the vitreous from the posterior lens surface. A PVD is *the*

major event leading to a retinal tear.[125] The 3 symptoms of a retinal break are flashing lights that last a split second (as opposed to ophthalmic migraines which last minutes to hours), new onset floaters or a curtain or veil of darkness (VOD).[125] When we are young, the vitreous is firm like Jell-O and pushes up against the retina. As we get older, the vitreous liquefies and tends to pull away from the retina, forming clumps in the vitreous which the patient sees as floaters. As it does this, it may tug on the retina causing lightning flashes or, worse still, tears in the retina. A PVD occurs in less than 10% of people under age 50 but occurs in 63% of people over age 70.[125] Retinal tears will develop into a complete RRD unless a laser retinopexy, gas injection or scleral buckle is performed. You can tell by the diagram above that the tension from the vitreous onto the retina is at the vitreal base, which straddles the ora serrata. **Your job is to:**

- examine the patient on an urgent basis,
- use the slit lamp to r/o vitreal blood and pigmented cells and the BIO to r/o tears
- do a scleral depressed DFE looking for any tears (© AAO 2014)

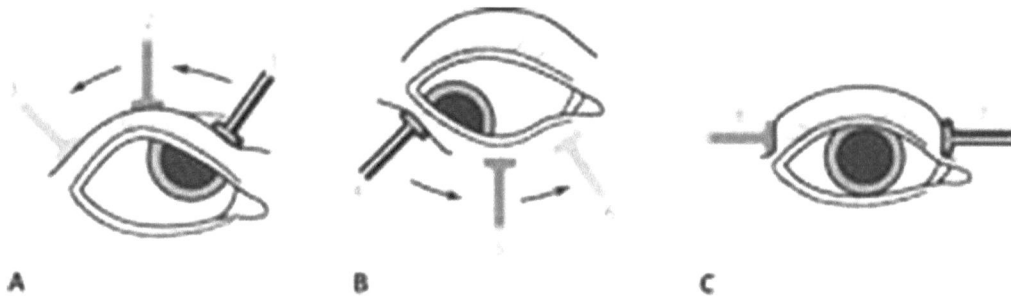

A   B   C

- refer out those patients that do develop tears urgently
- to see the patients without a retinal tear back in an appropriate time frame.
- I refer out all PVD patients that don't have tears but have **vitreal pigment** or **hemorrhage** to a retina specialist for a follow -up dilated fundus exam (DFE) in 1-2 weeks.

I follow active PVD patients every 1-3 months depending on the risk factors listed below. I gradually increase the follow up interval until there are no more lightning flashes. During the active/symptomatic PVD period, I tell the patient to return to the clinic (RTC) ASAP if any of these 3 symptoms of RD develop:

- hundreds of new floaters that represent red blood cells from bleeding
- the lightning flashes get worse
- if a veil of darkness (VOD) develops

Risk factors for the development of a rhegmatogenous retinal detachment (RRD) in a PVD patient include:[125]

- a retinal break that has retinal traction still present (such as a retinal tear)
- lattice: 25% of all RRD patients have lattice, 10% of people have lattice
- high myopia > 6.00 D
- family history of RD or history of the patient's fellow eye having a RD
- aphakia, not pseudophakia
- bleeding or pigment in the vitreous of an active PVD patient. **60% of PVD patients with vitreous hemorrhage have retinal tears, 11% of PVD patients without vitreous hemorrhage have a retinal tear**[125]

Document on your exam form the symptoms of the patient, the risk factors, and the presence or absence of vitreal pigment or hemorrhage. Document when you want the patient to RTC and to RTC ASAP if lots of new floaters/flashes occur or if a VOD develops. Also, document that you did scleral depression. **Refer out any patient with any of the risk factors above. Refer out ASAP for retinal tears, vitreal pigment, or vitreal hemorrhage.** Here is a photo of a retinal tear (© AAO 2014):

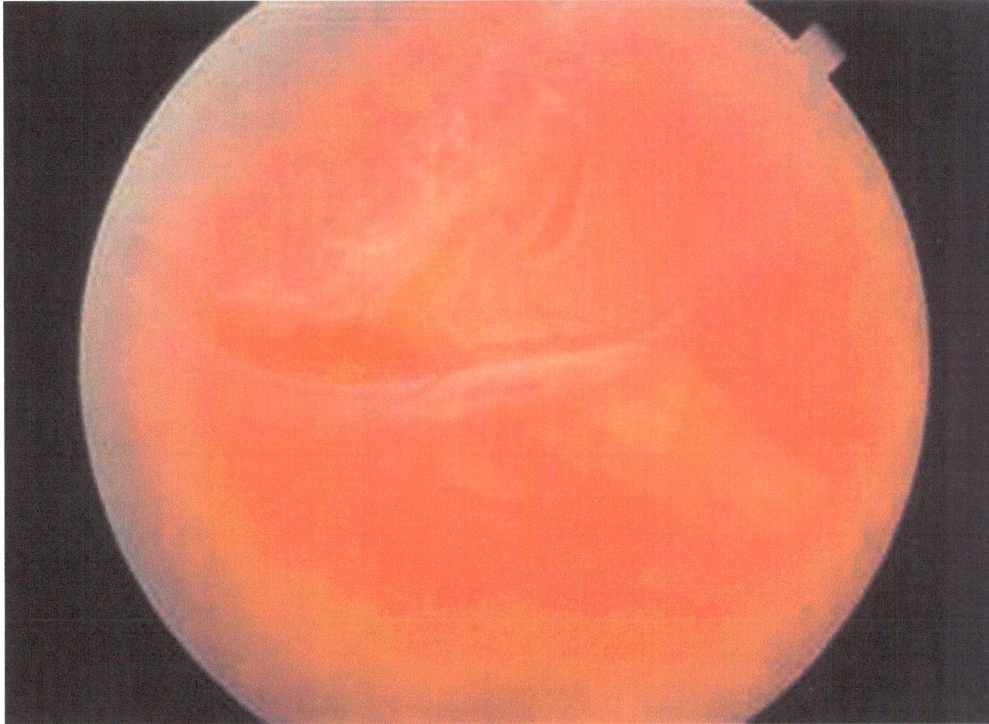

Here is a diagram showing a peripheral retinal tear and the use of laser to seal it down (© AAO 2014). Notice how the laser needs to be placed 360 degrees around the tear to properly insure that it won't spread into an RRD.

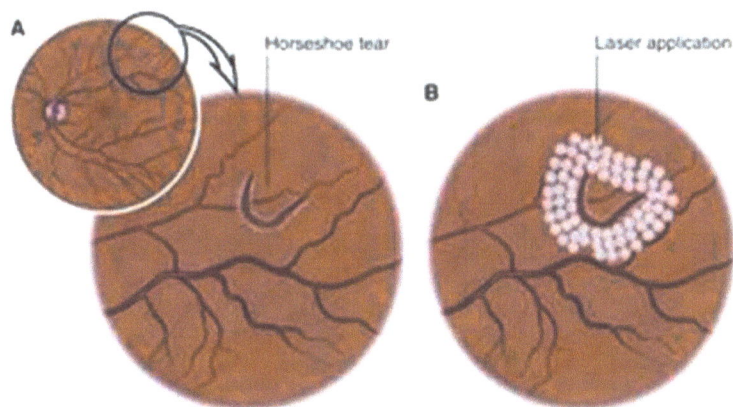

Here is a diagram of a large retinal tear being closed with air injected into the vitreous on the top photo rather than using a scleral buckle (© AAO 2014):

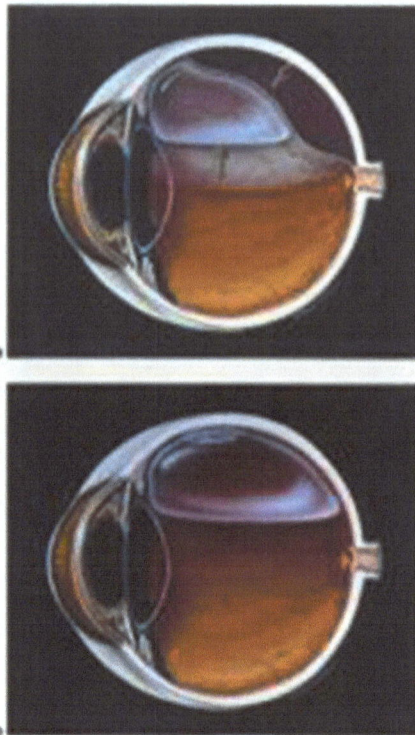

8. Diabetes:

Insulin facilitates glucose entry from the blood vascular system to inside the cells of our body so that the cell can use the glucose for energy. In type 1 diabetes, there is a lack of insulin production. In type 2 diabetes, there is a resistance to insulin facilitating glucose's entry into the cell. The lack of glucose getting inside cells results in those cells not getting the energy they need in order to function. The cells then start to die off. This culminates in a loss of support cells for the blood vessels of the eye (and the rest of the body as well), a thickening of the basement membrane of capillaries as well as an increase in the permeability of capillaries.[132] This causes retinal blood vessels to leak blood into the retina leading to a swelling of the retina and consequently, reducing the vision. The blood inside the retina eventually gets reabsorbed with the hemoglobin component of the blood getting reabsorbed sooner than the fat component of the blood. The fat components minus the hemoglobin are then seen as hard exudates (HE). HEs take a lot longer to reabsorb than the hemoglobin does. Over time, the capillaries occlude, causing the retina tissue they supplied blood to, to die. This is called capillary dropout. This leads to the retina growing new blood vessels in order to try resupplying the retina with nutrition and oxygen. These new blood vessels (neovascularization) break and bleed easily into the retina and vitreous, leading to reabsorption of the vitreal blood and traction between the vitreous and retina. This invariably leads to a tractional retinal detachment and blindness.

Here is a diagram of a normal foveal avascular zone (FAZ) (© AAO 2014):

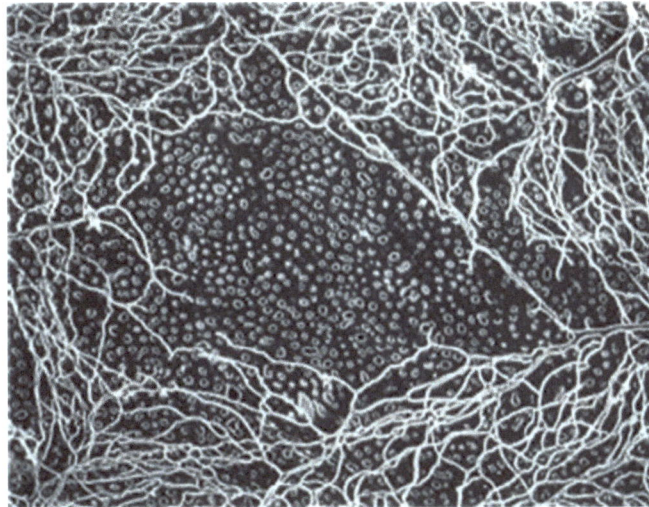

Here are 2 diagrams showing a damaged and enlarged FAZ from a BRVO(© AAO 2014):

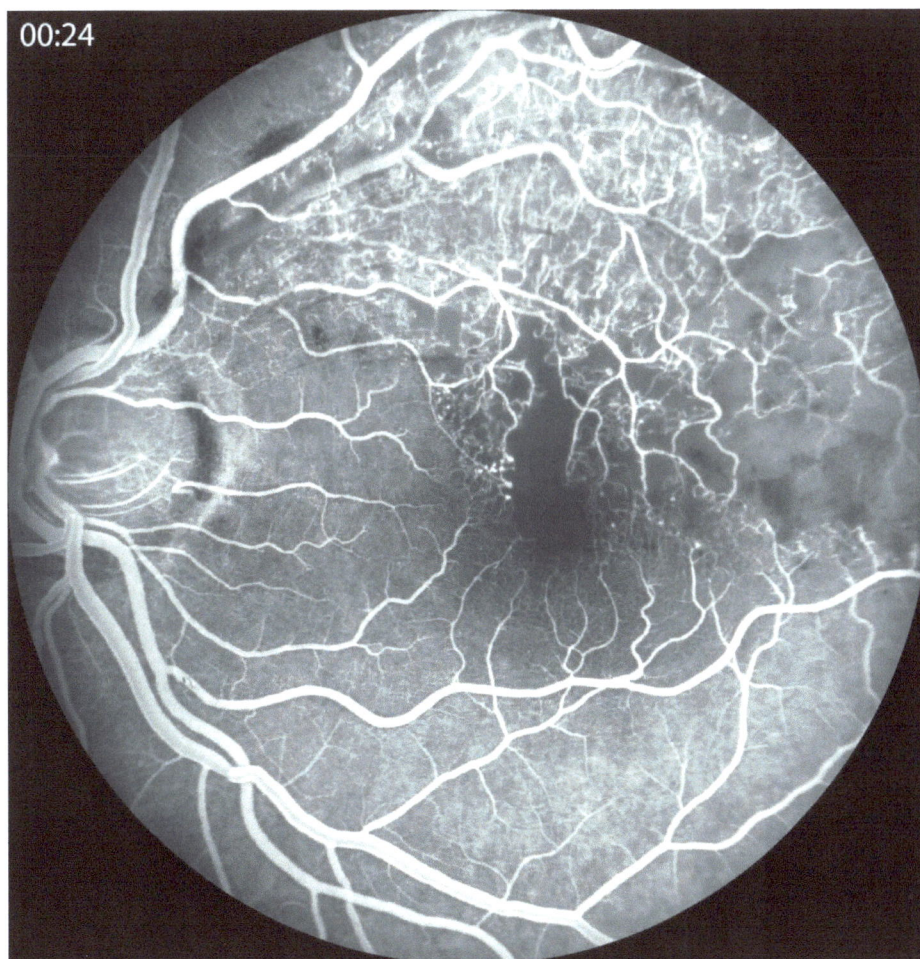

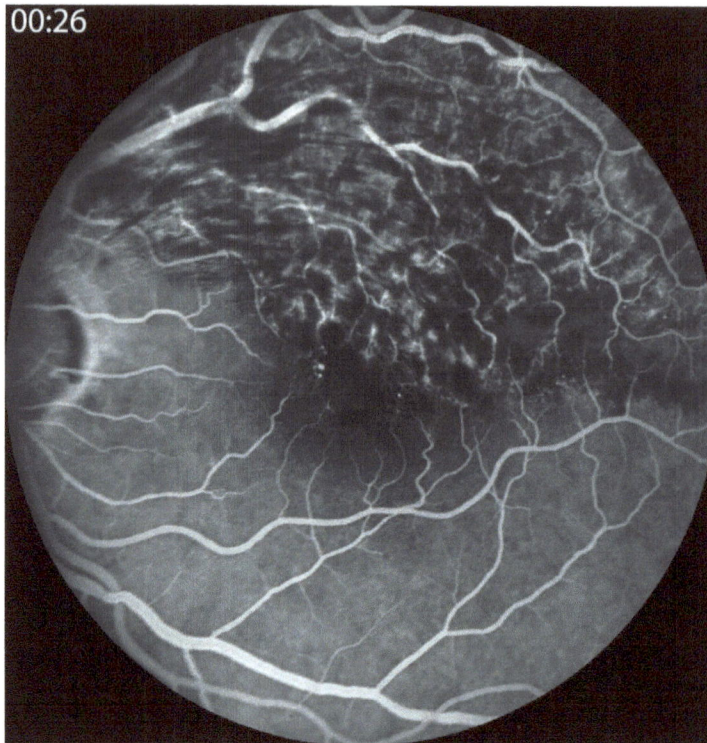

Here is a diagram showing retinal swelling on an OCT (© AAO 2014):

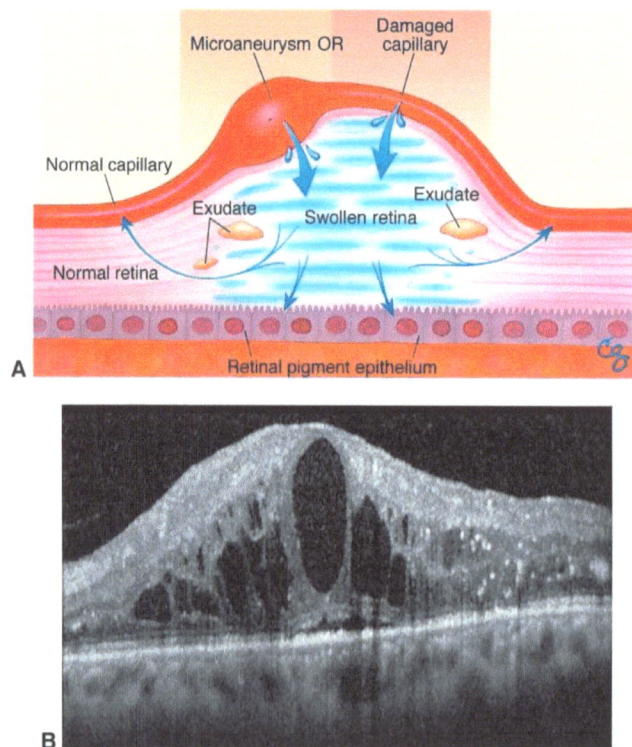

Significant hypoxia results in the hypoxic retinal area producing Vascular Endothelial Growth Factor (VEGF), which signals the retina to produce new blood vessels in order preserve the retinal tissue. These new blood vessels are called neovascularization (NV) and they look like a sea fan of small caliber blood vessels on the retinal surface. They can occur on or by the **d**isk where it is called NV**D** and **e**lsewhere away from the disk, where it is called NV**E**. Neovascularization leaks fluorescein on an intra-venous fluorescein angiography (IVFA) test. Normal blood vessels and shunt vessels do not leak, which is how NV is traditionally diagnosed in questionable patients. Due to the rare but serious side effects (anaphylactic death) of an IVFA, we usually use the visual presence of NVD or NVE on a DFE as enough of an indication to warrant PRP being performed. NVD and NVE eventually bleed into the vitreous. When this vitreal blood is reabsorbed, traction develops between the retina and vitreous, resulting in tractional retinal detachments (TRD) and eventual blindness.

TRDs start as white fibrosing strands along the arcades, eventually elevating and splitting the retina apart over time. To avoid NVD and NVE forming a TRD and eventual blindness, we kill off the retina producing the VEGF with PRP laser at the first sign of PDR. This is what PRP does. It destroys the ischemic part of the retina in order to save the functional retina. Once the decision is made to do PRP, full or complete treatment should be done. The DRS defines complete PRP as 1200 or more 500 um spots. This would look on your exam as burn marks from just outside the arcades all the way to the vortex veins all the way around the retina. PRP less than this amount may allow the further growth of NV. Sparing the 3 and 9 o'clock retina is an attempt to preserve the long ciliary nerves with its accommodation, and corneal sensation functions is occasionally done. It also saves a lot of pain for the patient during PRP. Here is a photo of a TRD in a diabetic patient. Notice the fibrosing bands along the arcades (© AAO 2014). The red arrow points towards NVE, the black arrow are PRP scars, the white arrow is a TRD, and the circle is a vitreous hemorrhage.

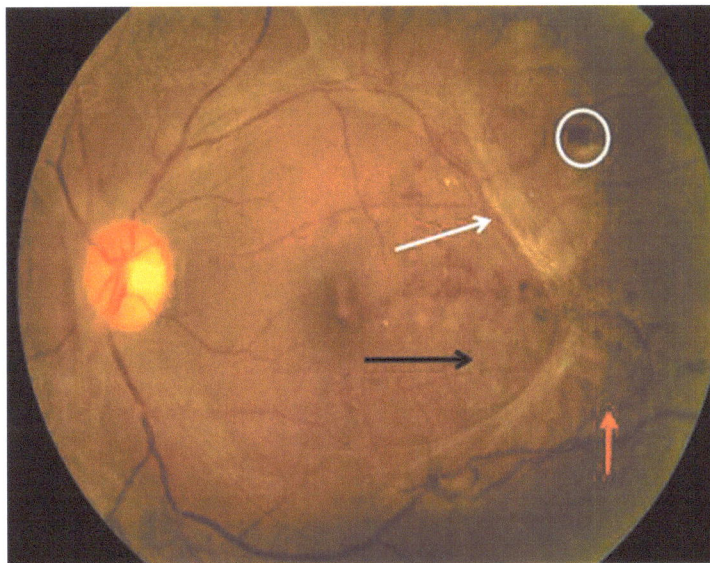

Your main concern, as the eye doctor, is to risk stratify the patient, to schedule appropriate follow-up DFE's, and to refer the patient for laser or anti-VEGF injections when the patient gets close to needing treatment, either for clinically significant macular edema (CSME) or proliferative diabetic retinopathy (PDR). You can think of diabetic retinopathy as having two main stages, non-proliferative diabetic retinopathy (NPDR), or background retinopathy, which does not have NV, and proliferative diabetic retinopathy (PDR), which has NV. NPDR can and will progress into PDR at some point, although it can take many years. Vision loss in diabetes occurs due to three main mechanisms,[126]

   - macular edema
   - capillary drop out in the macula which, unfortunately, is preferentially effected in diabetes
   - PDR leading to a Tractional retinal detachment (TRD) and blindness

Diabetic Retinopathy (DR) in order of increasing severity:

- no bleeding/no cotton wool spots (CWS) seen (microvascular damage and macular dropout can be present but not visible)
- blot/dot and flame hemorrhages, also called intra-retinal hemorrhages, deep and superficial respectively
- venous beading or sausaging
- cotton wool spots (CWS), which are micro infarcts of the nerve fiber layer (NFL)
- intra retinal micro vascular abnormalities (IRMA) which are a precursor to neovascularization and are the dilation of existing blood vessels. IRMA doesn't leak fluorescein, similar to ON shunt vessels in meningiomas. IRMA and shunt vessels both can look like neo (© AAO 2014):

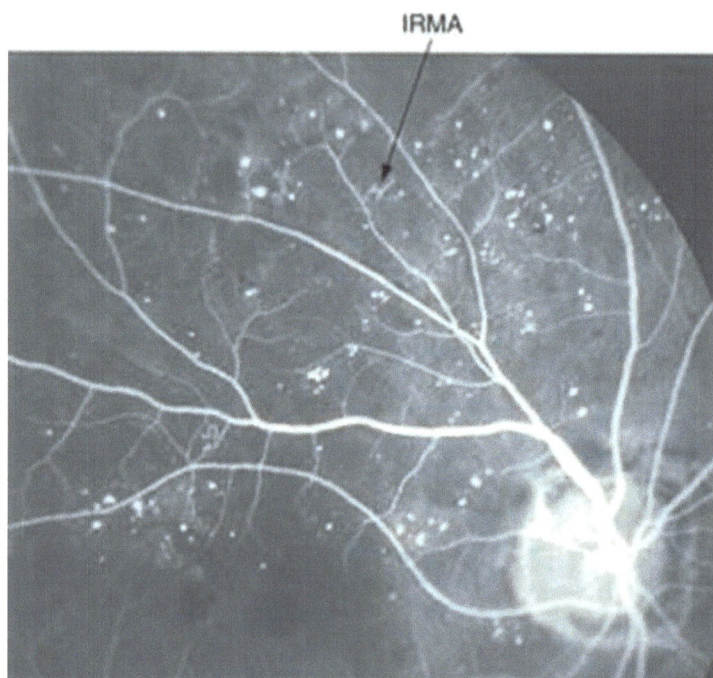

- pre-retinal or boat hemorrhages in the space between the retina and vitreous, usually from NVE/NVD
- vitreal hemorrhages which are from new blood vessels (PDR) bleeding into the vitreous. They look like ketchup in the vitreous. These new blood vessels leak fluorescein on an IVFA and break and bleed very easily. The re-absorption of this vitreal blood leads to traction between the vitreous and the retina, causing traction retinal detachments (TRD) and blindness. This whole cascade of events is halted with properly timed PRP.

The Early Treatment of Diabetic Retinopathy Study (ETDRS)[127] showed that focal macular laser (FML) treatment applied to the areas of the macula with clinically significant macular edema (CSME) showed a significant reduction in future visual loss. FML laser should be avoided in ischemic areas since it may worsen the ischemia and permanently reduce the VA. The new microburst laser avoids this complication and can be applied on top of the fovea. Anti VEGF treatment is also revolutionizing diabetic retinopathy treatment, especially macular edema. CSME is diagnosed on the exam and is defined as:[127]

- having retinal edema within 500 microns (1/3 of a disk diameter) of the foveal center
- having hard exudates associated with retinal edema (ME could be further than 500 um from the fovea) within 500 microns of the foveal center
- having edema 1 disk size in diameter within 1 disk diameter of the foveal center

The ETDRS categorizes "high risk" NPDR with the 4-2-1 rule:[128]

- diffuse intra-retinal bleeding (> 20 in total) in the 4 quadrants of the retina
- venous beading in any 2 of the quadrants
- IRMAs in any 1 quadrant

Here are some diagrams showing blot/dot hemorrhages, CSME and the use of laser to treat them (both © AAO 2014):

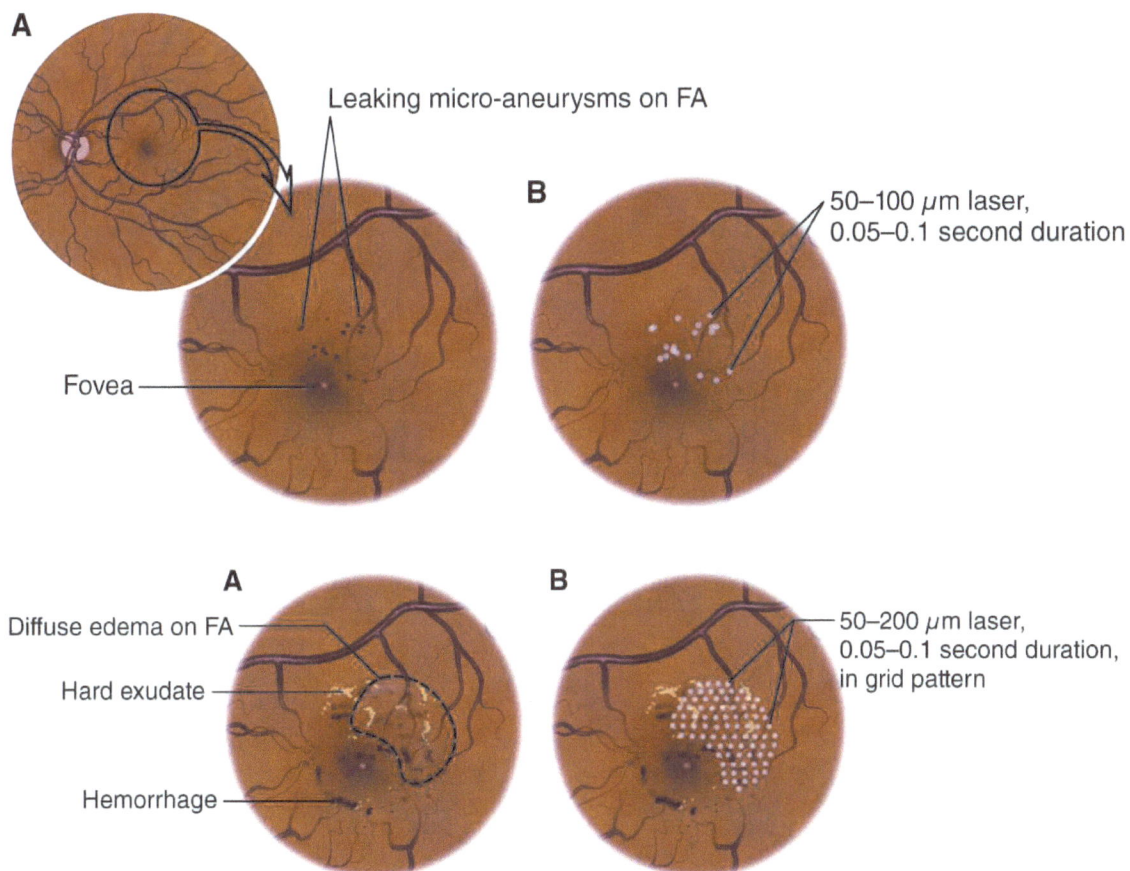

Here are some photos of a BRVO, showing the use of an FML in the treatment of macular edema, similar to what is done in diabetic macular edema (© AAO 2014):

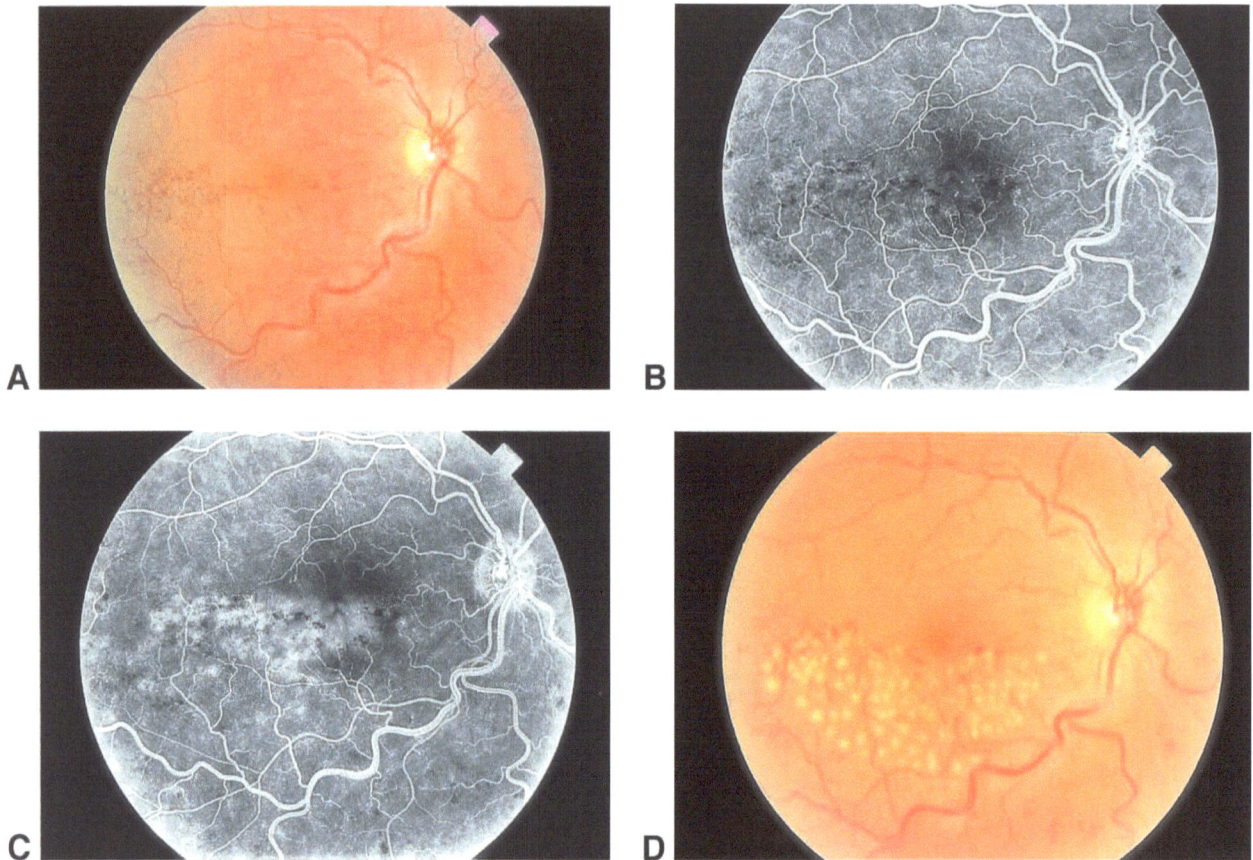

Here is a patient of mine that shows old FML laser spots OU:

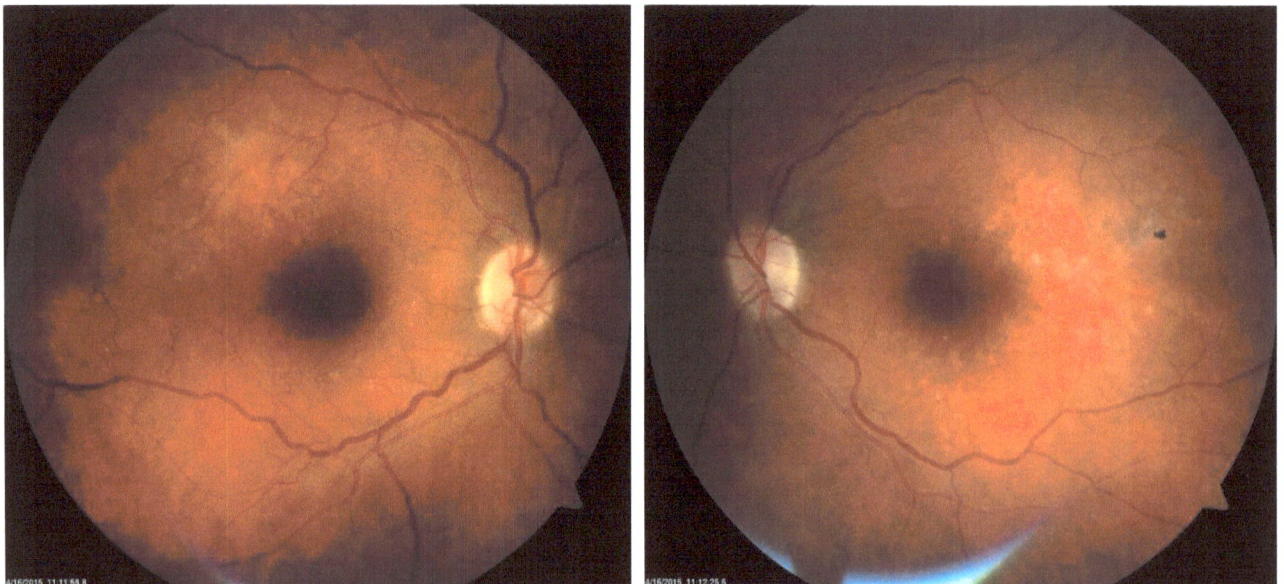

The ETDRS showed that having 1 of these 4-2-1 categories carries a 15% chance of developing "high risk" PDR, with high risk requiring PRP within 1 year. Having 2 categories carries a 45% risk of needing

PRP within 1 year. Therefore, patients having 1 or more of the 4-2-1 categories should be considered for early PRP[128] despite not meeting the requirements of the DRS. This is partially due to the risk of losing the patient to follow up and developing complications without treatment. Also, patients today are getting treatments earlier due to the development of anti-VEGF injections, OCT and micro-pulse lasers, which were all developed since the DRS/ETDRS studies were done. Also, OCT testing has supplanted IVFA in the diagnosis of diabetic retinopathy since an IVFA carries a small but significant risk of anaphylactic death from the injection of the dye into the veins. So, don't be too surprised if your patient does not get an IVFA before an FML or vitreal injection or seems to be getting anti-VEGF or laser treatments before the ETDRS/DRS would normally call for them.

Anti-VEGF injections have changed many of the treatment decisions for many of the diseases of the retina. Leucentis or avastin (antibodies to VEGF that bind VEGF) have been miracle drugs for us. Contact your local ophthalmologist to determine when they like to initiate treatments and work with them. If they know that you know these terms and use them correctly, they should be willing to work and support your care to the patient until it is time for laser treatments or injecting anti-VEGF drugs into the eye.

PDR patients are referred out immediately. CSME and high risk NPDR patients should be seen for DFE every 3-4 months depending on the severity. Moderate NPDR are seen every 6-8 months. None and very mild diabetic retinopathy patients are seen for DFE every year. If a patient has both CSME and PDR, refer ASAP. Most surgeons treat the CSME first and then do PRP unless the PDR is really bad. This is to try and preserve visual acuity as much as possible. Diabetes is the number one cause of blindness in Americans aged 20-64.[130]

Gestational diabetics with no NPDR are not seen again during pregnancy. Moderate NPDR are seen Q trimester and severe NPDR are seen every month looking for CSME or PDR.[129]

In a large survey of diabetics (n= 996 for type 1 and n= 1370 for type 2),[131] the following results were found:

Chance of Retinopathy:		Diabetes <5years	Diabetes >15 years
	Type 1:	17%	**98%**
	Type 2:	29%	**78%**
		Diabetes <10 years	Diabetes >35 years
Chance of PDR:	Type 1	1.2%	67%
		Diabetes <5 years	Diabetes >15 years
	Type 2	2.0%	16%

The DCCT study[133] showed that tight control of a type 1 diabetic's blood sugar significantly reduced the complications from diabetes for the kidney, nerve, and eyes. The incidence of retinopathy increased slightly during the first 18 months and then decreased by 74% at a mean follow up of 6.5 years. There were also more severe hypoglycemic episodes associated with tight control as well but these complications were considered worth the overwhelming benefit to the eyes, kidneys and nerves. A similar result was shown in type 2 diabetics in a large European study.[134]

Ophthalmic studies subdivide PDR into early, high-risk, and advanced PDR but with the development of anti-VEGF injections, most clinicians don't follow this as much. Many ophthalmologists now treat any ME with Leucentis and/ or micro pulse laser and any NV with anti-VEGF injections and then use laser without strictly following the ETDRS/DRS guidelines. The Diabetic Retinopathy Study[135] showed that PRP reduces the chance of visual loss by more than 50% if the patient develops "high risk" PDR, which is defined as:

NVD ¼ of a disk or greater
NVD associated with pre-retinal hemorrhage (PRH) or vitreal hemorrhage (VH)
NVE ½ a disk diameter when associated with PRH or VH
NVI or NVA

Here are some diagrams showing PDR and the use of PRP to treat it (© AAO 2014):

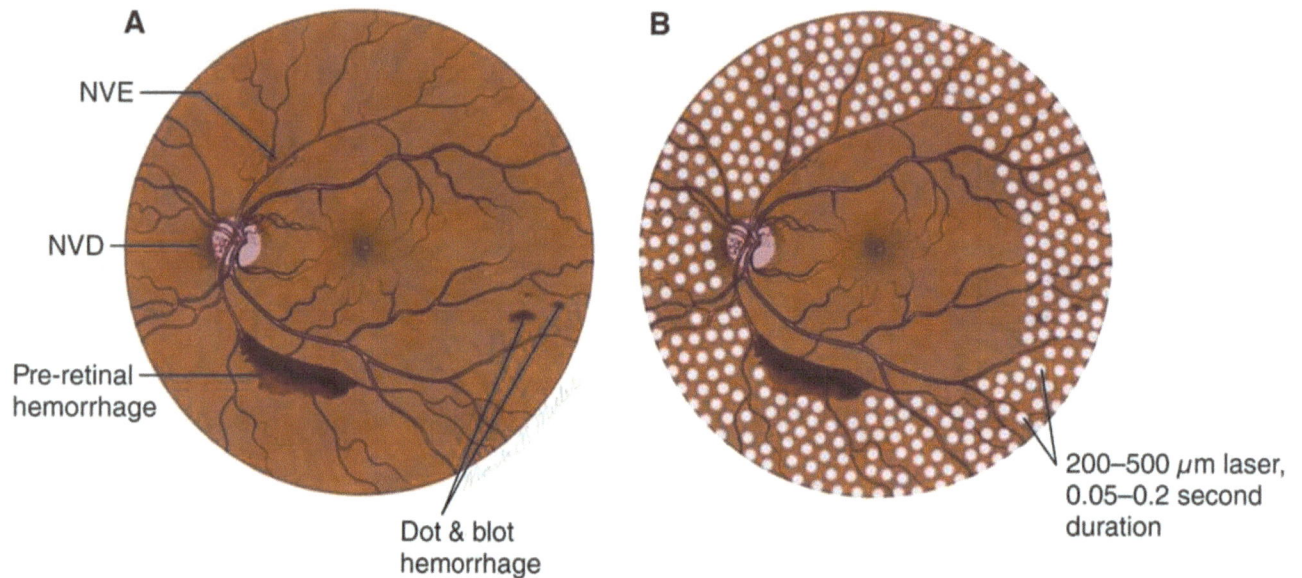

Here is a patient of mine with CSME:

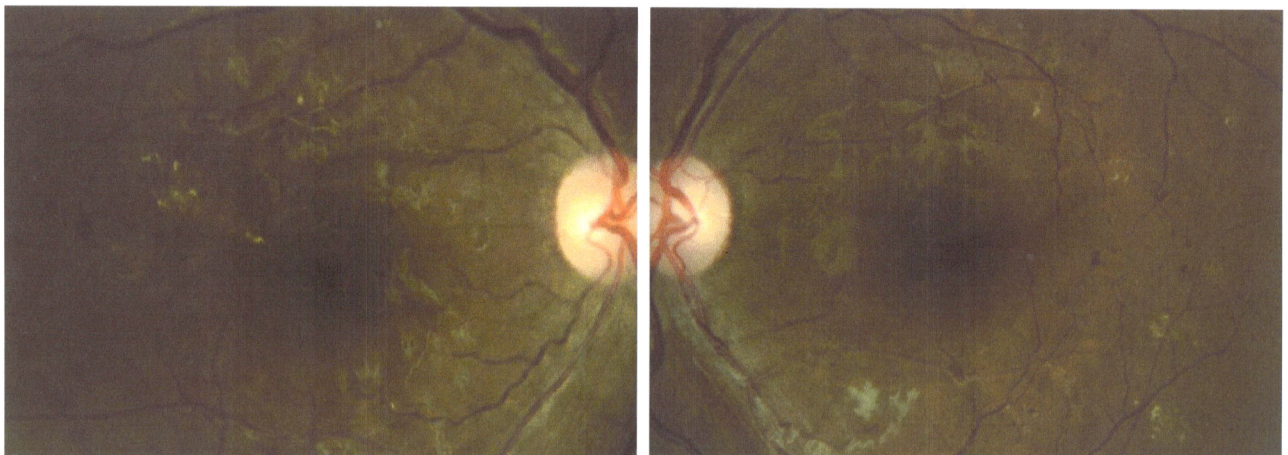

Another with moderate BDR that developed CSME over 2 years:

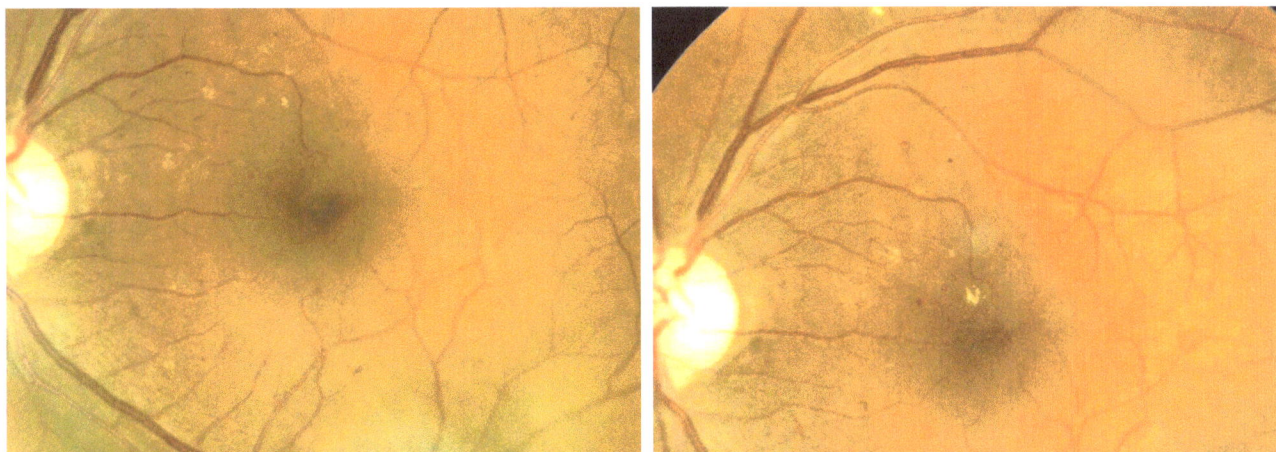

Here is a non-compliant diabetic that had a significant reduction in BCVA over a 1 year period due to non-compliance with his diabetic regimen. The second photo is 1 year after the first photo and shows severe macular swelling:

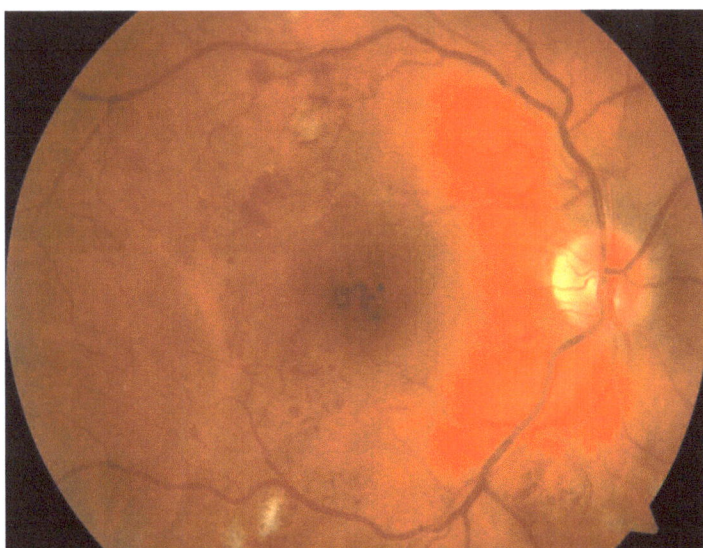

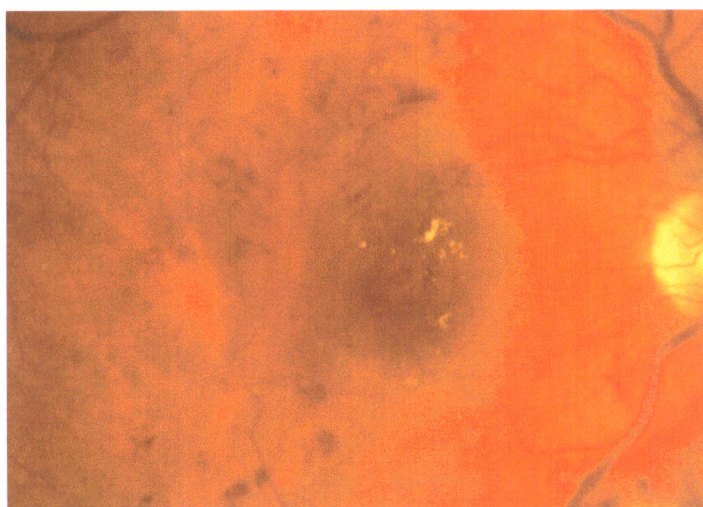

Be careful not to call myelination a cotton wool spot! Here is a patient with what might look like a CWS but is actually myelination. You can tell the difference by waiting 4-5 months. A CWS will go away, myelination won't. These photos are 1 year apart:

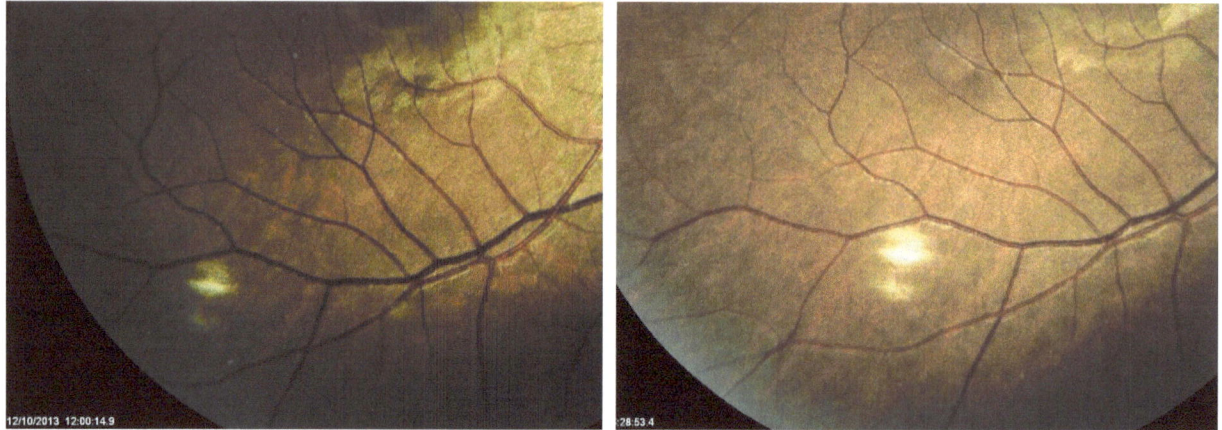

Here are photos of a real CWS taken 3 months apart. The photo on the left shows the faint CWS between the artery and vein. The right photo doesn't have the CWS present. I wasn't sure if this was a CWS but its disappearance 3 months later confirmed the diagnosis. I ordered an A1C, carotid doppler, cardiac echo, 24 hour monitor, Protein C and S, anti-thrombin 3, and lupus anti-coagulant levels, all of which were normal.

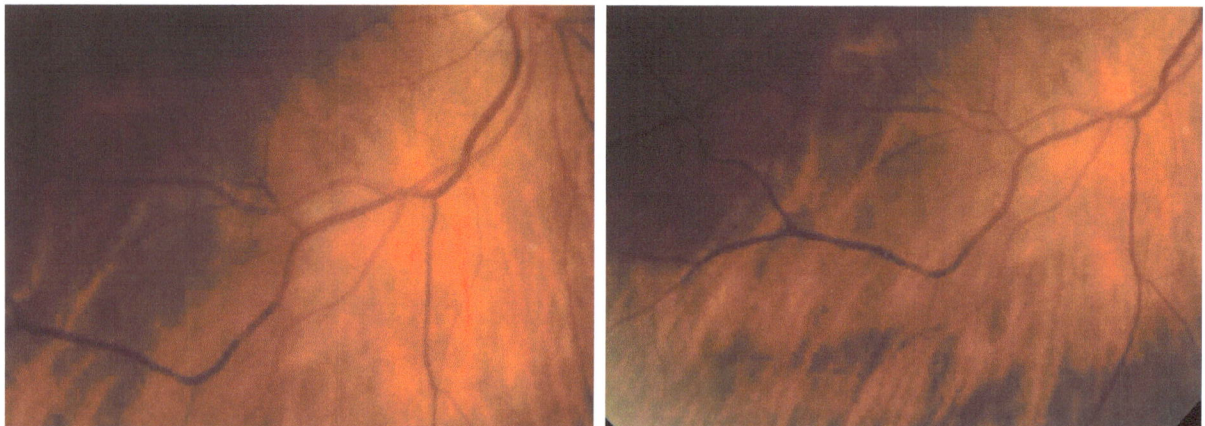

Here is another CWS in a known diabetic in photos taken 12 months apart:

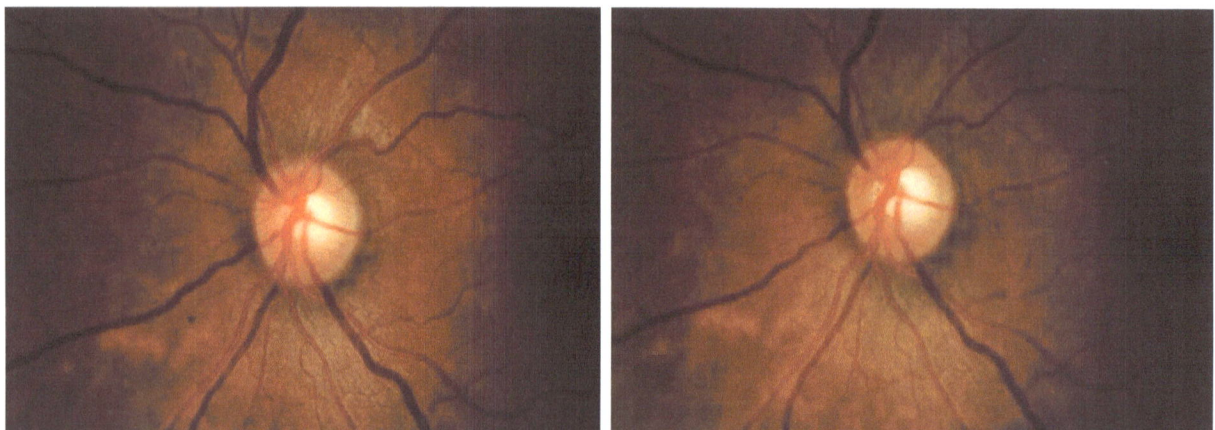

Here is another CWS from a patient with metastatic breast cancer undergoing chemotherapy:

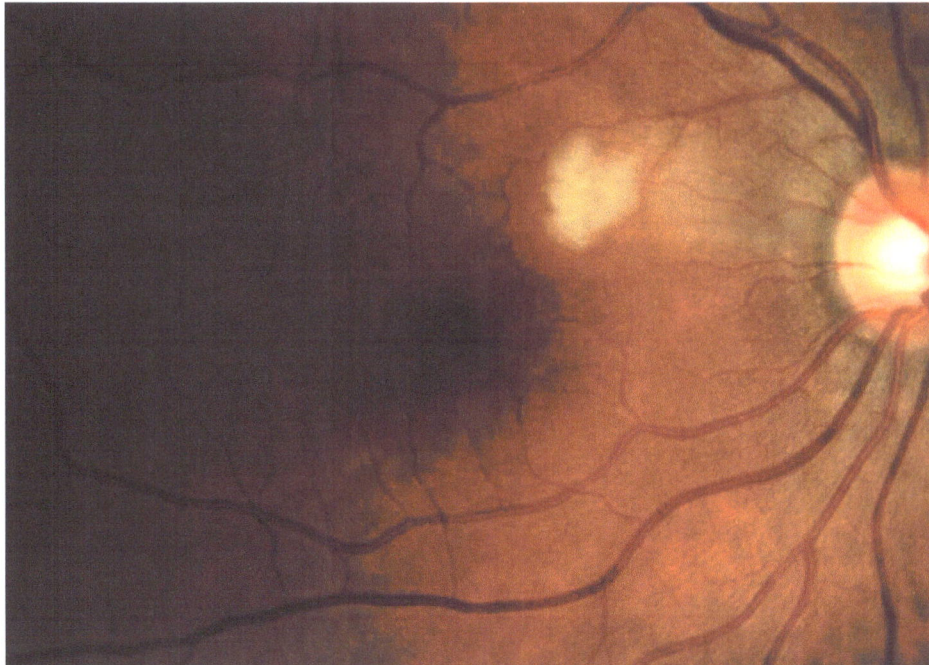

Anti VEGF injections are changing the way CSME and PDR are being treated. In a study combining the RIDE and RISE data, two randomized, prospective, multi-center clinical studies, 759 patients with CSME and no PDR were randomized to monthly anti-VEGF injections of two different concentrations for 24 months versus a third group receiving saline ("sham") injections. At 24 months of treatment, the saline group was also injected with anti-VEGF medicine due to the obvious benefit of the treatment. 707 patients completed the 36 months of follow up. The risk of developing PDR is graphed below. © Elsevier 2014. Anti-VEGF injections reduced the risk of development of PDR significantly (P < 0.0001). The risk of developing PDR in the saline group (sham) was reduced to the same rate as the treatment group after anti-VEGF treatments were started at the 25 month point (the vertical dotted line). Note how 36% of the sham group developed new onset PDR during this study and that at least 15% of the treated group still developed PDR at the end of 36 months of anti-VEGF injections. [193]

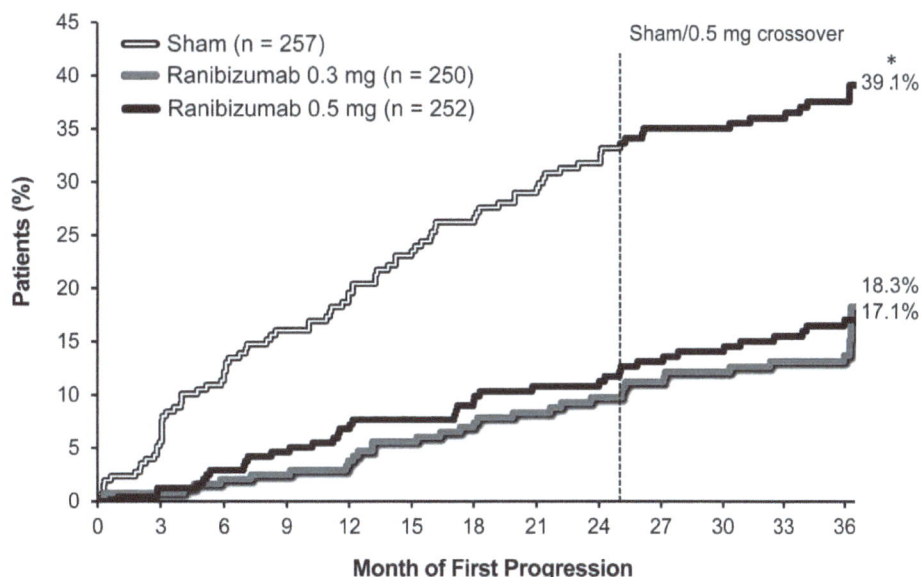

What else can look like Background Diabetic Retinopathy?[136]

-CRVO: The bleeding from diabetes occurs mostly inside the macula and is deep in the retina tissue which is seen as blot and dot hemorrhages. The bleeding from a CRVO occurs both inside and outside of the macula. The bleeding from a CRVO is more superficial in the retinal tissue where it presents as flame hemorrhages. It also has dilated and tortuous veins and tends to be unilateral. Diabetics tend to have similar retinopathy between the two eyes. Remember, we treat both eyes of all CRVOs for glaucoma, in addition to watching for NVG which leads to PRP.

-Ocular Ischemic Syndrome (OIS): OIS is due to an occluded carotid artery reducing the blood flow to that side of the head. Bleeding from OIS occurs in the periphery of the retina and rarely inside the macula and does not have HE. You listen for bruits of the carotid artery. Document that you did this. Also, refer to a cardiologist or to the patients PCP for carotid duplex testing. Here is a diagram showing the areas of the macula (© AAO 2014):

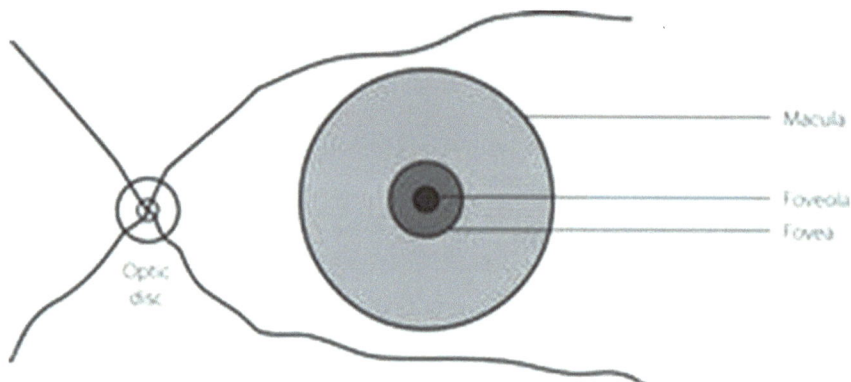

-Hypertensive retinopathy: flame shaped hemorrhages with arterial narrowing.

Here is a photo of a macro-aneurysm showing the classic hard exudate ring. This shouldn't be diagnosed as diabetic retinopathy but an A1C is appropriate to get anyway:

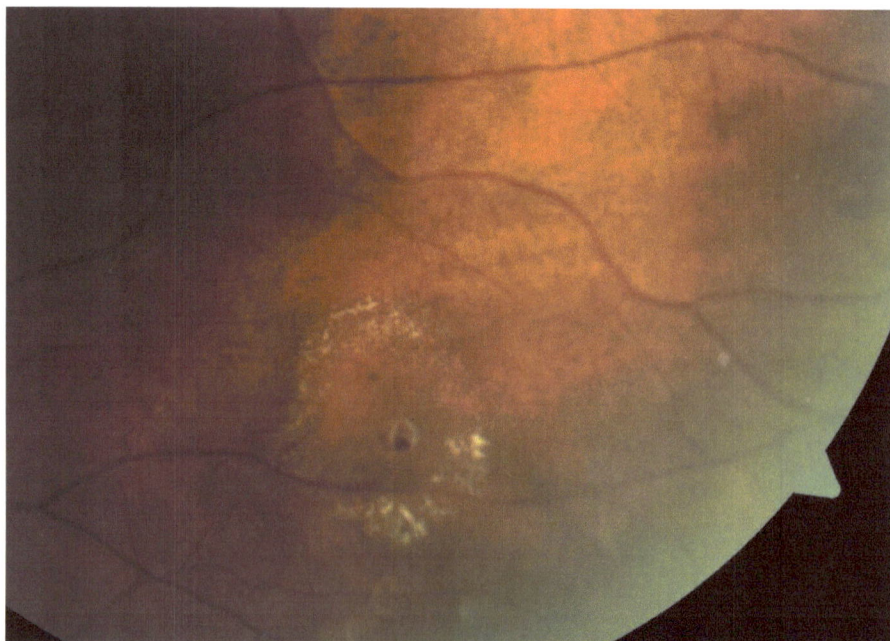

What else can cause retinal NV?[136]

- CRVO/CRAO/BRVO
- Sickle Cell Retinopathy: has a sea fan NV in the far periphery. Seen in African Americans.
- sarcoid: with HE along BVs called candle wax drippings
- Talc Retinopathy from IV drug use, which has multiple bright emboli in arteries
- hyper coagulation states

9. Uveitis/Keratoconnus:

Inflammation of the uveal tract is called uveitis. The uvea has 3 parts: the choroid, ciliary body, and the iris. Inflammation of these parts is called, respectively, choroiditis (posterior uveitis), pars planitis (intermediate uveitis) and iritis (anterior uveitis).

Uveitis within 5-7 days after intra-ocular surgery is an immediate referral to rule out endophthalmitis. This usually occurs within the first 5-7 days after cataract surgery or at <u>any time</u> after a trabeculectomy. Late endophthalmitis does occur and is suspected in all patients that have had surgery inside an eye as the #1 diagnosis to rule out in uveitis. I have seen a late onset endophthalmitis, occurring years after cataract surgery, diagnosed as non-infectious uveitis even by an NIH "uveitis specialist".

We treat the presence of cells, not flare, with steroids. Uveitis is graded as 1+ with about 10 WBCs in a high-powered field (HPF) on SLE, 2+ with 20 cells per HPF, 3+ with 30 cells per HPF and 4+ with 40 cells per HPF. A hypopyon is 4+ by definition even if you do not see cells floating in the AC. Iritis may spill over into the vitreous and vice versa. Identify which area of the uvea has the most cells to determine the type of uveitis present (iritis, pars planitis or posterior uveitis). Your job is to classify the type of uveitis, identify treatable causes of uveitis, and to minimize the inflammation before posterior synechia, CME, glaucoma, or other complications occur.

Collections of cells on the back surface of the cornea are called keracitic precipitates (KPs). The come in small and large or mutton fat KPs. I don't really care what size the KPs are, all sizes need to have the same work up performed. Here is a photo of mutton fat KPs (© AAO 2014):

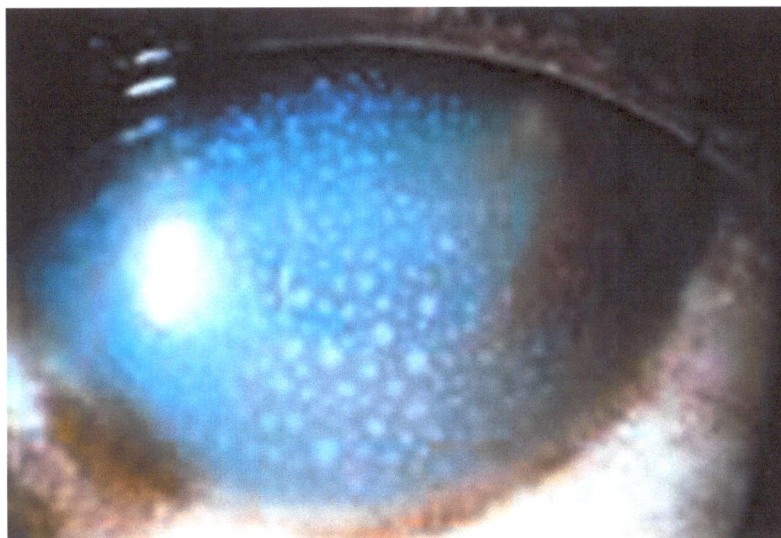

A more typical presentation of mutton fat KPs in a sarcoid iritis (© AAO 2014):

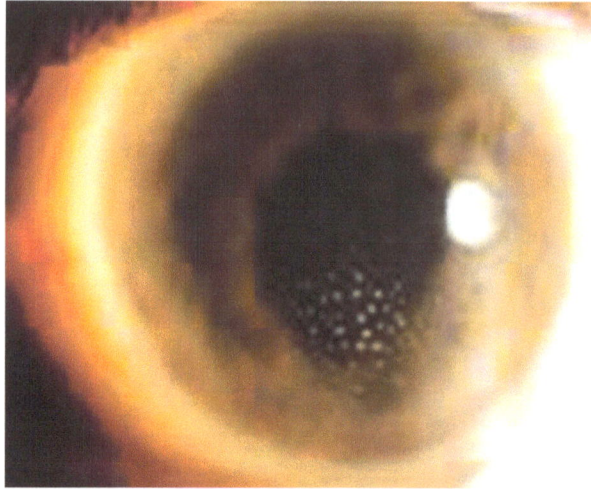

This is what posterior synechia from an old case of iritis looks like. (Anterior synechia are between the iris and the TM):

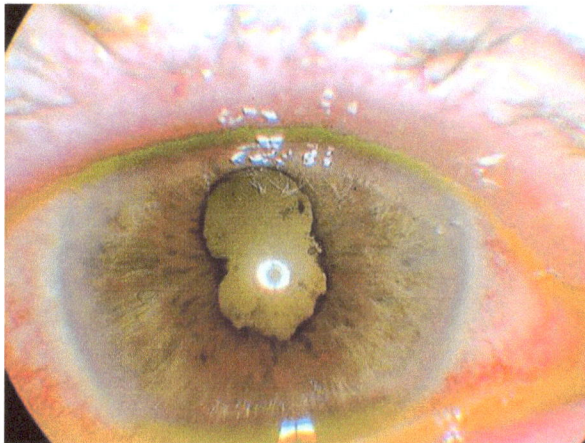

My mnemonic for remembering an iritis W/U is:

1. Iritis work up:

Just	**J**RA, found in young girls that have a history of JRA, refer them out
Better	**B**ehcet's with mouth ulcers (98% of Behcet's patients have these)[137] and genital ulcers, found on the medical history taking
Get	**G**onorrhea with a painful penile discharge. GC is typically asymptomatic in females/ **G**laucomatous crisis also called Posner-Schlossman, treat with PF
Help	**H**erpes history or seen on exam, topical PF and PO acyclovir
Understanding	**U**GH seen with AC IOLs, has glaucoma, uveitis (iritis), and a hyphema
The Friggin	**F**uch's iridocyclitis has iris atrophy causing a lighter iris in the affected eye, fine KPs occur over the entire cornea and fine B's traversing over the TM. May not respond to or need topical PF. Rare.
Labs	**L**yme labs in California may have better western blot results

Here is a photo of the hard to see, fine ANV sometimes present in Fuch's heterochromic iridocyclitis that typically do *not* cause NVG (© AAO 2014):

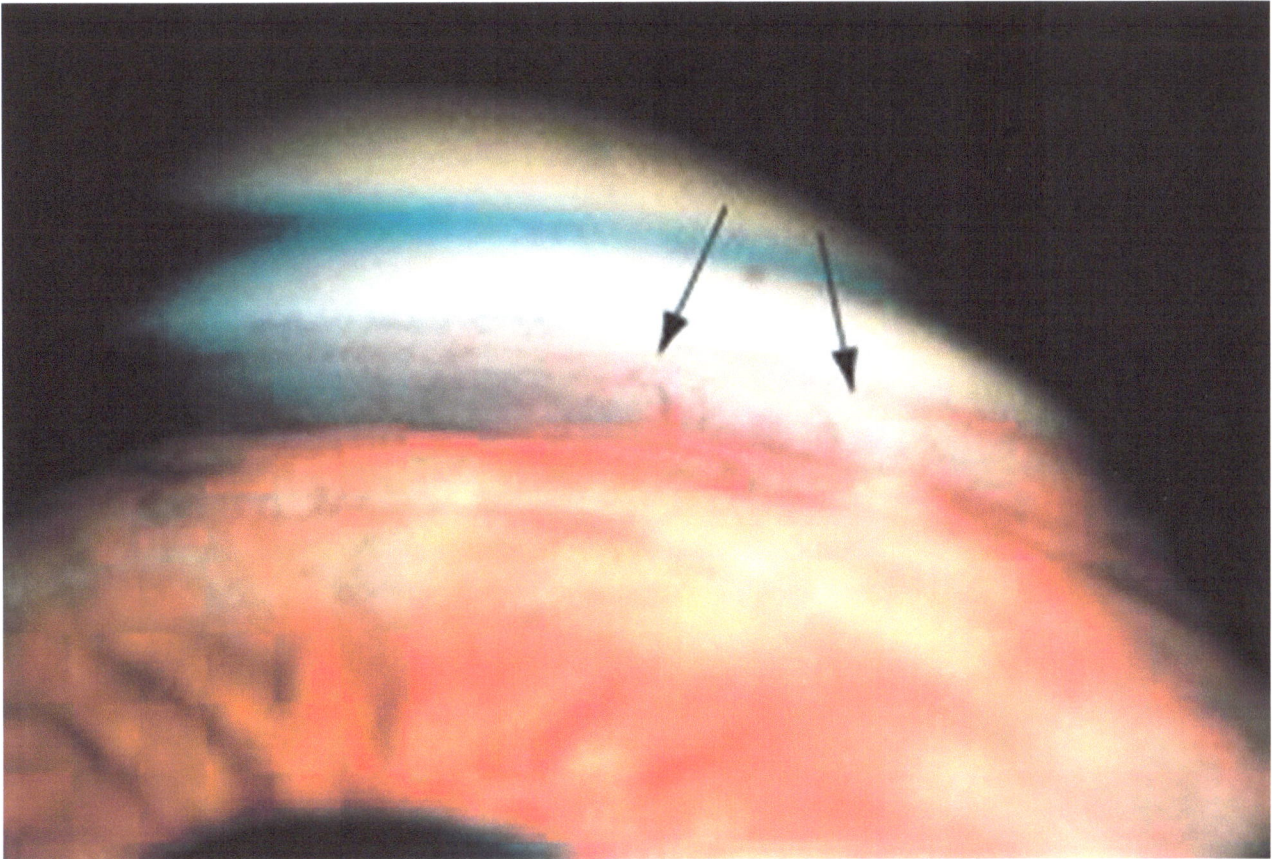

I also get a *HLA B27* blood test for the 4 diseases associated with HLA B27 positive uveitis entities. These are **PAIR** for:

- **p**soriasis · · · · · · · · · · · · · · · · · with skin lesions
- **a**nkylosing · · · · · · · · · · · · · · · · spondylitis with chronic lower back pain
- **i**nflammatory bowel disease · · · Crohn's and ulcerative colitis, both have chronic bloody diarrhea
- **R**eiter's syndrome · · · · · · · · · · with chronic penile discharge, needs chlamydia treatment too

2. Posterior uveitis or choroiditis, found in the *3 "S"s and 3 "T"s*

- **S**arcoid. Get an ACE and CXR (very useful, look for hilar adenopathy). Look for candlewax drippings around BVs. Up to 50% of sarcoid patients are anergic and don't respond to PPD.[138]
- **S**ympathetic Ophthalmia which has a severe vitritis in the *non*-involved eye usually less than 1 year after penetrating trauma or vitreal surgery to the other eye. #1 diagnosis to R/O in an uninjured eye with pan uveitis as it can lead to bilateral blindness. Refer out for PO steroids. I've never seen a case.
- **S**yphilis gets an RPR and FTA-abs blood test. Refer out all positive responders
- **T**oxoplasmosis seen on DFE, #1 cause of posterior uveitis,[139] get titers and refer out if positive since treatment requires strong PO drugs
- **T**oxocaria seen on DFE
- **T**B tested for with PPD with anergy panel to insure normal immune reactions

3. Intermediate uveitis or *pars planitis* has all of these and multiple sclerosis as well. Pars Planitis has snowballs or snow banking seen inferiorly on exam as well as CME. This is how you diagnosis it. You ask for symptoms of MS and consider an MRI of the brain if some are found. The #1 symptom of MS is reduced vision from optic neuritis. It can also present as a tingling or loss of motor function in the extremities.

In my experience, if you treat uveitis inflammation very aggressively in the beginning, you shorten the duration of treatment. I use brand name Pred Forte (PF) 1% every 30 minutes while awake (WA) and homatropine BID to avoid posterior synechia in the effected eye/s for the first 2-4 weeks. (The brand name PF gets inside the eye better than the generic does and I insist on brand name PF as does the Will's Eye Manuel, calling it "necessary"[140]). Keep in mind that some patients are steroid responders and need to have their IOP checked every so often. Both uveitis and PF can cause high IOP so treat the uveitis with PF aggressively and then the glaucoma, if it occurs with glaucoma drops. Don't worry about trying to determine if the uveitis or PF is causing the high IOP, use the PF aggressively and treat the high IOP with drops if it occurs. I taper off the PF *slowly* or a rebounding of the uveitis may occur. I use PF for a number of weeks or months before stopping it altogether. We do not find a cause in 50% of the patients with iritis. Some patients require long term QD dosing of PF to keep recurrent attacks from occurring. Refer out posterior uveitis patients as topical PF will not control the uveitis and these patients may require a sub-tenons, retrobulbar or intra vitreal injection to control the posterior uveitis. Refer out patients with a significant choroiditis (swelling seen on exam) or any retinitis (retinal bleeding and retinal bumps) as lymphoma now enters into the differential diagnosis. Any retinal bleeding or vasculitis becomes a retinitis diagnosis (which can also have a vitritis associated with it); refer it out.

Keratoconus:

This is a form of corneal ectasia that usually enters the differential diagnosis when the phoropter BCVA is less than 20/20 and the pin hole is 20/20. Corneal ectasia is not uncommon in LASIK patients. They usually had forme fruste Keratoconus that was missed on the pre-LASIK screening. Corneal topography is the best way to screen for this but the reimbursement is so low, I use the keratometer to screen for keratoconnus. Not being able to superimpose the mires all the way around, 360 degrees on the keratometer, can be graded as 1-4 plus. I find that regular and scleral gas perm lenses are the best initial option for post-LASIK patients. Hybrid contact lenses have a lot of hypoxia associated with their use.

A recent long term study of corneal cross-linking (Riboflavin with ultraviolet A) showed an improved BCVA (P = 0.002), no loss of endothelial cells, and a reduction in corneal curvature (P < 0.001) over a mean follow up of 131.9 months in keratoconnus patients. [199] This was a retrospective study involving 34 eyes. I believe that all keratoconnus and post-LASIK ectasia patients should be offered the option of corneal cross-linking in an attempt to stabilize the cornea from future worsening of their ectasia.

Circulating auto antibodies are thought to be the cause of TED. These auto antibodies can affect the thyroid gland and/or the EOM's. TED is a self-limiting disease that on average lasts 1 year in non-smokers and 2-3 years in smokers, so encourage patients with TED to stop smoking.[141] 5-10% of patients with TED reactivate at some point in their lifetime.[141] TED patients can be hyperthyroid (90%), hypothyroid (4%), or euthyroid (6%).[124]

TED can have acute, severe episodes that need to be differentiated from an orbital pseudo tumor episode. Both have acute proptosis with a very red eye. A CT of the orbits shows no EOM tendon involvement in TED and tendon swelling in pseudo tumor. This is a CT of the orbits showing sparing of the tendons seen in TED. (© AAO 2014).

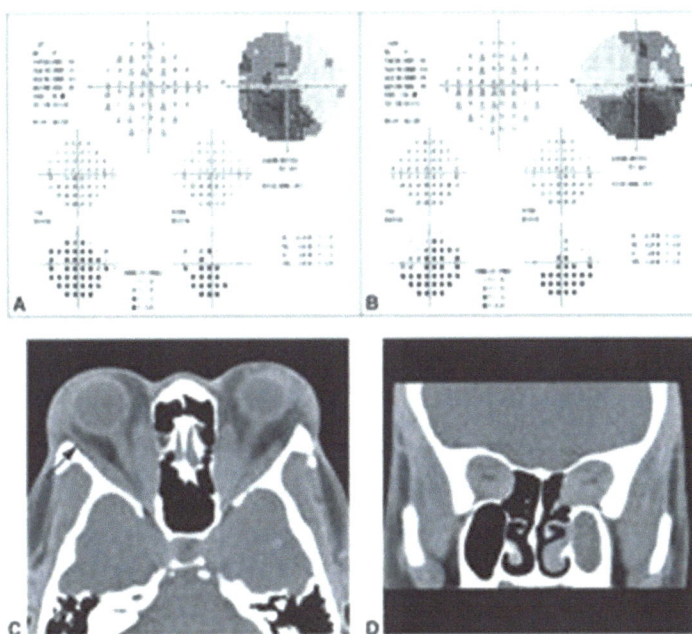

TED is suspected when you notice scleral show either in the upper lid or lower lid in 90% of cases at some point.[142] The upper lid also can have more scleral show in down gaze. This is known as *lid lag* and is called *Von Graefe's* sign. Here is a photo showing scleral show above and below the cornea (© AAO 2014):

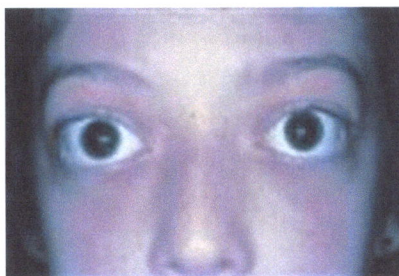

Non-TED causes of scleral show include:[143]

- a severe ptosis in one eye with significant stimulation of the levator to raise that eyelid which also raises the other eyelid due to Herring's law of innervation, giving a false scleral show in the non-ptotic eye. This is called a false Von Graefe sign. You raise the ptotic lid and see if the other upper lid comes down or not. Or, just measure for true proptosis with a Hertel exophthalmometer.
- 3rd CN aberrant regeneration causing a false Van Graefe sign, usually in one eye.
- Rarely Parinaud's syndrome that has a reduced upward gaze, dilated pupils with a light-near dissociation, and convergence-retraction nystagmus. I've never seen one.

A deep dull orbital pain is the most common symptom in TED, occurring 30% of the time[141] so think of TED when you are presented with this symptom. True proptosis is measured with the Hertel exophthalmometer. This measures the distance from the corneal apex to the lateral orbital rim. A Hertel reading of 22 and below is normal for non-Africans, 24 and below is normal for Africans, 2mm or less of a difference between the eyes is normal. Readings above this warrant a TED work up. For some unknown reason, myasthenia gravis is associated with TED in 1% of patients[141], so look for intermittent ptosis (the opposite of scleral show) and intermittent diplopia complaints (unfortunately also found in TED).

Your biggest fear in TED is optic nerve impingement from swollen extra ocular muscles (EOM). Although it occurs in only 5% of TED patients, it must be ruled out at each visit with HVF 24-2 testing.[144] Counter-intuitively, optic nerve compression tends to occur with minimal or no proptosis but is associated with restrictive strabismus.[144] A **TED work up** includes:

- full annual eye exam with Hertel exophthalmometer, checking for an APD, VF defects on formal 24-2, D15 color testing, IOP in straight ahead gaze and in up gaze.
- TSH, T4, T3, TSI (thyroid stimulating immunoglobulins), Thyroid antibodies (thyroglobulin antibodies and thyroid peroxidase antibodies), TBII, and acetylcholine receptor antibodies in patients with MG symptoms (intermittent symptoms worse in the PM).
- are they a smoker? Inform them of the 2.5 times higher risk of developing TED in smokers[141]
- a CT of the orbits, thin cuts, axial, and coronal/no contrast with unilateral or significant proptosis. I get an initial CT in most cases of proptosis to R/O something else pushing the eye forward and to look for enlarged EOMs pushing on the ON. The tendons of the EOMs are not involved with TED but they are with pseudo tumor. I don't repeat the CT very often. Order a CT of the orbits (thin cuts, axial and coronal) only if the Hertel readings or ocular symptoms become significantly worse.

Primary care providers and orbital surgeons want to know if the HVF shows impingement of the ON. If it does, maxillary sinus surgery, EOM surgery to address the resulting potential diplopia and ptosis surgery, in this order, needs to be done. This is why your HVFs are so important. Treat dry eyes as you would normally. Refer out acute TED episodes, all HVF losses, diplopia complaints, and abnormal TFTs.

11. Corneal FB:

If you want to take out corneal FBs, use a 30 gauge diabetic needle aimed obliquely to the cornea and try to pry it out. Don't approach the cornea straight on since you might penetrate the eye if the patient suddenly moves forward and create an open globe. The rust ring can usually be pried out with the same needle but a burr may be used as well. Don't go too deep with the burr as the cornea will be soft in the area of the FB and you might burr into the AC. This possibility should really scare you. Always check beneath the upper lid for foreign bodies left over from the original insult (both photos © AAO 2014):

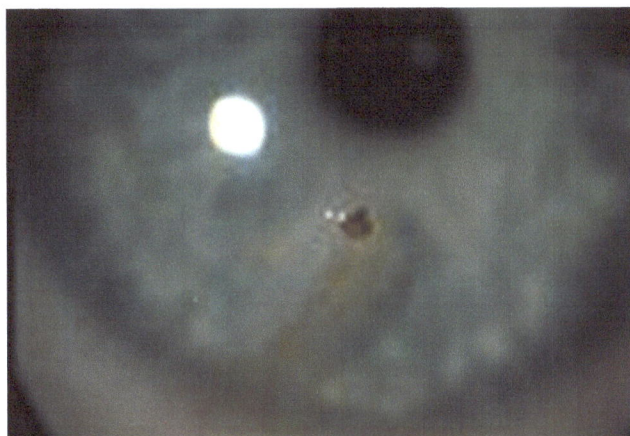

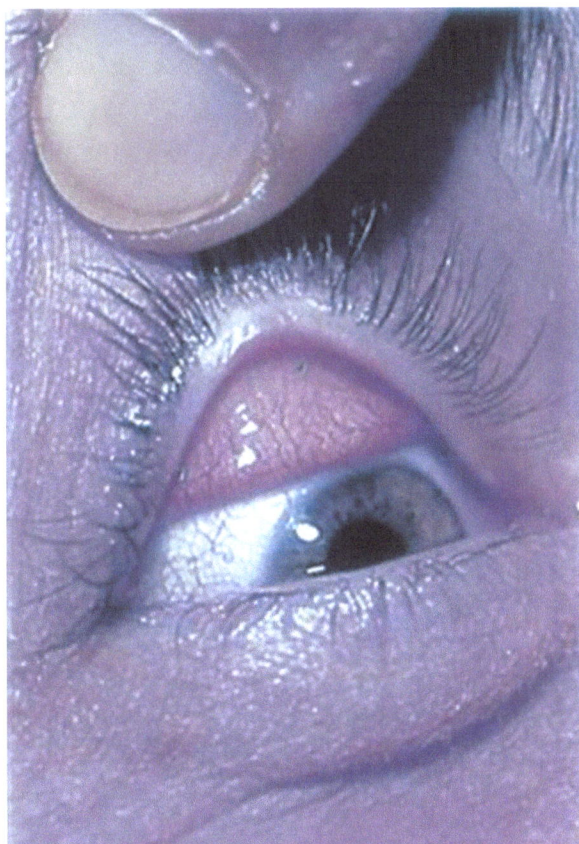

Use topical anesthesia and a topical antibiotic drop. I don't use a lid speculum but I do tell the patient to keep their eye open and not to blink, otherwise I might poke inside their eye or scratch it badly. Make sure the patient's head is tight up against the forehead rest. You don't need their head coming forward just when you are up against the cornea with your needle. You can use a consent form if you wish and do a separate operative note for the insurance company when they come looking for an excuse not to pay.

I agree with *Abelson's triad:*

> If it itches, it's allergic conjunctivitis
> If the lids are sticking together, it's bacterial conjunctivitis
> If it burns or has a sandy feeling, it's dry eyes

Let's add one: treat for Demodex with any chronic blepharitis patient
Here are some photos of the demodex mite and the tree tea oil available for lid wipes Q day for 6 days:

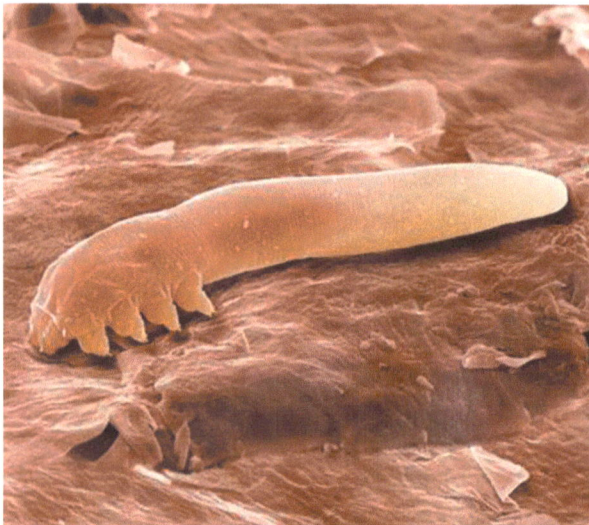

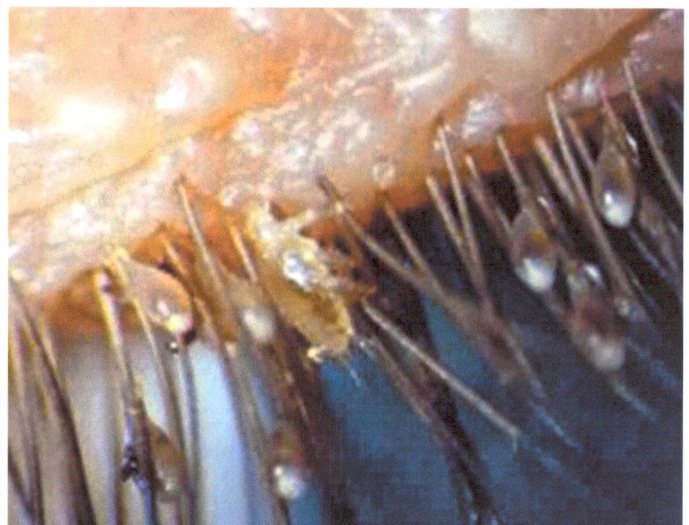

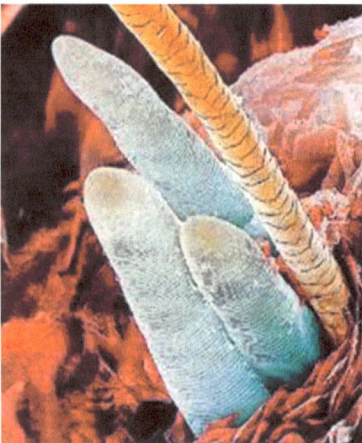

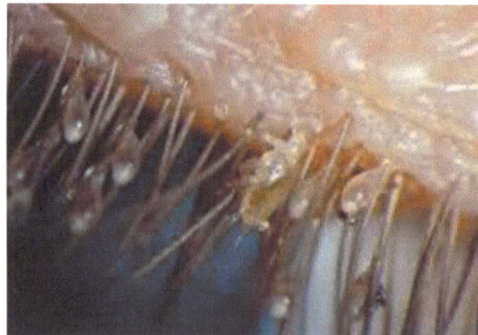

Let's make this as simple as possible. There are a lot of diseases that can cause a red eye but let's approach this topic like I do in my clinic. The first question I have for the patient is "do you wear SCLs?" If they do, invariably the cause of the pink eye is SCL-induced sterile sub-epithelial infiltrates (SEIs). SEIs are a non-infectious collection of WBCs in the cornea that you can see on high power (we only use high power on the SLE right?). SEIs are faint and you can see through them. Corneal ulcers are dense and you cannot see through them. SEIs do not look like ulcers. Ulcers are an infection inside the corneal stroma so a break in the corneal epithelium is required for their development. SEIs have an intact corneal epithelium.

There is a lot of confusion and mis-diagnosis surrounding SEIs and I don't know why. The best etiology I have read for SEIs is that they are an allergic reaction to the protein buildup on the SCL. It is rare for this to occur in HCL GPs. I treat SCL induced SEIs by having the patient stop wearing the SCLs for at least 6 weeks and have them start a topical antibiotic drop QID for 4 days. I tell the patient that if they wear the SCLs for even 5 minutes, they will restart the 6 week stop clock again. I do not give PF drops as I want the WBCs in the SEIs to go away on their own so they stay away. SEIs treated with PF can come back once the PF is stopped.

I see the patient back in 6 weeks, ASAP if a worsening red eye or decreased vision occurs and I document that I told the patient this in the chart. If the SEIs have not gone away in 6 weeks, then I might consider using PF Q 1 hour WA for 4-7 days, see them back and then start a quick taper. If the patient is on PF longer than 10 days, I see them back to check the IOP. Once the SEIs have resolved, the patient has to improve whatever wearing/cleaning regimen they were using so the SEIs don't return. It is typical for the patient to have a long history of recurrent red eyes with SCL use by the time they finally come to see you for the condition. Whatever wearing and cleaning schedule the patient is using needs to be changed so that a *cleaner lens is on the patient's eyes for fewer hours*. I think that changing to the H_2O_2 (hydrogen peroxide) cleaning system or changing to a daily SCL regimen (a new SCL on every day) is the most effective wearing schedule change.

Let's cover some of the more common causes of a red eye.

Corneal ulcers. You will read a lot of different strategies when it comes to ulcer treatment; some of it is not accurate. Corneal ulcers, especially in SCL wearers, are serious and need to be treated aggressively. Any red eye in a SCL wearer needs to be seen that day. Central, big, vision threatening, or ulcers threatening corneal perforation are admitted to the hospital for fortified (concentration higher than normal) drops of vancomycin (or cefazolin) and fortified tobramycin (or gentamicin) Q 30 minutes *around the clock* (ATC). Patches are never used as they can make the infection worse. Patches were used for pain control in corneal abrasions before topical NSAIDS were developed, but never for corneal ulcers. You will not be involved in the treatment of patients that should be admitted to the hospital. Your job and what I do as well, is to recognize serious ulcers (large, central or having a risk of perforation due to the depth of the ulcer) and refer them out. The ulcers that you will treat will be the vast majority of non-vision threatening ulcers. These are ulcers that are not large and do not pose a threat of perforating the cornea (not deep). Sleeping in SCLs increases the risk of an ulcer 10 to 15 times, with the longer the extended wear time, the higher the risk of developing an ulcer (AAO Quality of Care Secretariat, Hoskins Center for Quality Eye Care, May 2013).

The cornea is unique in that we don't use the blood system to get antibiotics to the infection. The microbiology lab measures resistance of organisms to antibiotics by using disks embedded with the antibiotic spread over a blanket of the bacteria in question and seeing if the bacteria grow close to the disk. If the bacteria grow close to the disk, then the bacteria are resistant to that antibiotic. When we treat an infection of the cornea, we can reapply drops to the cornea right on top of the infection and increase the exposure to the antibiotic, thus overcoming what the lab measures as bacterial resistance to the antibiotic. So, tell the patient to stop SCL use and have them apply antibiotic drops Q 15 to 30 minutes WA with or without an antibiotic ointment QHS until there is an improvement in the ulcer, and then for 3-5 more days Q 60-120 minutes WA. Once the infection is significantly decreased in size and density and the *eye is not red*, I will step down the dosing to 6x/day for 3-7 days (depending on how bad the ulcer was in the beginning), then 4x/day for 1-2 more weeks. Go longer than you think is necessary so the ulcer doesn't come back, all the time not wearing SCLs.

I use this plan on all of my corneal ulcers that don't need admission to the hospital. Again, admitted patients need fortified antibiotics applied ATC. I tell the patient that improvement is not expected for 2 days, and that we are okay as long as the infection does not get worse during these first 2 days. I also stress that there is a risk that if they do not do what I ask them to do, they will lose the eye and go blind. I document that I told them this.

You have to see corneal ulcer patients every day until the infection starts to get better. No choice here. You could have a pseudomonas infection which can perforate the cornea in 1-2 days. I use Ocuflox Q 30 minutes WA. You can make a good case to use a newer flouroquinilone if you wish. Besifloxacin or moxifloxacin is expensive but is probably better defended in court should an ulcer get worse with treatment than ocuflox is. However, I have never had an ulcer not resolve using ocuflox.

Most doctors fret over which antibiotic to use QID rather than using any antibiotic drop Q30 to 60 minutes WA, missing the big picture altogether. Just use whatever antibiotic drop you chose Q 15 to 30 minutes WA, with or without an antibiotic ointment QHS. You give up the tremendous advantage you have in the treatment of corneal ulcers by using BID or QID dosing. I never use topical PF for ulcers, especially since the prospective, randomized, multicenter Steroids for Corneal Ulcer Trial (**SCUT**) showed no benefit in BCVA (P = 0.39) nor in size of the scar (P = 0.69) after using PF, 48 hours of moxifloxacin use (prospective, randomized, multicenter trial, n = 399 patients followed for 12 months)[145] and since PF is *contraindicated* for fungal, acanthamoeba, norcardia, and HSV infections. The point is, how do you really know which organism is causing the ulcer in the beginning? You are never sure so *don't use topical PF for corneal ulcers*.

I don't culture the cornea. Most community-acquired corneal ulcers are treated successfully empirically without cultures (PPP 2013, Bacterial Keratitis). I recommend that you not do so either as you may perforate the cornea when you scrape it for a culture, turning the ulcer into an open globe with probable loss of the eye. If the ulcer gets worse during any one of the first 3 days, refer the patient out. Aminoglycosides can cause an irritation of the cornea/conjunctiva if used for more than 7 days and therefore should not be used chronically. If the red eye comes back, refer out for a potential chlamydia chronic infection.

Don't treat infections that look like fungal (with satellite lesions) or a ring ulcer that is very painful (acanthamoeba) with antibiotics, refer those patients out immediately.

The first photo is a fungal ulcer with feathery borders and the second photo is an acanthamoeba infection with the ring ulcer (both © AAO 2014):

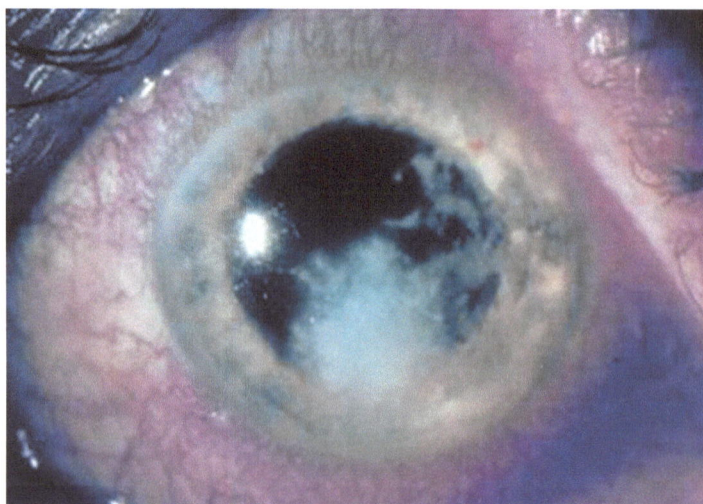

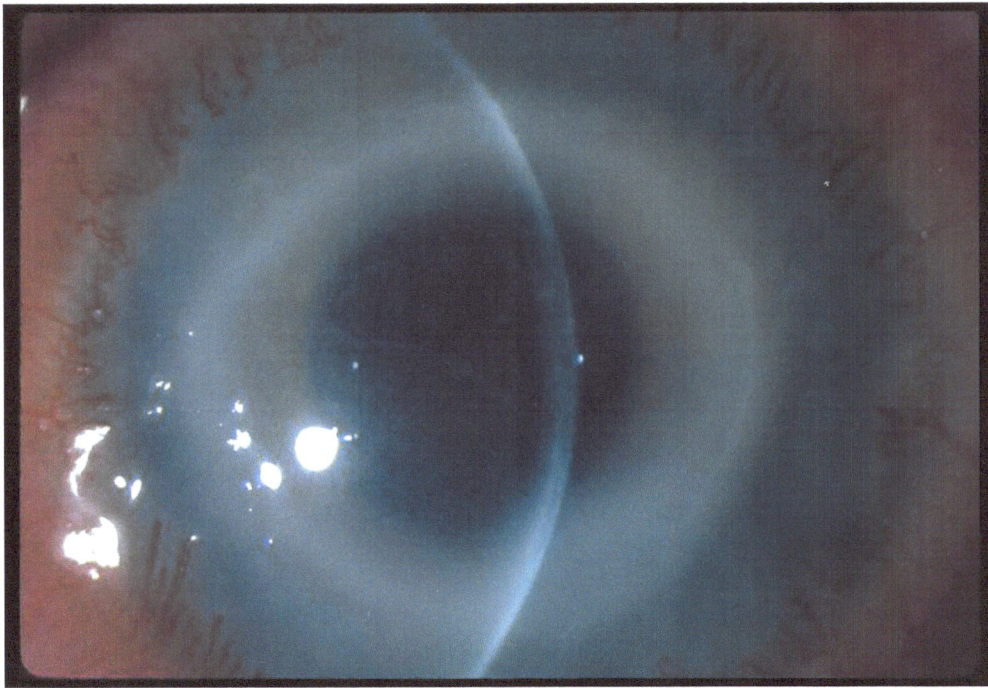

Refer out corneal ulcer patients with significantly reduced eyelid closure abilities (severe bells or ectropion patients) as they won't heal normally and may require a bandage SCL.

Viral conjunctivitis: Caused by adeno, the herpes viruses, or the rare EKC infection, which is covered below. Herpes simplex virus (HSV) and herpes zoster virus (HZV) can cause conjunctivitis or a keratitis. Stain the cornea looking for dendrites and refer out if HSV or HZV is suspected. Adeno with significant reduction in the BCVA is treated with topical PF once HSV has been ruled out. Generic PF is okay as the drug does not have to get inside the eye. SEIs in some adeno infections may remain permanently and reduce the BCVA. **Make sure there isn't HSV present before you use PF for a pink eye.**

The first 3 photos show an HSV infection with classic non-dermatome skin eruptions and dendritic keratitis. The fourth and fifth photos are of an HZV infection with a classic V1 dermatome eruption and corneal dendrites as well (all 5 © AAO 2014):

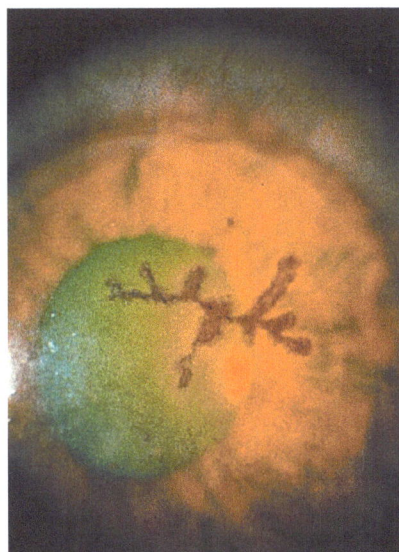

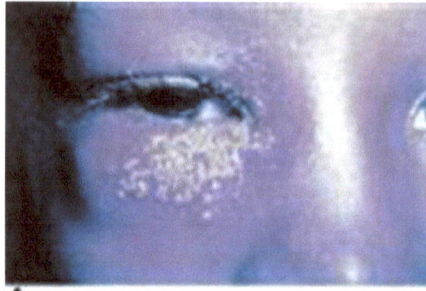

A

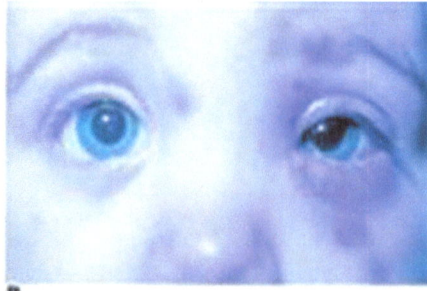

B

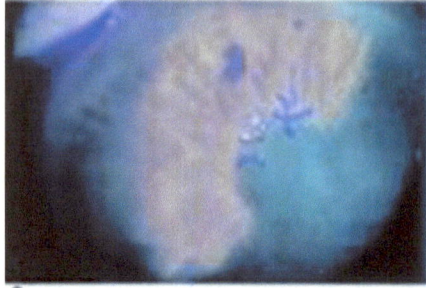

C

D

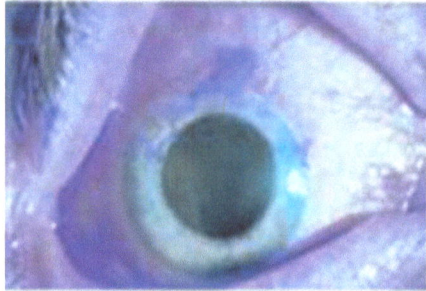

E

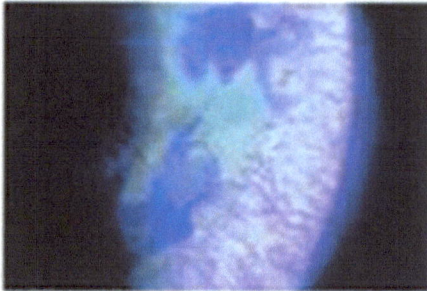

F

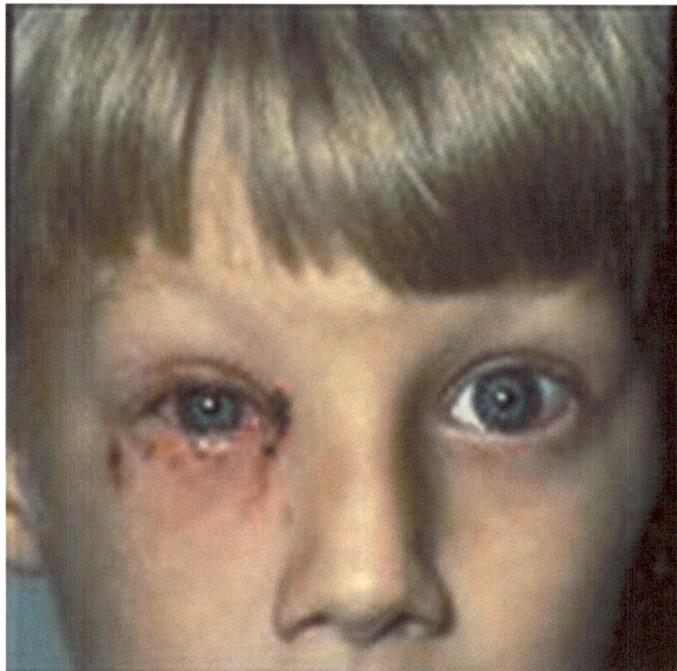

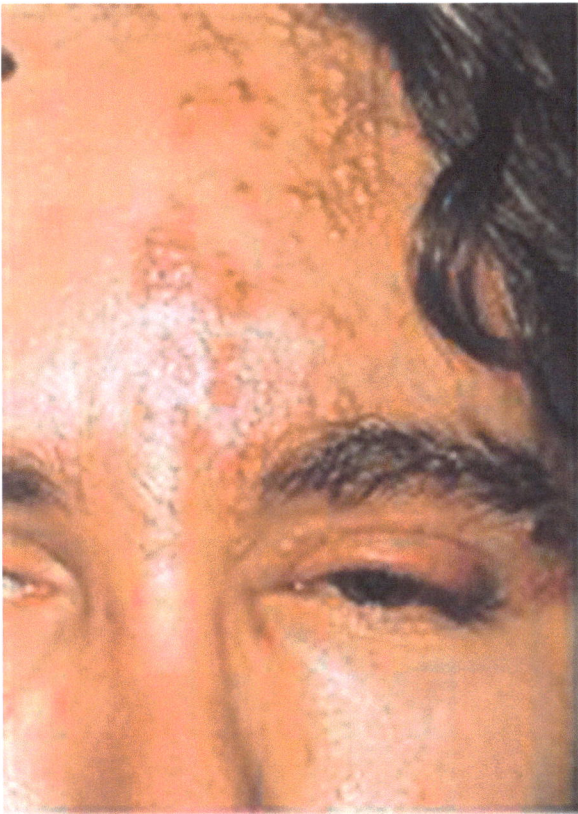

A

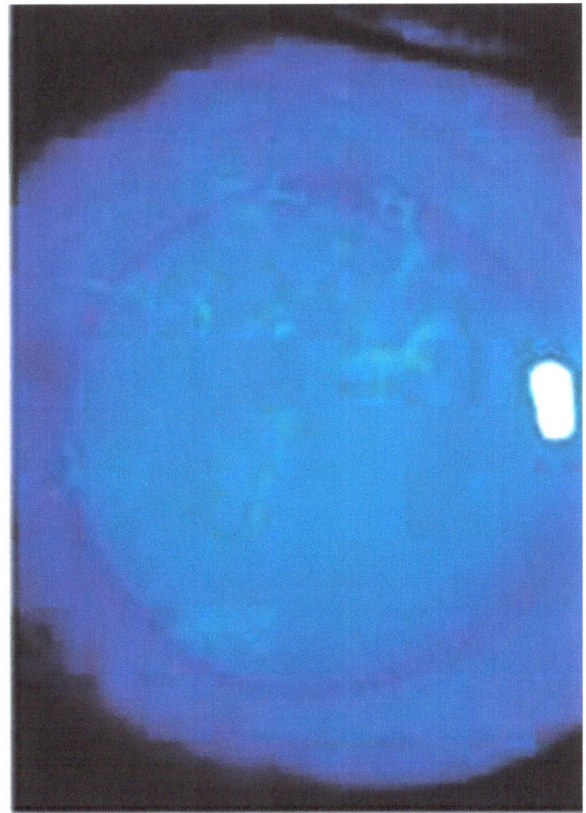

B

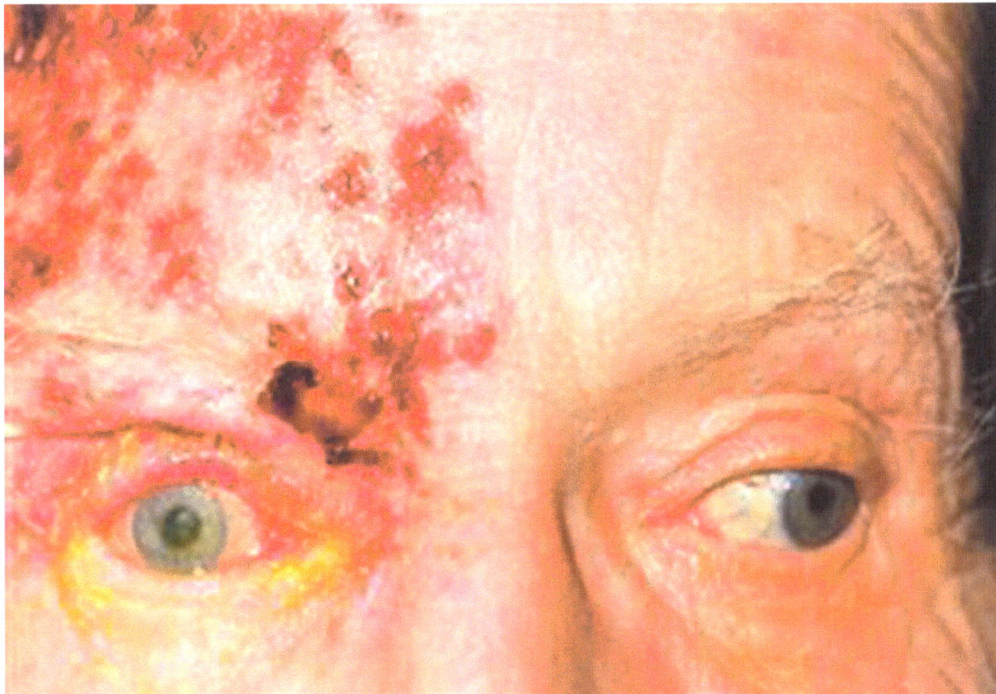

EKC. Unless you have experienced an EKC outbreak, you have not seen EKC. I can diagnose an EKC case from across the room. EKC has severe SEIs that usually permanently reduce BCVA along with severe swelling and redness of the eyelids so that it looks like pre-septal cellulitis. You have to do a 5% Betadine wash OU in the clinic and then ganciclovir 5 times a day and topical (generic is okay since it does not need to get inside the eye) PF Q 1 hour WA. All hands on deck for this disease in order to avoid a penetrating corneal transplant (PKP). The most important thing you can do is the Betadine wash, which you will probably not get paid for. Here are two photos of the SEIs from an EKC patient (both photos © AAO 2014):

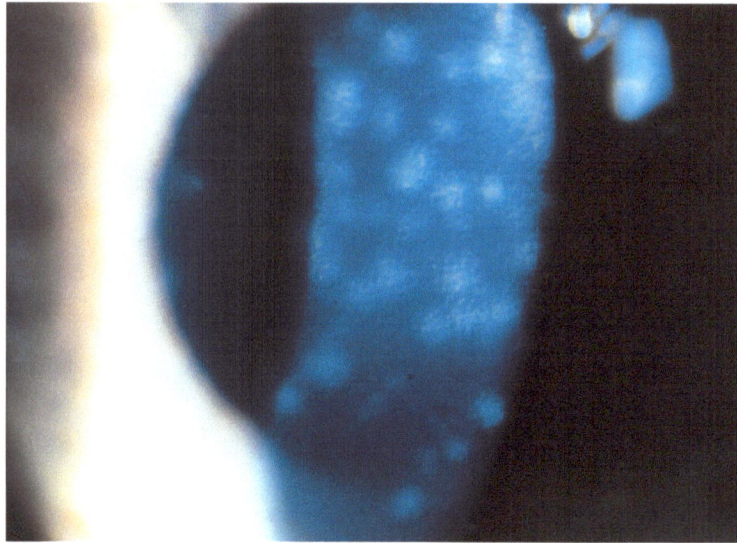

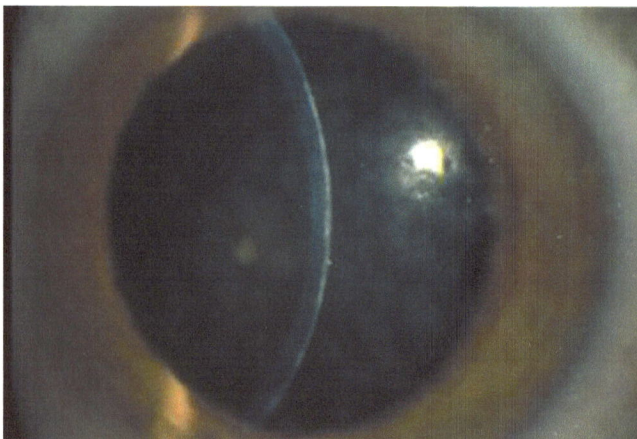

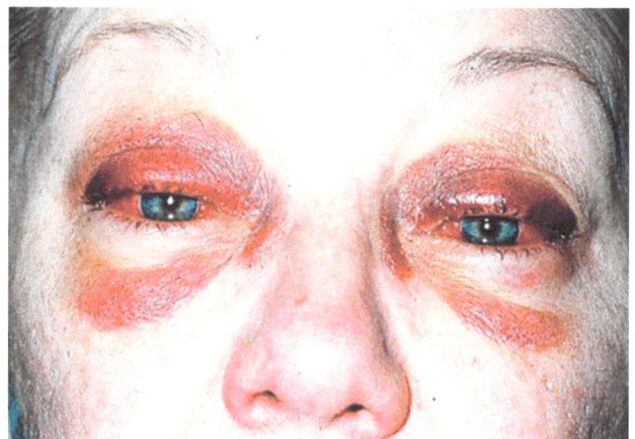

Iritis or uveitis is next. Look for cells in the AC and/or vitreous in all pink/red eye patients. PCPs use tobradex QID for all pink/red eye patients and if they don't respond, they send them to you. They don't have a slit lamp, you do. Use it to R/O uveitis in all pink eye patients. This is what the PCP is concerned about, the patient having something that they cannot see and is not being resolved with Tobradex. Remember, we treat cells, not flare, in uveitis. Use brand name PF (remember, brand name gets inside the eye better than generic) Q 30 to 60 minutes WA with homatropine 5% BID for the first 7-10 days to prevent posterior synechia. Do you notice that I use drops Q 30 minutes in the beginning for a number of different treatments? I find that if I gain control a disease quickly, I shorten the total treatment time. It kills me when I see besifloxacin used initially BID for a corneal ulcer or PF initially BID for iritis. Those patients are receiving significant under treatment of their disease, with the risk of the disease getting worse despite receiving treatment. I get referrals for this a lot.

Allergic Conjunctivitis: I have patients use OTC alaway BID, sometimes for the whole year, to relieve itchy eyes. Alaway is acidic and burns a little but I think that it gets inside the conjunctiva better than Patanol because of its acidity. If it is an allergic conjunctivitis, alaway will relieve the symptoms within a few days. Tell the patient to RTC if the itching is still there after a week of using alaway for dry testing. Don't miss meibomianitis as a cause of mild itchy eyes at your initial visit! I rarely use generic PF for allergic conjunctivitis, but I will for a short period of time to get a bad episode of itching under control so Alaway can resume control. I have the patient use alaway at the same time as PF.

Bacterial Conjunctivitis: I use Ocuflox QID OU for 10 days. Treat *both* eyes even if only one is affected as the other one will probably get it eventually. Most bacterial conjunctivitis patients respond well within 4-5 days and then the patients want to stop the drops. Tell them to complete the entire 10 day course or the infection will come back. If it is a really bad infection, consider reducing the bacterial load with a 5% betadine conjunctival wash. I use the betadine lollipop stick, 1 drop in each eye after numbing the eye with topical tetracaine. Betadine lollipops are inexpensive and one lollipop is good for both eyes.

Episcleritis is my waste-basket diagnosis for pink eyes when I am not sure what is going on and in the absence of any of the above. Make sure that a scleritis is not present with its deep boring pain and potential for perforation of the eye. I treat episcleritis with topical PF (generic is okay since it does not need to penetrate into the eye) Q 1 hour WA and I expect to see dramatic improvement in 1-2 days. Have the patient RTC in 4-5 days and, assuming there has been a dramatic improvement in the redness, step the dose down quickly to nothing in 10 days. The patient is asked to call back if the redness ever returns for a re-evaluation.

Sub conjunctival hemorrhage: Read page 120-121 of the Will's eye manual to see what you should do if the hemorrhage is associated with proptosis or decreased motility. (CT of the orbit with and without contrast to rule out a retrobulbar lymphangioma). Frequently, these patients are taking blood thinners (aspirin, Plavix, Coumadin, ...) and no further work is needed.

13. Retinal Conditions:

Lattice degeneration. Lattice is a thinning of the inner peripheral retinal tissue that occurs in about 10% of the population.[146] 25% of rhegmatogenous retinal detachments have lattice[146] so lattice presents a 2.5 fold increase in the lifelong risk of RRD. <u>**Lattice tends to progress to RRD in young myopic patients without symptoms**</u>[146] so see all lattice patients yearly for DFE and document that you have reviewed the 3 symptoms of RRD with the patient and told them to RTC ASAP if they occur. Here are some photos of lattice:

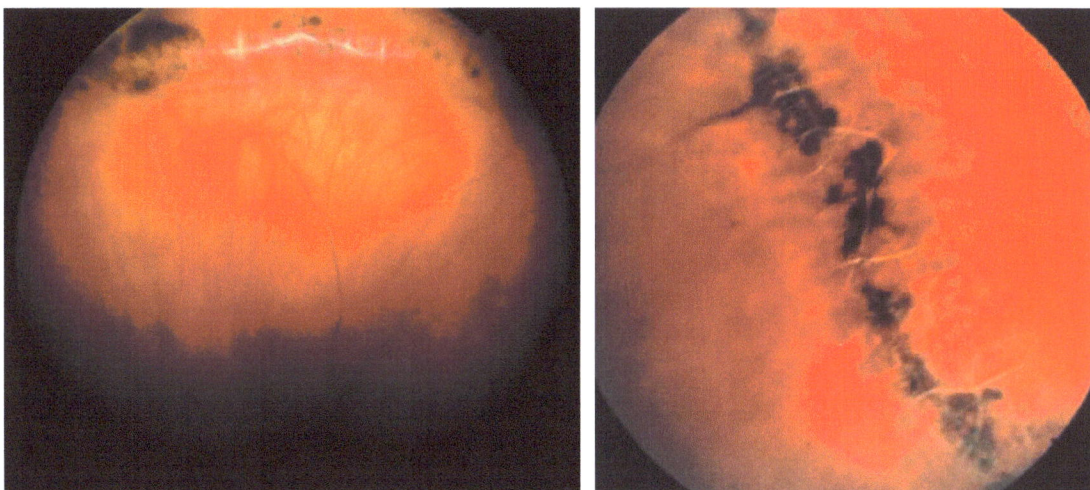

Here are some optos photos of lattice on the left and retinal tears and retinal holes on the photo to the right:

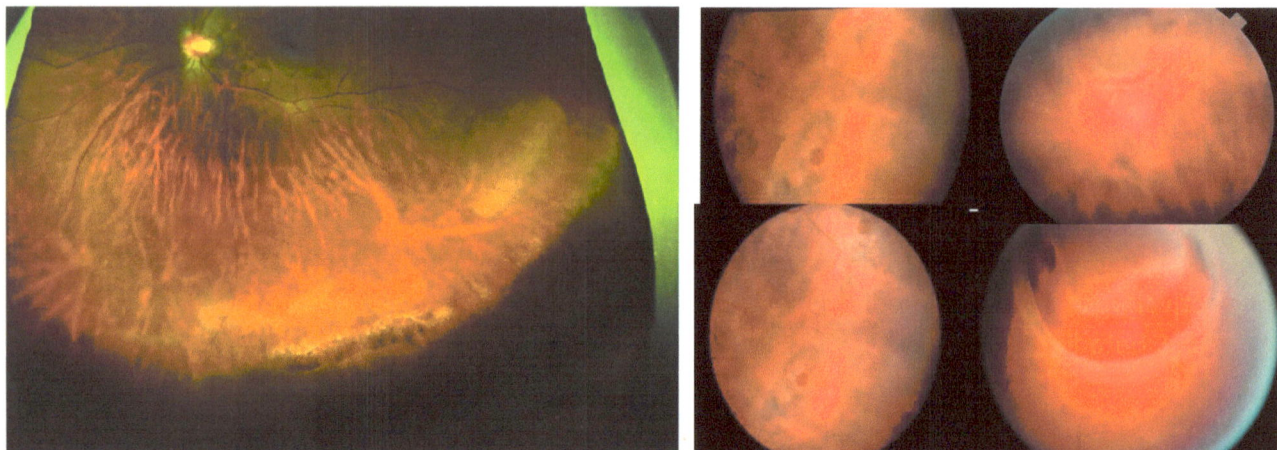

Here is a photo of lattice that has developed a localized retinal detachment:

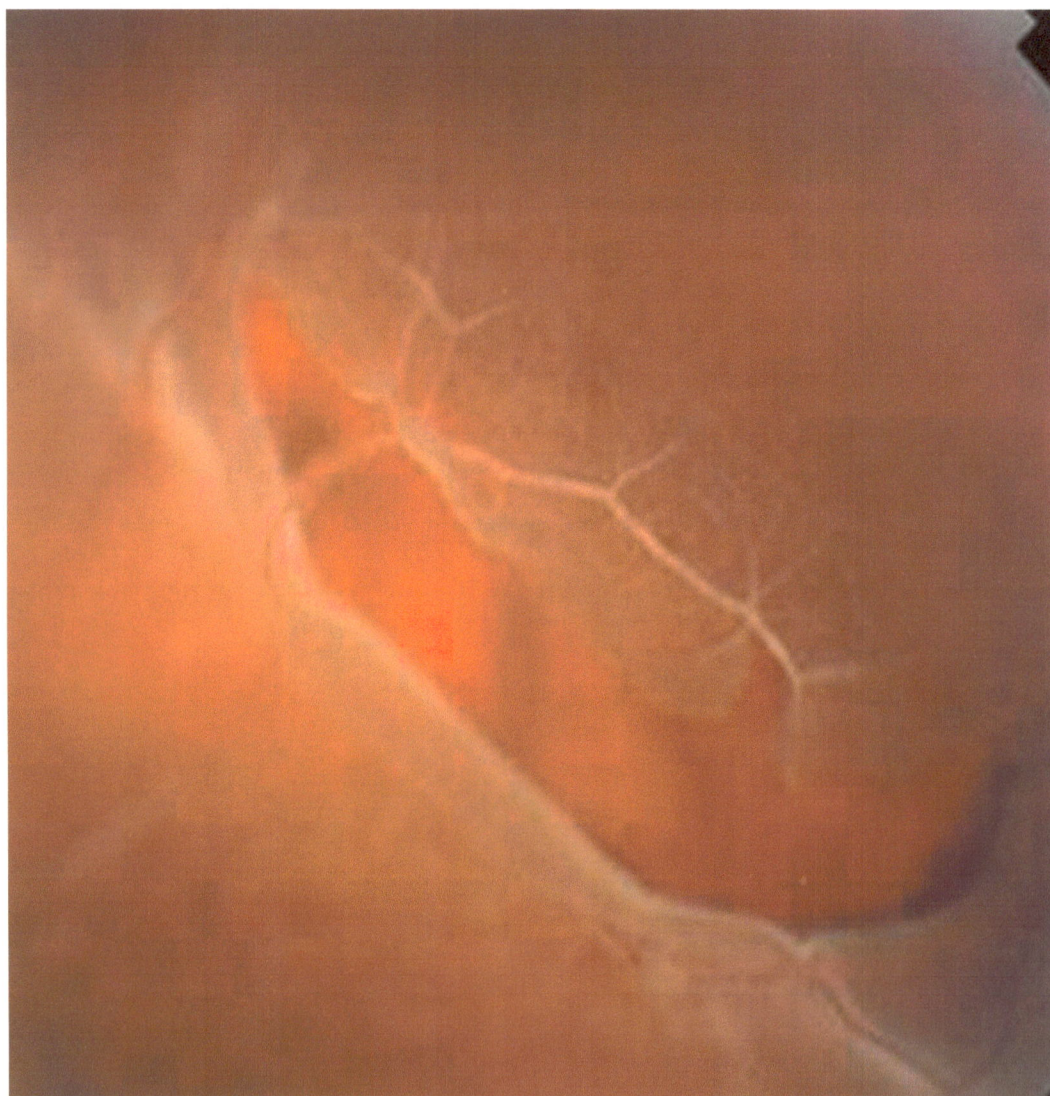

Age Related Macular Degeneration (**ARMD**): I have used the 2011 PPP for ARMD extensively for this section. ARMD comes in two forms, dry (90%) and wet (10%).[147] Wet is defined as the presence of a choroidal neovascularization membrane or an RPE detachment. Here are 2 diagrams showing the flow of nutrition between the choroid and the retina and the retina (RPE)/Bruch's membrane/choroid anatomy. Notice that drusen is in Bruch's membrane:

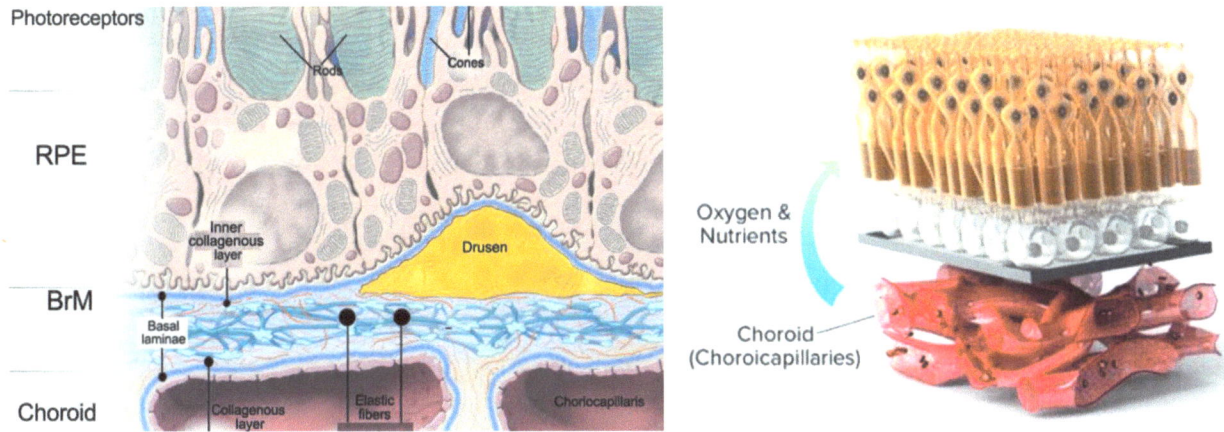

Prior to the development of Leucentis, we watched wet ARMD patients go blind despite out best efforts. Leucentis (and Avastin) have been miracle drugs for us. Both have been shown to be, by far, the most efficacious treatment available for wet ARMD. Two prospective, randomized, blind, multicenter studies confirmed the efficacy of anti-VEGF treatments. The **ANCHOR**[200] study looked at *classic* wet ARMD (n = 420 who completed the 2 year follow up, p < 0.0001 better BCVA with treatment) and the **MARINA** study [201] looked at *occult* wet ARMD lesions (713 completed the 2 year follow up, p < 0.0001 better BCVA with treatment). [The difference between classic and occult wet ARMD have little meaning in the anti-VEGF period.

A retrospective study looked at injections monthly or every 2 months for 7 years (n = 44 eyes). [202] The graph below breaks out the gain of letters for good, fair and poor BCVA at entry of the study. You can see that the worse the entry BCVA is, the more letters are gained with treatment and that BCVA is maintained very well over the 7 year follow up period.

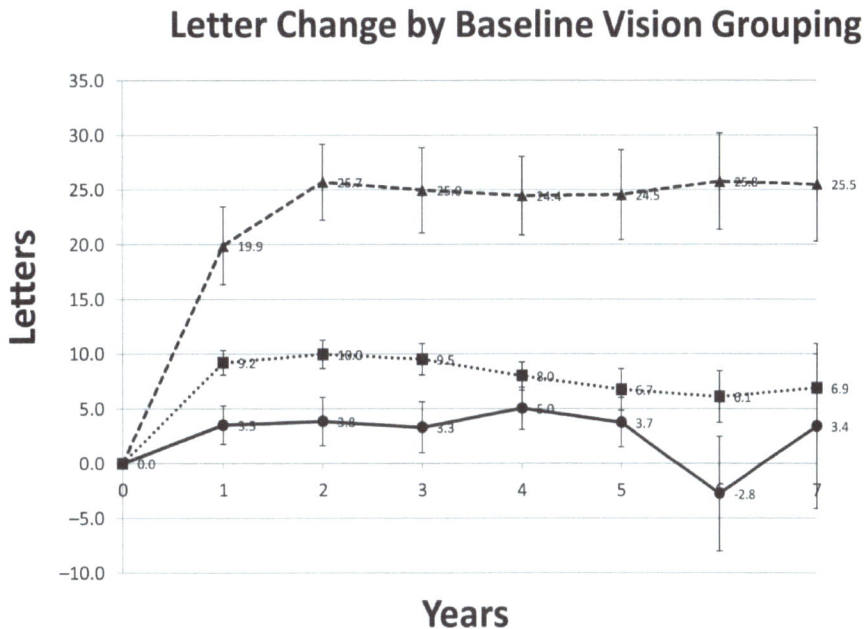

Letter Change by Baseline Vision Grouping

Graph showing the mean change in Early Treatment Diabetic Retinopathy Study (ETDRS) letter score at each yearly time point for eyes with 20/40 vision or better, 20/50 to 20/100 vision, and 20/200 vision or worse at baseline. Vertical standard error bars are shown.

—●— Good vision (≥20/40); ··■·· impaired vision (<20/40, >20/200); —▲— blindness (≤20/200)

___

Due to the cost of the drug and clinic time involved in monthly injections (for today's drugs, this is forever), various alternatives have been looked at for the frequency of long term anti-VEGF injections. A study of ABCHOR, MARINA and HORIZON patients looked at maintenance of vision improvement after 7 years of follow up (the **SEVEN-UP** study). [203] The frequency of anti-VEGF injections was dropped from monthly in the "initial study" period to an "as needed" basis. The SEVEN-UP study showed not only a loss of the BCVA gain seen during the initial 2 year Q monthly injections (seen at the 2 year mark in the graph below) but a worsening of BCVA from baseline after 7.3 years of follow up. There doesn't seem to be any way around it, Avastin and Leucentis anti-VEGF injections need to be given Q month or every 2 months forever. Newer anti- VEGF injections promise to need fewer injections per year. Eylea is an example of one of these newer drugs.

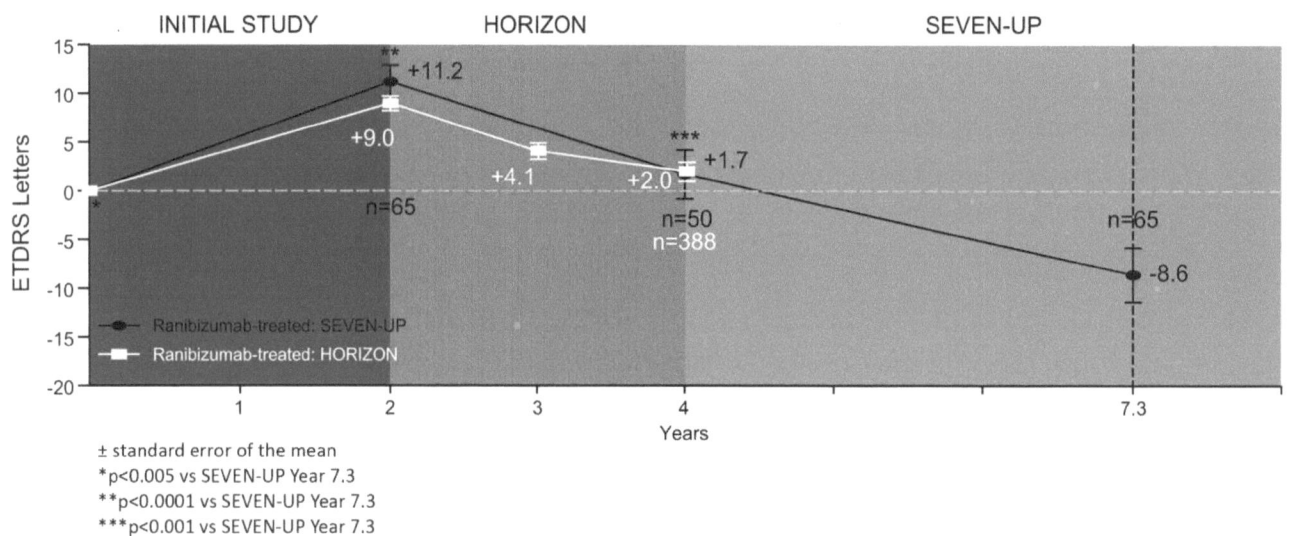

± standard error of the mean
*p<0.005 vs SEVEN-UP Year 7.3
**p<0.0001 vs SEVEN-UP Year 7.3
***p<0.001 vs SEVEN-UP Year 7.3

Anti-VEGF injections are used to treat a host of other hard-to-treat ocular diseases as well. The list of the new uses for avastin seems to grow almost monthly. Leucentis and Avastin are antibodies that bind to the naturally produced VEGF molecule, preventing it from promoting the formation of new blood vessels. Both drugs have been shown to be equally effective at reducing retinal neovascularization, although the cost difference is huge. They are typically used intra vitreally long term for a number of diseases. "Because neither drug eliminates neovascularization, treatment continues indefinitely for most patients."[148]

Major risk factors for ARMD that have been consistently identified in numerous studies include (PPP Age Related Macular Degeneration 2008):

- Age above 70, it is rare to have ARMD develop below 55 years of age. This is the #1 risk factor for advanced ARMD and is why it is called "age related" in the first place.
- Caucasian race

- Smoking. Smoking doubles the risk of ARMD with a higher pack year history increasing the risk. Smoking cessation after 20 years approximates the risk of never having smoked for the development of ARMD.
- Family history, hypertension and low blood levels of antioxidants have also been weakly linked to the development of ARMD.

Your main job is to identify the degree of ARMD, place patients on Age Related Eye Disease 2 Study vitamins (**AREDS 2**), a prospective, randomized, multicenter study (n = 3,694, average follow up 5 years)[151], encourage cessation of smoking, start home Amsler grid use when appropriate, take a photo and refer out all the patients who might be getting close to developing wet ARMD. The AREDS 2 vitamins are in addition to a MVI taken every day. Lutein and Zeaxanthin were not included in the **AREDS 1** study, a prospective, randomized, multicenter study (n = 3,597, average follow up 6.3 years)[149] as they were not commercially available at that time.[149] The zinc in the AREDS 1 formula increased hospital admissions for prostate issues and the vitamin A has been linked to lung cancer in smokers and ex-smokers. Because of these risks, the AREDS 2 study was done to see if substituting lutein and Zeaxanthin for vitamin A and reducing the zinc content would keep the efficacy in the ARMD formation. It did.[151] Again, substituting Lutein and Zeaxanthin and/or omega 3 long chain fatty acids (DHA and EPA) for vitamin A (beta carotene) did not reduce the effectiveness of AREDS 1 vitamins in reducing the chance of developing advanced ARMD, as shown in the AREDS 2 study.[151] I have intermediate patients RTC Q 6 months for DFE. The risk of progressing from intermediate to advanced ARMD in the AREDS 1 study was 18% over 5 years.[152]

So, only recommend the AREDS 2 formulation and not the AREDS 1 formula. Document that you told the patient to do the above as well. A large group of retinal specialists under the Beckman Initiative have attempted to standardize the classification of ARMD. Their classification is similar to the AREDS one but I'll use the Beckman Initiative below:[150] Here is the photo of drusen that was used to define the sizes of drusen for the AREDS ARMD studies.[206]

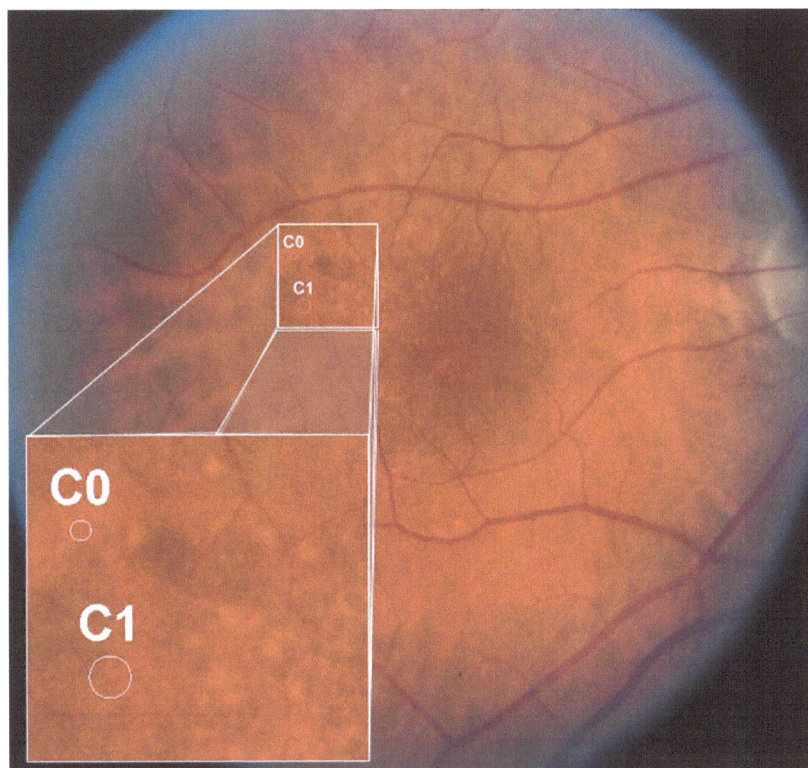

Figure 1. In an eye with multiple drusen variants, the Age-Related Eye Disease Study drusen grading circles C0 (63-m diameter) and C1 (125-m diameter) are superimposed for size comparison. Small drusen are smaller than the C0 circle (drupelets). Lesions larger than C0 but less than C1 are considered medium drusen, and lesions larger than C1 are large drusen. Within the inset, drupelets and medium drusen are seen. Faint reticular drusen also may be seen in the superior macular region.

No ARMD consists of a few small, < 63 um, hard drusen only, also called drupelets, that represent no increased risk of developing late ARMD. I place these patients on AREDS 2 vitamins even though it hasn't been shown to be beneficial in "no" and "early" dry ARMD over 5 years, it might over 15 to 20 years. Tell them to stop smoking and RTC for DFE Q year. I get yearly macular photos on all of my ARMD patients. I consider letting patients who refuse to sign the ABN for retinal photos, go see other doctors for their eye care.

Early Dry ARMD has at least one medium sized, ≥63um and ≤ 125um, druse *without* RPE abnormalities. I treat them the same as above. Early ARMD has 1.3 % risk of progressing to advanced *ARMD* over 5 years.[149]

Intermediate Dry ARMD has at least one large druse, >125 um (or the size of a large vein at the disk margin) or pigmentary abnormalities associated with at least one medium druse. Add the Amsler grid done *monocularly* to the AREDS 2 and cessation of smoking treatments above and tell the patient to RTC ASAP with any changes in the straight lines (metamorphosia) or if scotomas develop. AREDS 1 vitamins have been shown to reduce the chances of intermediate ARMD progressing to advanced ARMD by (n= 3,597; average follow up 6.3 years):[149]

    25% with antioxidants and zinc
    21% with zinc alone
    17% with antioxidants alone

Advanced (usually wet) ARMD has a choroidal neovascular membrane that is seen as a gray-green area in the fovea, a retinal pigmented epithelium detachment (RPE detachment), subretinal fibrovascular scar formation with/without bleeding or geographic atrophy. Once these patients bleed out from the NVM, the horse is out of the barn and it is very difficult to get their vision back. Our job is to get them onto Avastin before this happens. The risk of the fellow eye also developing advanced ARMD is 22% over 5 years so all patients with advanced ARMD in one eye should be followed by a retinal specialist. You do not need the liability exposure. Not much can be done for geographic atrophy except as described in the treatment for intermediate ARMD above. I learned to classify ARMD as dry (mild-moderate-severe) or wet. Similar to diabetic retinopathy. I still tend to use that classification today.

Soft and confluent drusen (drusen whose borders touch each other) may have a higher risk of progressing to advanced ARMD.[152]

OCT has replaced IVFA for most patients in the screening of intermediate and advanced ARMD patients. Although rare, anaphylactic reactions resulting in death do occur with IVFA and cases of IVFA induced death reverberate around the retinal community like a tidal wave. Here are some examples of ARMD in patients of mine. I would classify these two eyes as moderate to severe dry ARMD:

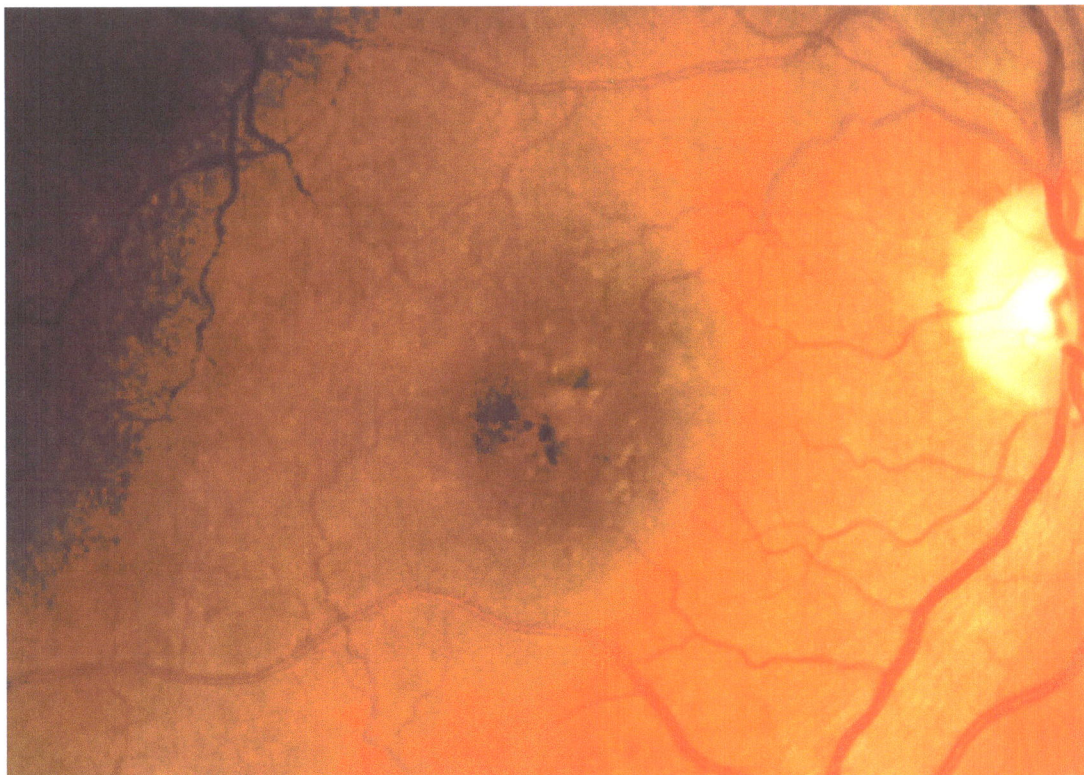

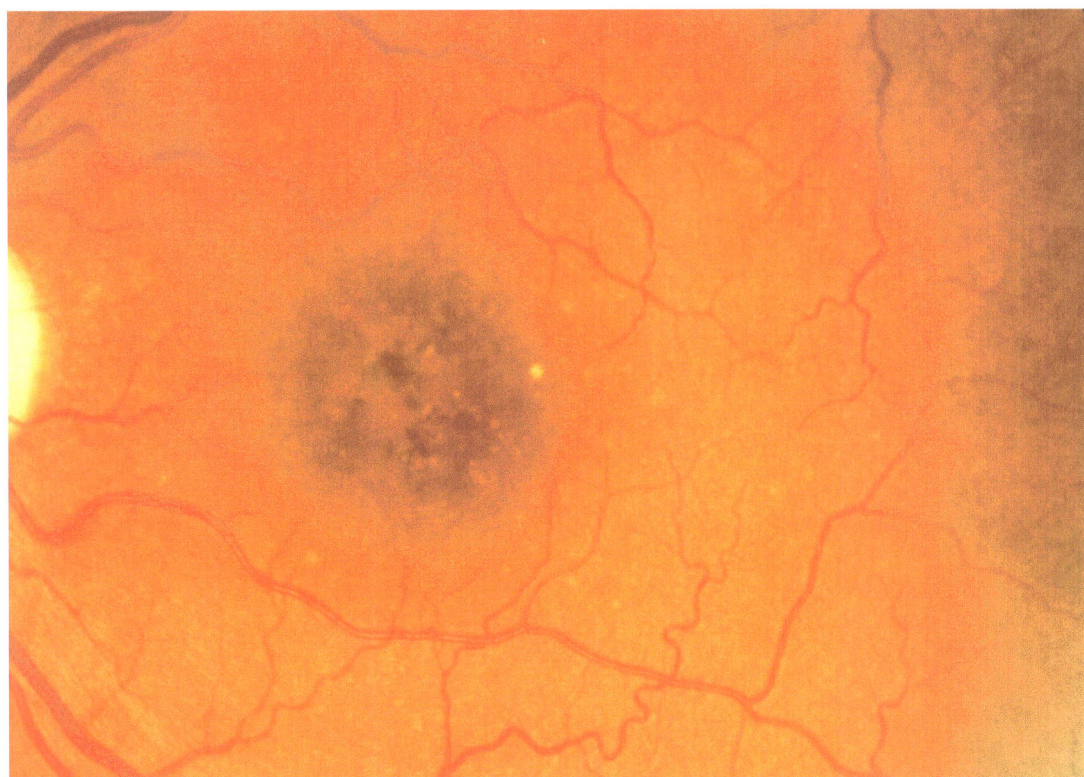

Here are more. These 2 photos are from the same patient's OD, 7 years apart. Despite no smoking and taking AREDS 2 vitamins BIB, she moved from mild dry to severe dry/ borderline wet ARMD:

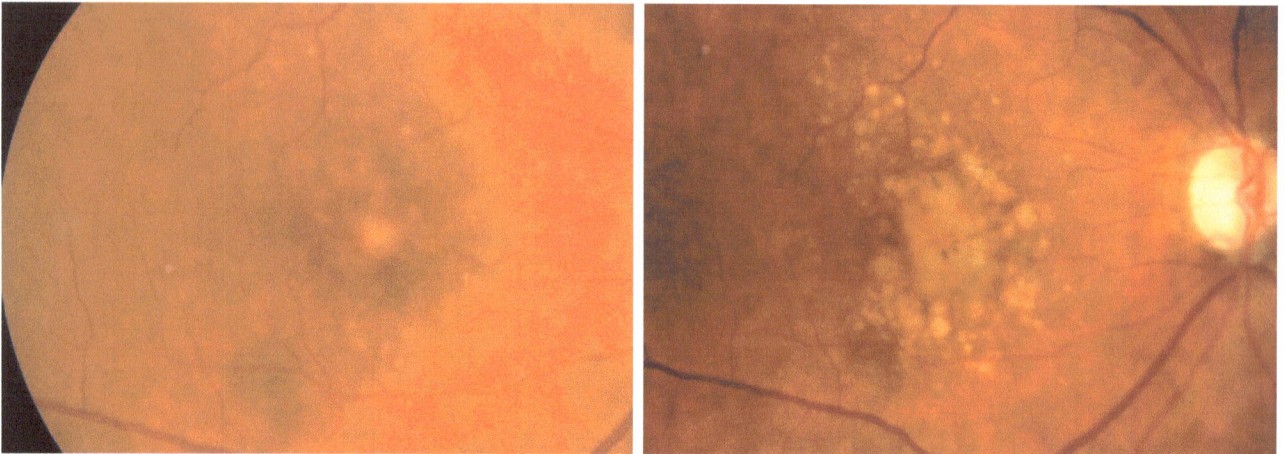

Here is another patient that showed worsening of his geographic atrophy AEMD over a 4 year period despite no smoking and taking AREDS 2 vitamins BID:

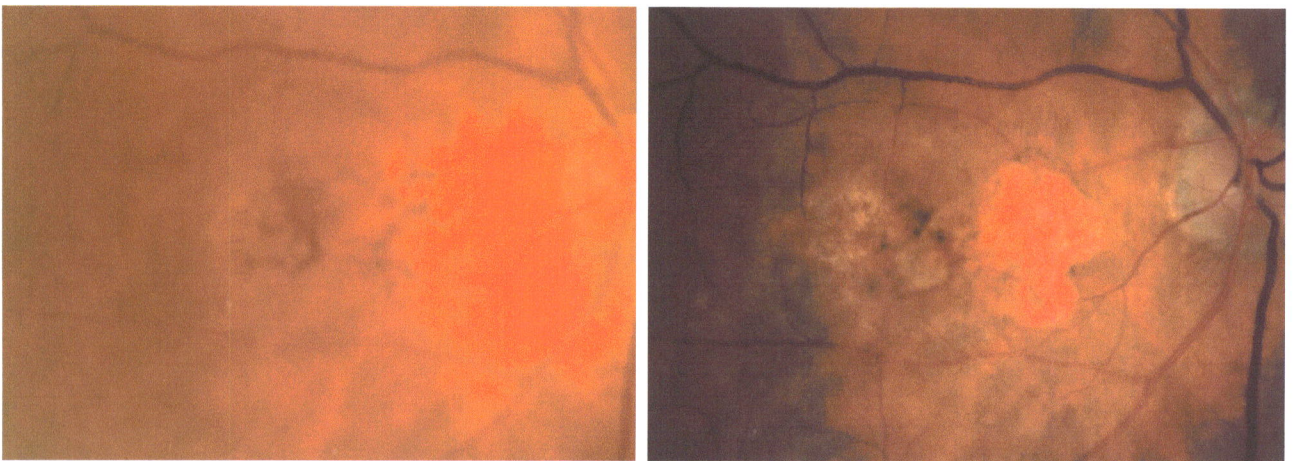

Another patient that converted from mild dry to wet in 2 years despite no smoking, AREDS 2 vitamins BID and a normal Amsler grid. She started receiving intra-vitreal shots, obviously:

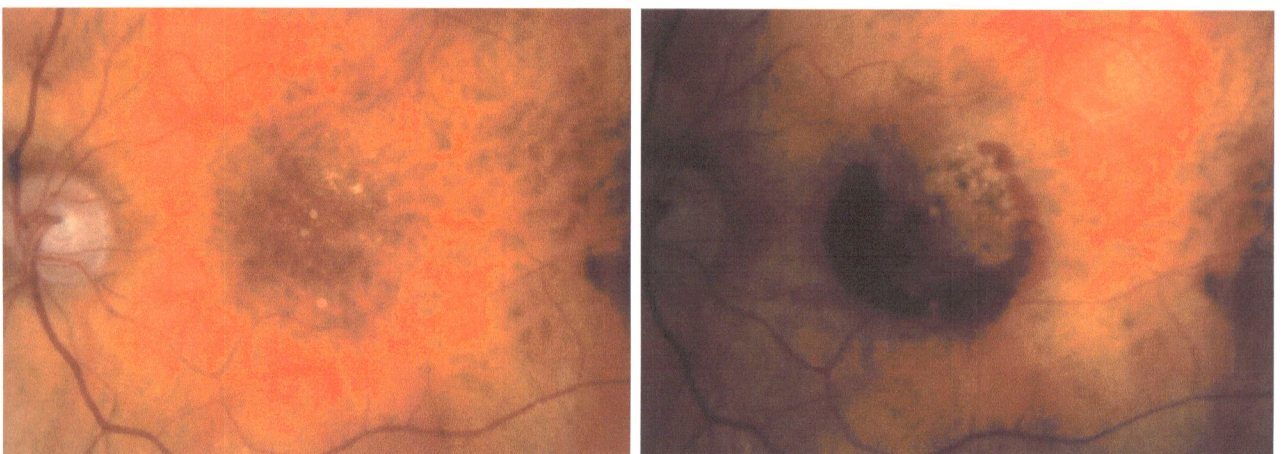

Here is her photo OS after 2 years of intra-vitreal shots:

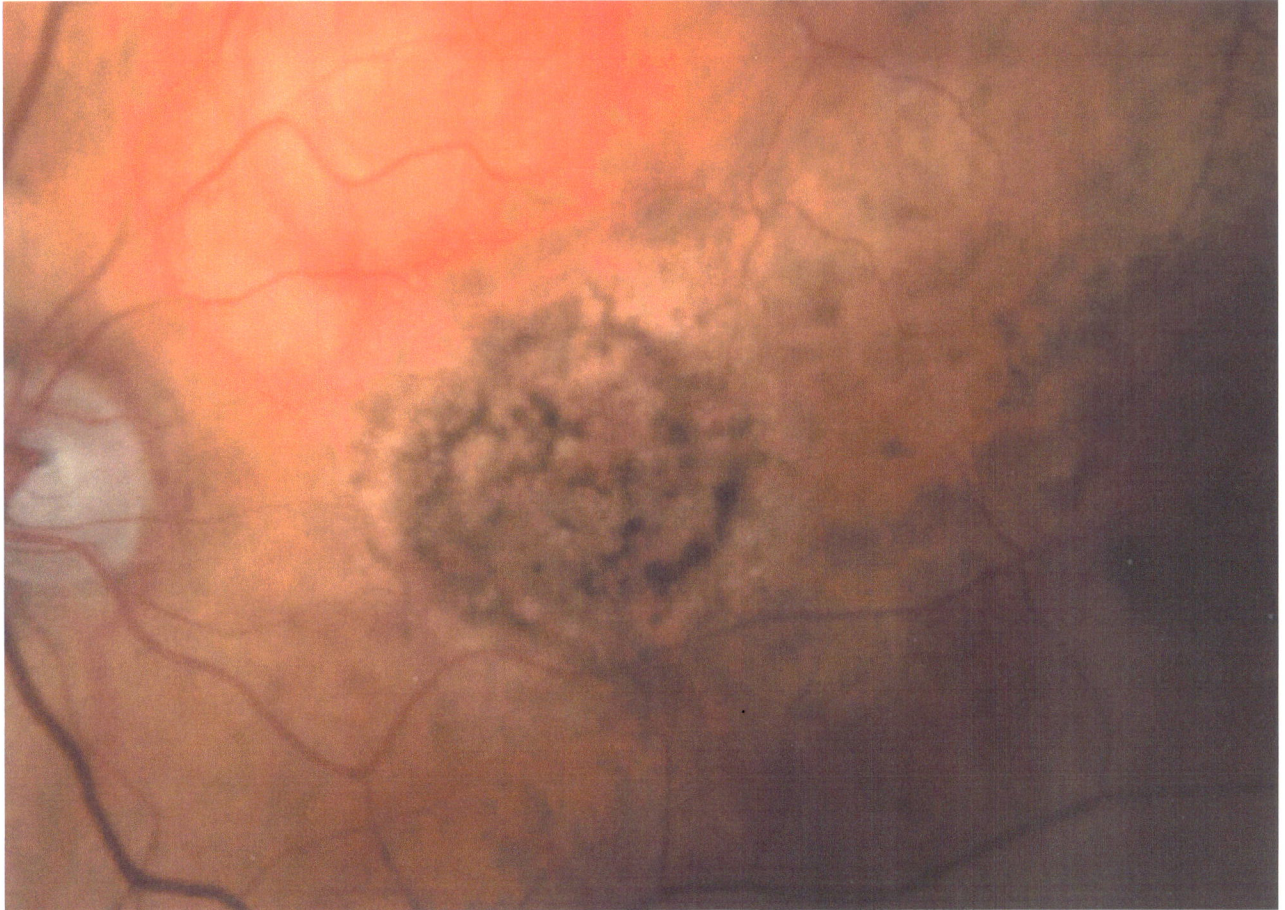

Here is a patient with mild dry ARMD OD, mild to moderate dry OS:

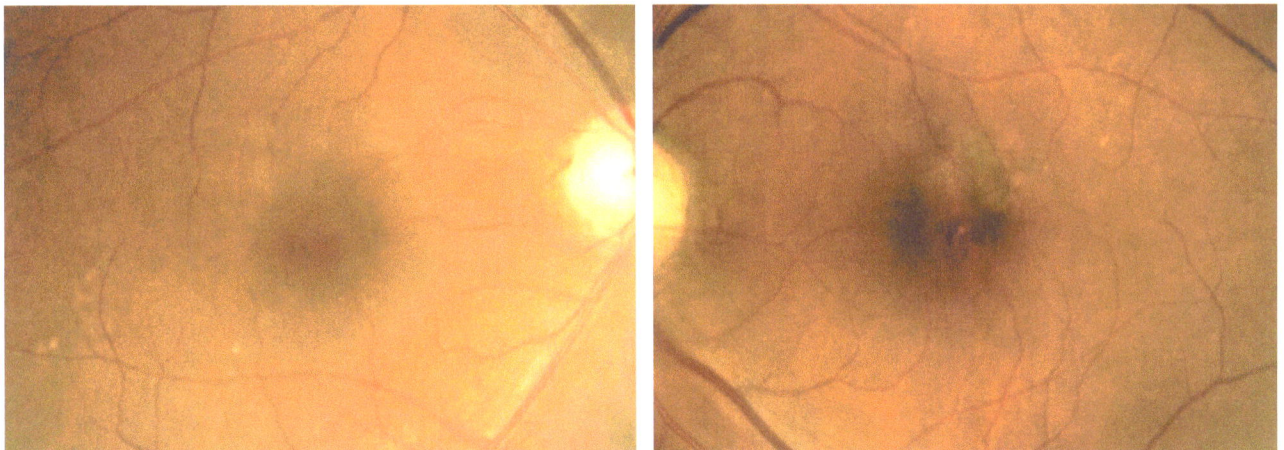

Another with moderate dry ARMD:

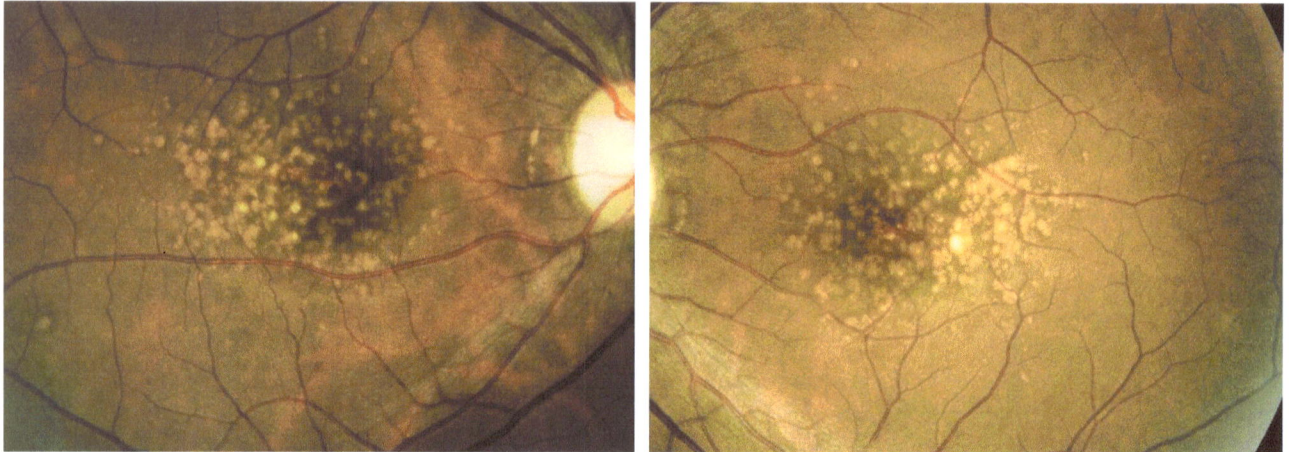

Another with mild dry changing to probably severe dry ARMD in 4 years despite no smoking and AREDS 2 vitamins BID:

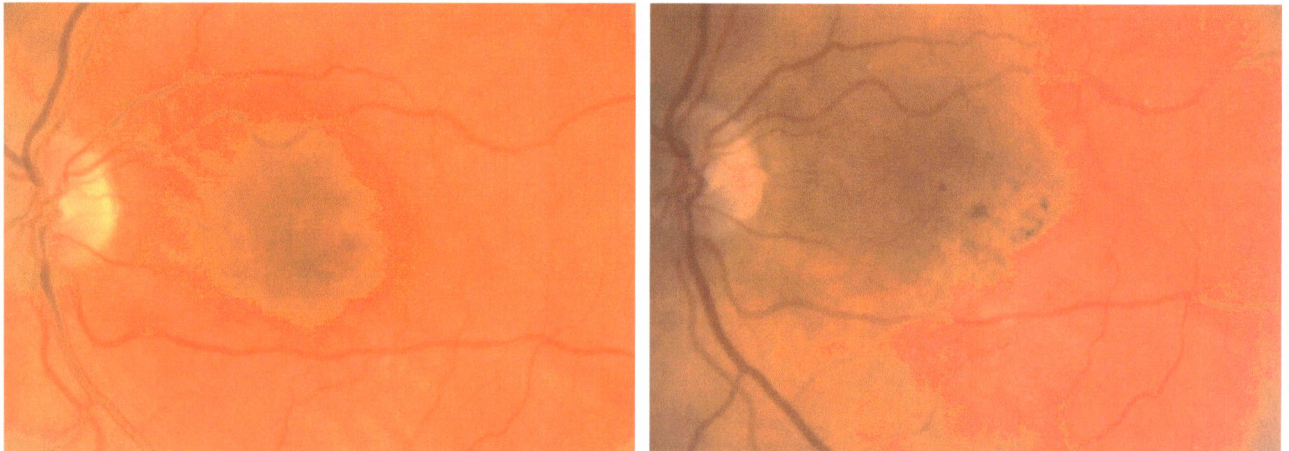

Here is another patient with geographic atrophy OD (BCVA 20/25-, she is taking AREDS 2 BID, HA QD and doesn't smoke) and wet ARMD OS in pre-anti VEGF days (BCVA CF).

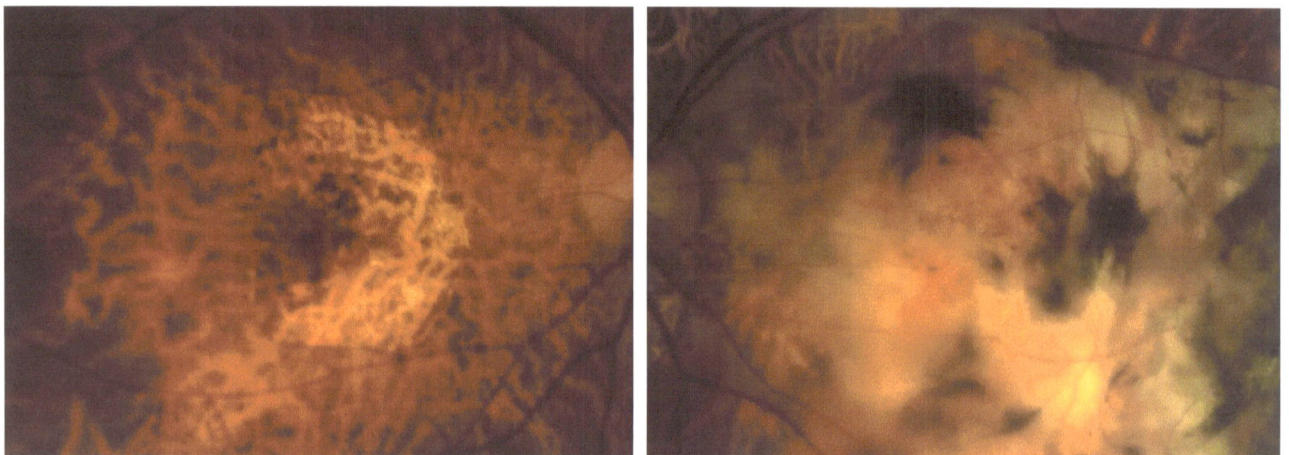

Retinal deposits: this is a term I use for drusen seen in younger patients where ARMD is clearly not a cause. I have the patients followed like in the no ARMD category above. Here is a photo of a patient with deposits outside of the macula. Something that I see a lot of:

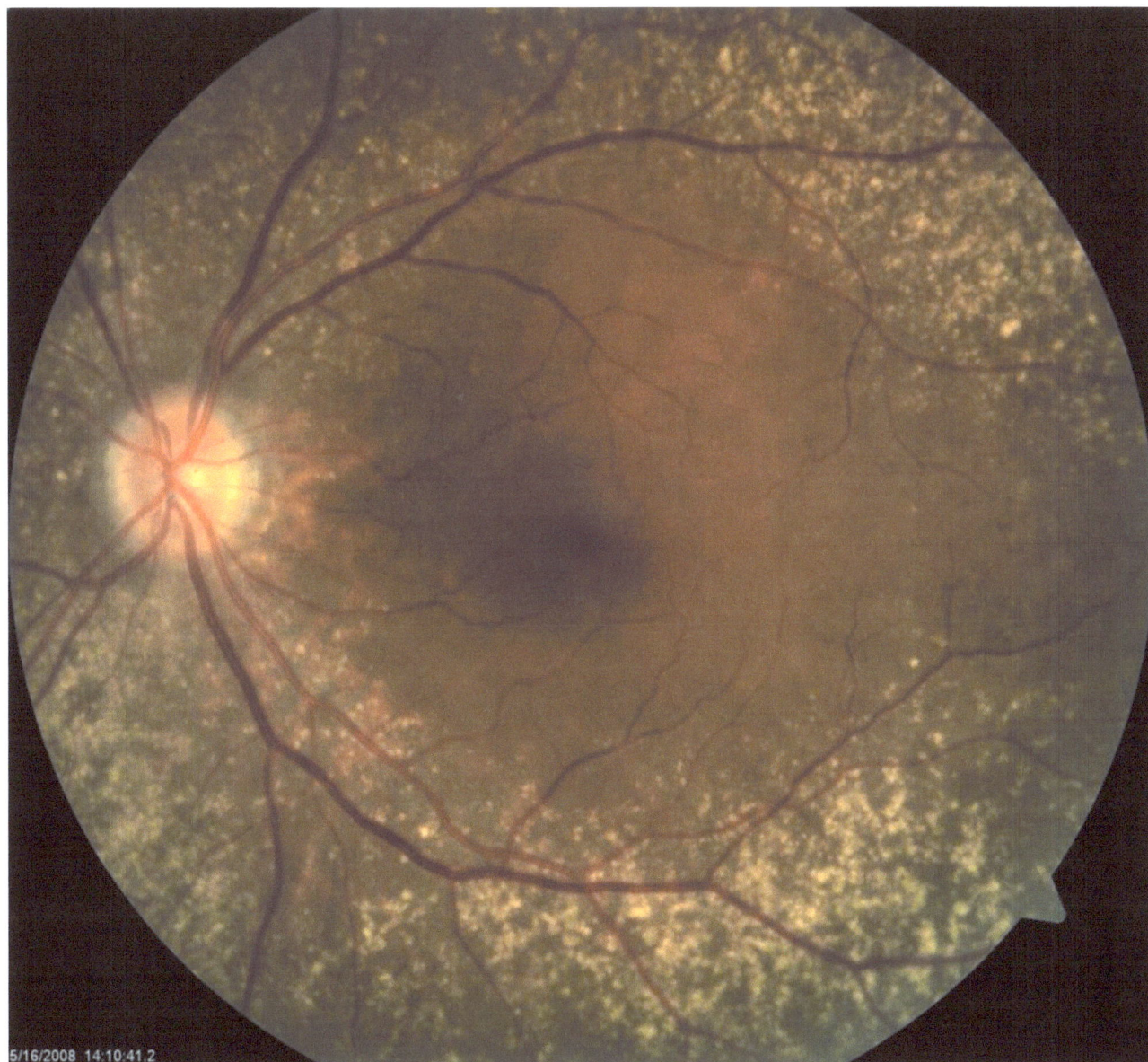

5/16/2008 14:10:41.2

Here is another with "fleckled retinal syndrome" per retina. No work up was necessary:

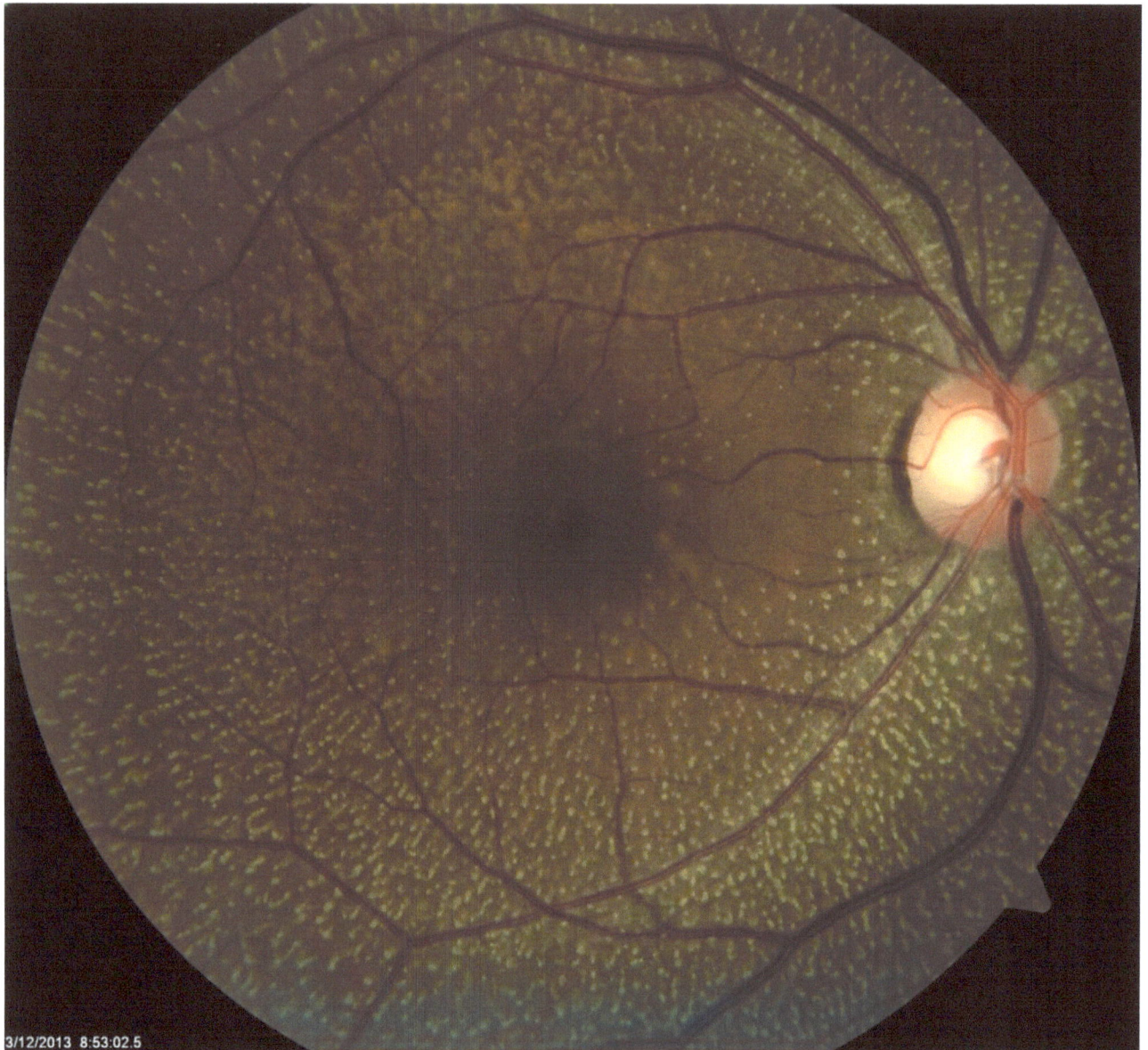

Here is a 22 yo smoker patient of mine with a blonde fundus and mild deposits OS>OD. I placed her on AREDS 2 and asked her to stop smoking:

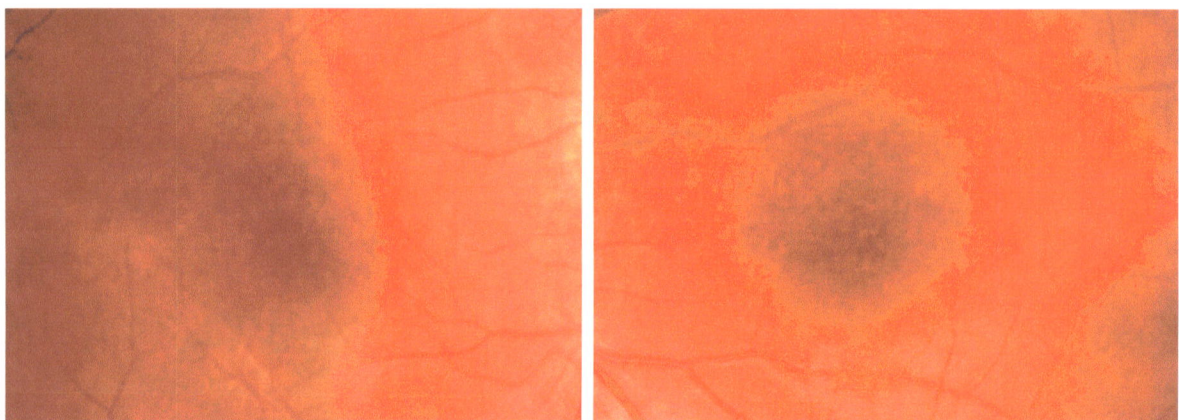

CSR:[153] Central Serous Chorioretinopathy is described as occurring in 25-55 year old white, asian, and hispanic, males more than females, especially with type "A" personalities. These are the borderline obsessive-compulsive personalities that fret over almost everything in their life. You will see patients with this disorder, many showing the pigmented scars of prior episodes. Active episodes look like a fluid filled skin blister in the macular area and they have a very specific OCT and IVFA pattern. 90% of cases reabsorb spontaneously with mild to no residual BCVA loss. A 1974 study using a red laser showed a more rapid resolution of CSR compared to no laser but the final BCVA was the same with or without laser treatment.[187] Due to these two facts, most cases get no treatment. The advent of the micro burst laser may change this. Photo dynamic treatment (PDT) is sometimes used for recurrent and recalcitrant cases. Stress-relieving techniques are recommended.

Here is a photo of CSR (© AAO 2014):

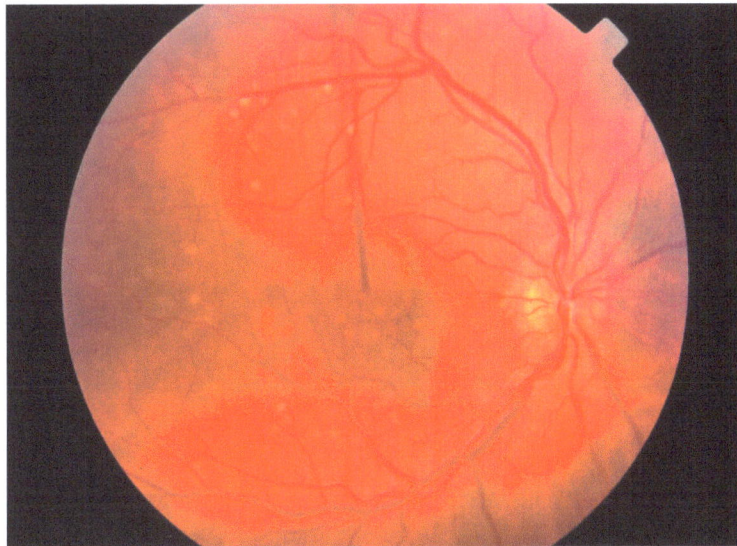

The first two photos are of patients of mine who presented with c/o mild distortion in their vision with BCVA around 20/40. You can see the round edge of the large CSR elevated area if you look closely. In the third photo, the CSR lesion is smaller but the retina shows signs of prior episodes of CSR. This patient had normal BCVA. I believe that a lot of younger patients with similar pigmentary abnormalities in their retina represent prior CSR episodes that were asymptomatic, making CSR a more common entity than is generally reported.

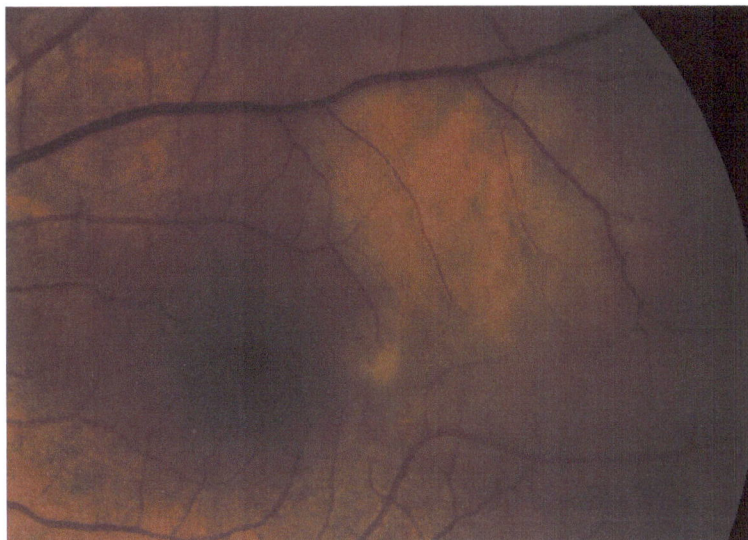

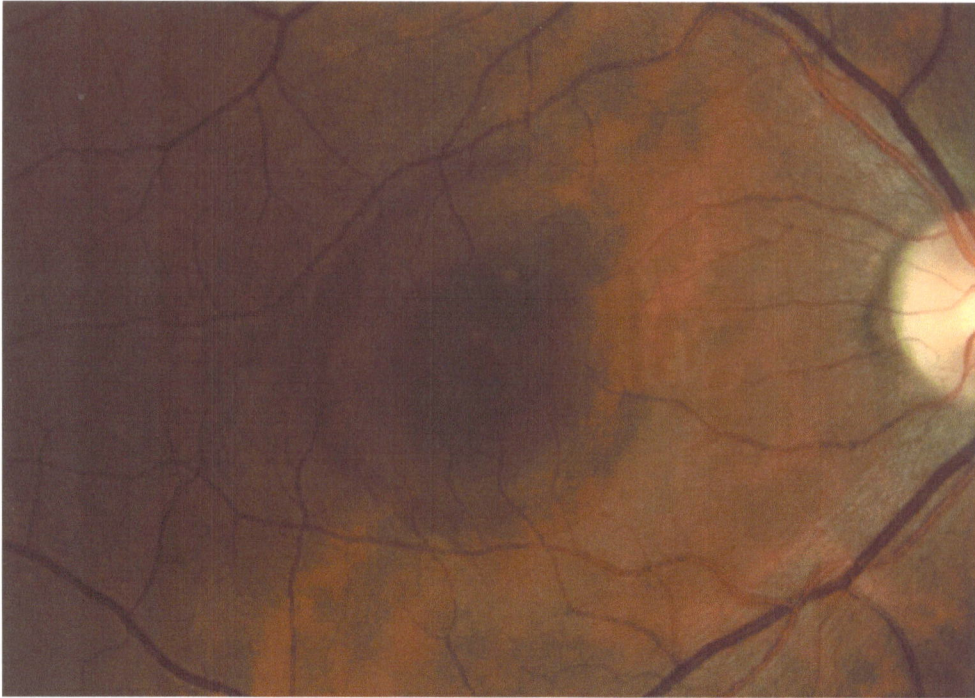

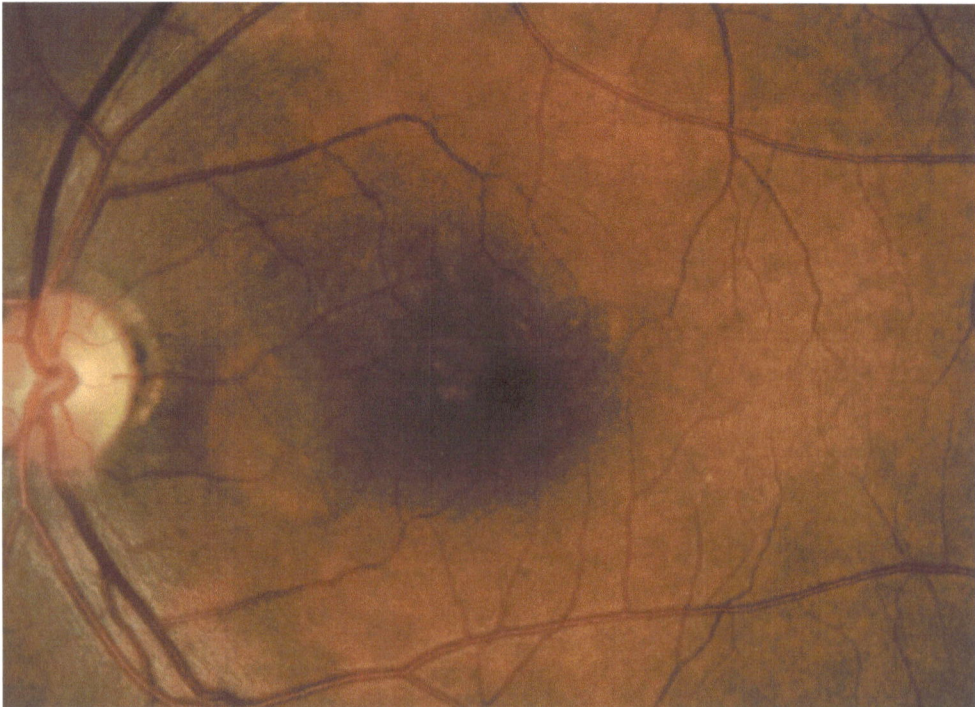

Vitreo-retinal traction (VTR) syndromes:[154] The vitreous attaches anteriorly to the lens. A separation there is called an anterior vitreal detachment. The vitreous also attaches to the optic disk, blood vessels, macula, and ora serrata. A separation there is called a posterior vitreal separation (PVD). Typically, the vitreous stays attached to its "base" in a circumferential band straddling the ora serrata. Stress along the vitreal base from a PVD can cause the peripheral retinal tears that we are so afraid of in an active PVD. Vitreal traction on the retina can cause a number of conditions. The most common, a PVD, has been covered in #8 above.

Persistent macular adhesion from a partially detached vitreous (PVD) is thought to be the cause of a number of syndromes, including a macular hole, as well as the progression of an ERM, diabetic retinopathy, ARMD and CME among others.[195] Until recently a vitrectomy, surgically removing the vitreous from the retina, has been the only treatment for removing this traction. The **MIVI-TRUST** study, a prospective, multi-center, double blind study with n = 652 eyes injected with Ocriplasmin or saline, showed that injecting a truncated form of the human serum protease, Plasmin significantly resolved VTR (P < 0.001) as well as closure of macular holes (P < 0.001). [195] The FDA has approved Ocriplasmin when 2 conditions are met. Symptoms consistent with vitreal-retinal traction (metamorphopsia, blurry vision, relative blind spot, etc.) and an OCT that shows VRT. A macular hole without VRT would not qualify for Ocriplasmin injections.

Here are OCT results showing the vitreo-retinal traction that give rise to various disease states. [189]

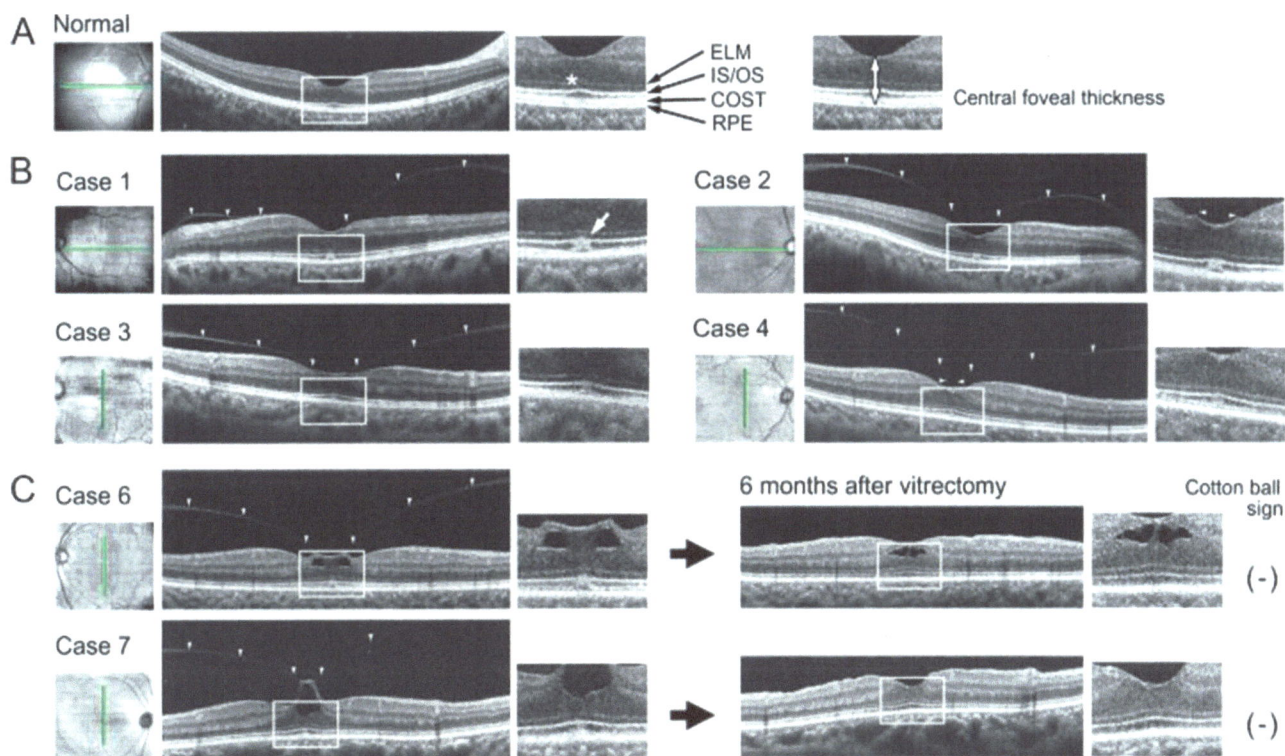

Epiretinal Retinal Membrane (ERM): Continued traction from a PVD on the macula is thought to give rise to the membrane seen in an epiretinal membrane (ERM). An ERM is seen as a glistening sheet on the surface of the retina causing cystic changes in the retina and causing the curved blood vessels of the retina to straighten out, sometimes reducing the BCVA. The ERM has a whitish glisten to it and sometimes a hole in the ERM forms over the fovea, causing a pseudo macular hole, not a true macular hole. 20% of autopsy patients 75 years old and older have an ERM. The only treatment is to surgically remove the vitreal attachment and traction as well as to peel the ERM from the retina. Most retina specialists won't attempt surgery unless the BCVA is reduced to 20/50 or less, since a lot of mild ERMs resolve spontaneously and the surgery has a mixed rate of success. I refer out ERMs with a reduced BCVA of 20/40 or less that I estimate is reduced due to the ERM.

Here are two photos of the same patient taken 2 years apart. You can see how the ERM developed in the second photo. This patient experienced a reduction in his BCVA from 20/30 to 20/70 between the two photos.

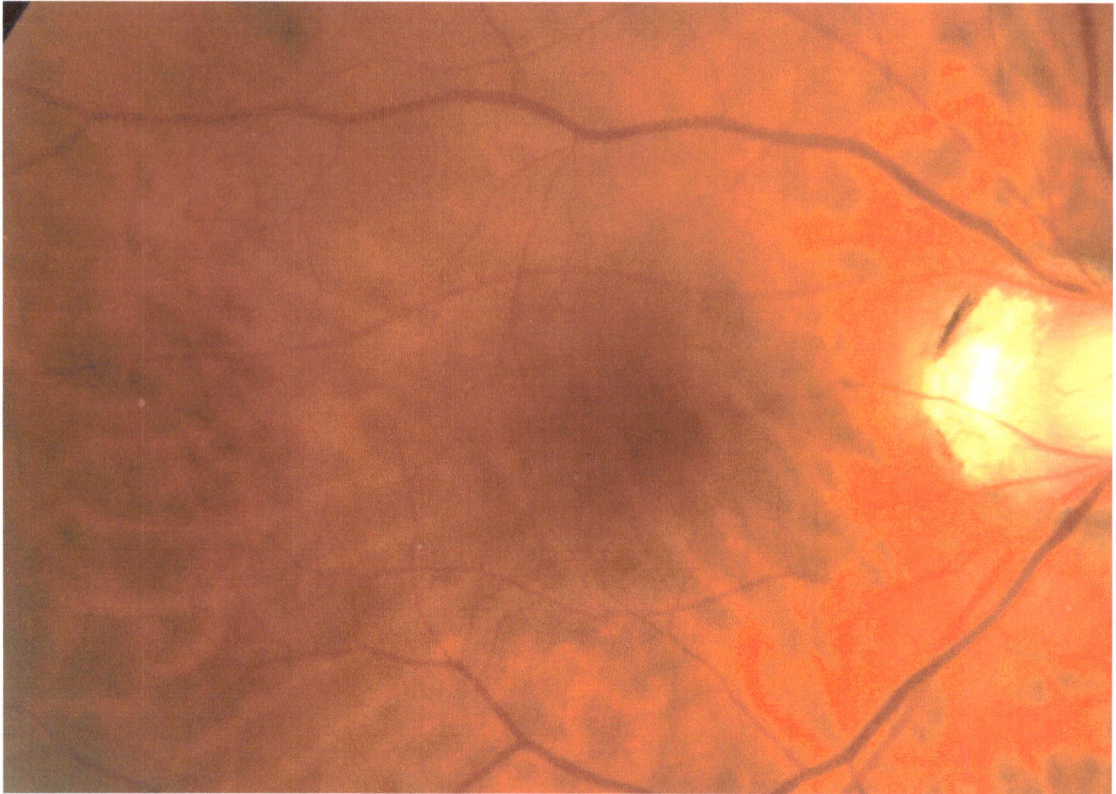

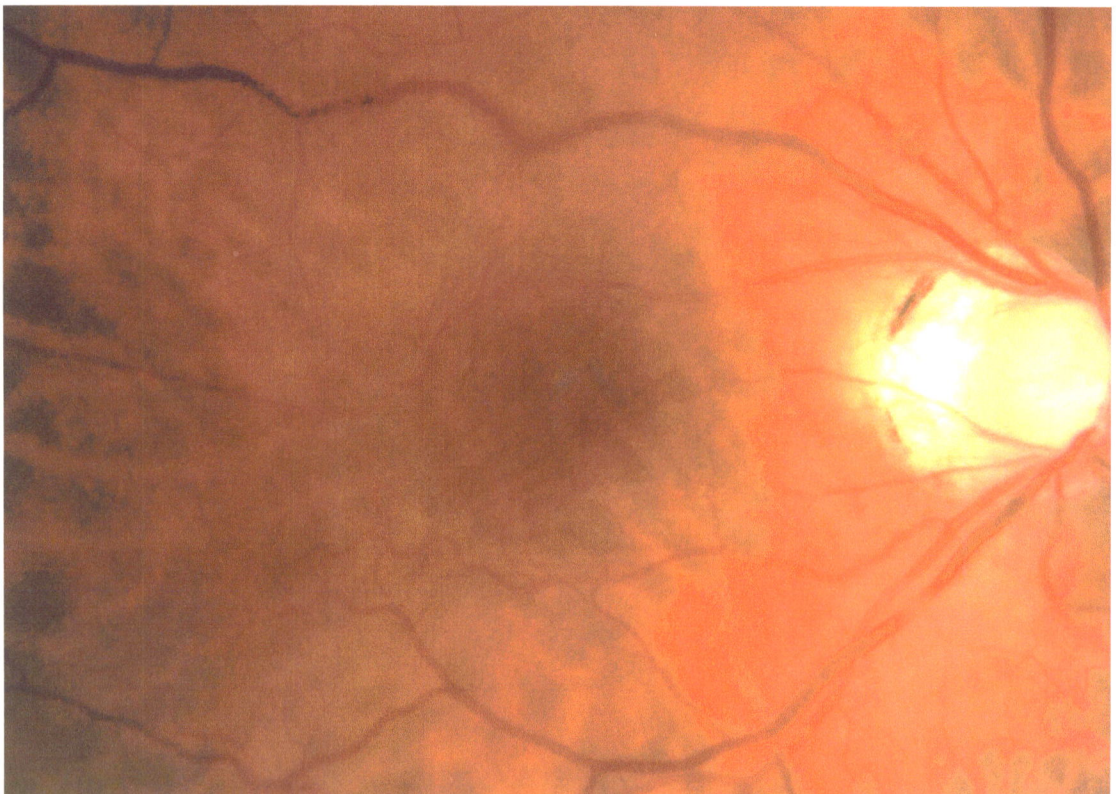

Some ERMs can get severe. Here is one of my patients' ERM that had a retinal hemorrhage too. Her A1C was normal:

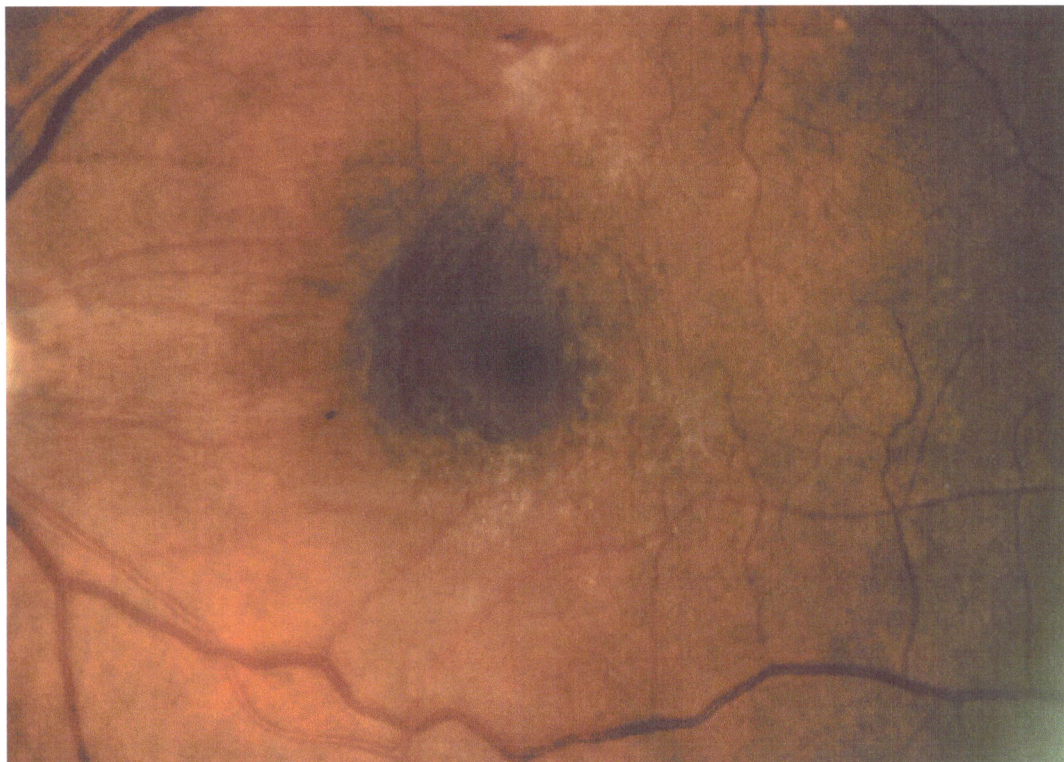

This patient of mine also had optic nerve drusen OU in addition to his ERM OU. The first 2 photos are of his ON drusen, subtle OD and obvious OS. The third photo is centered on his ERM OD:

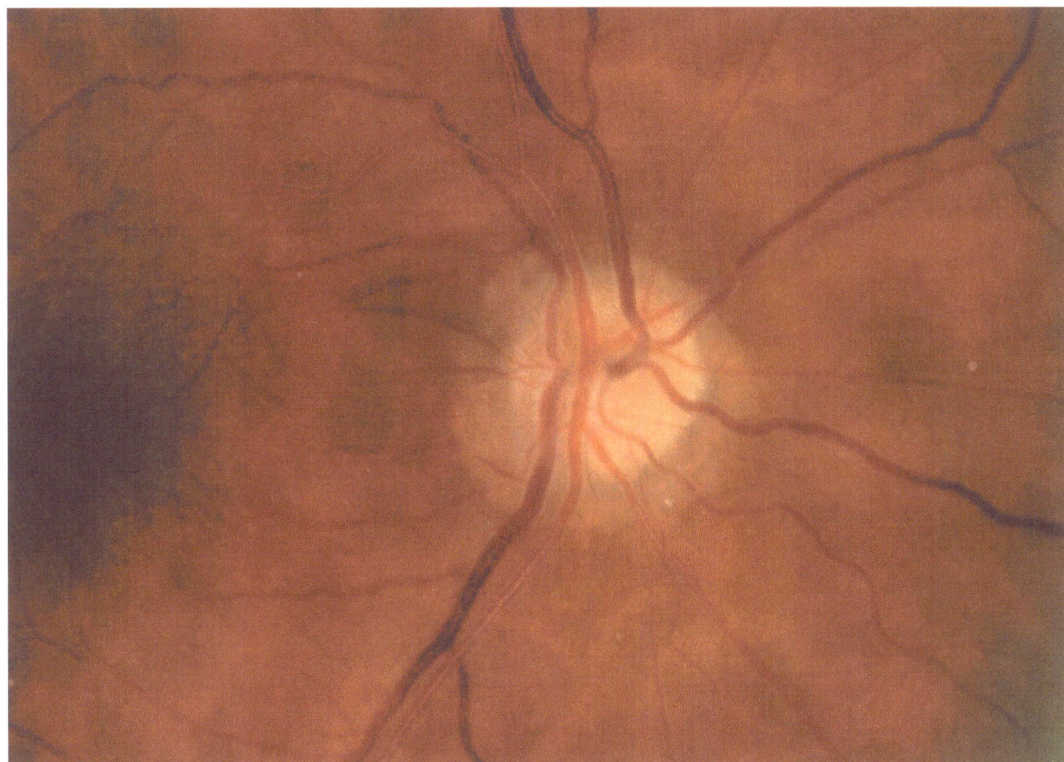

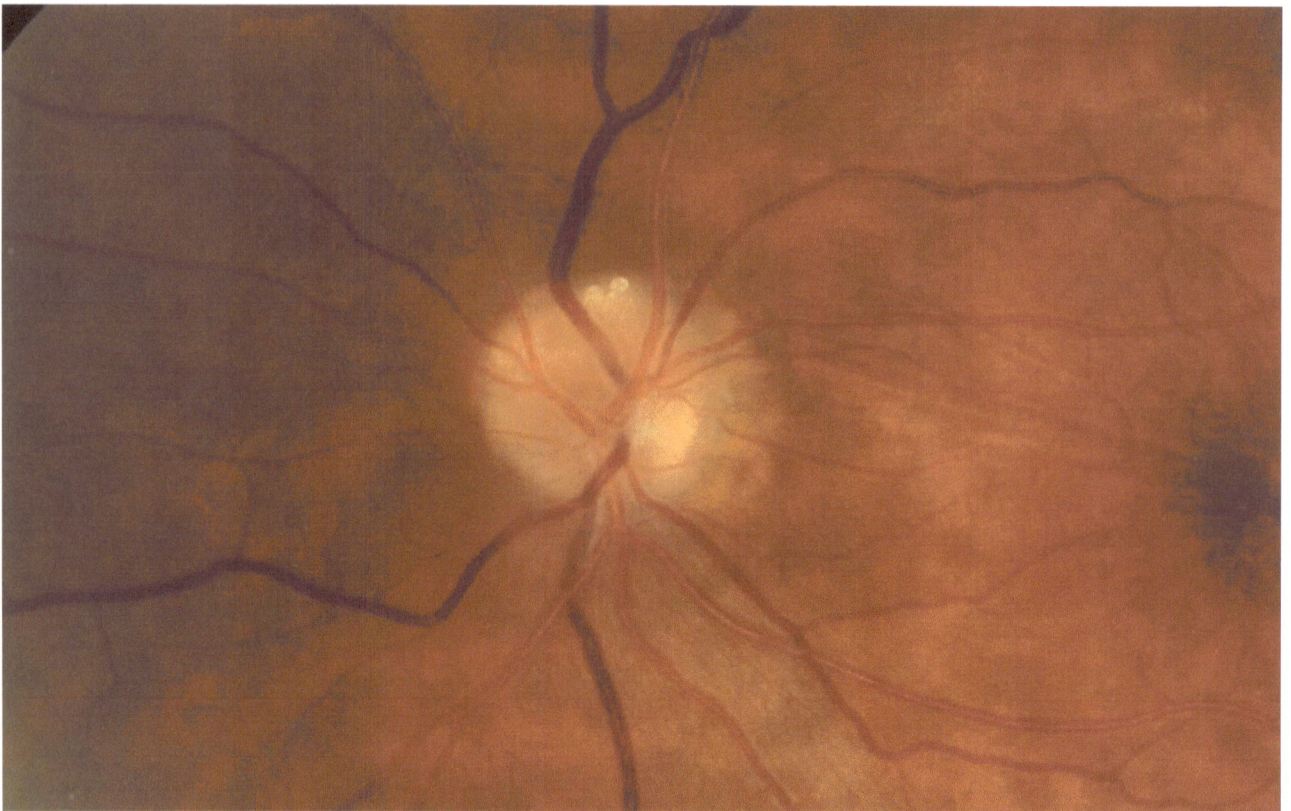

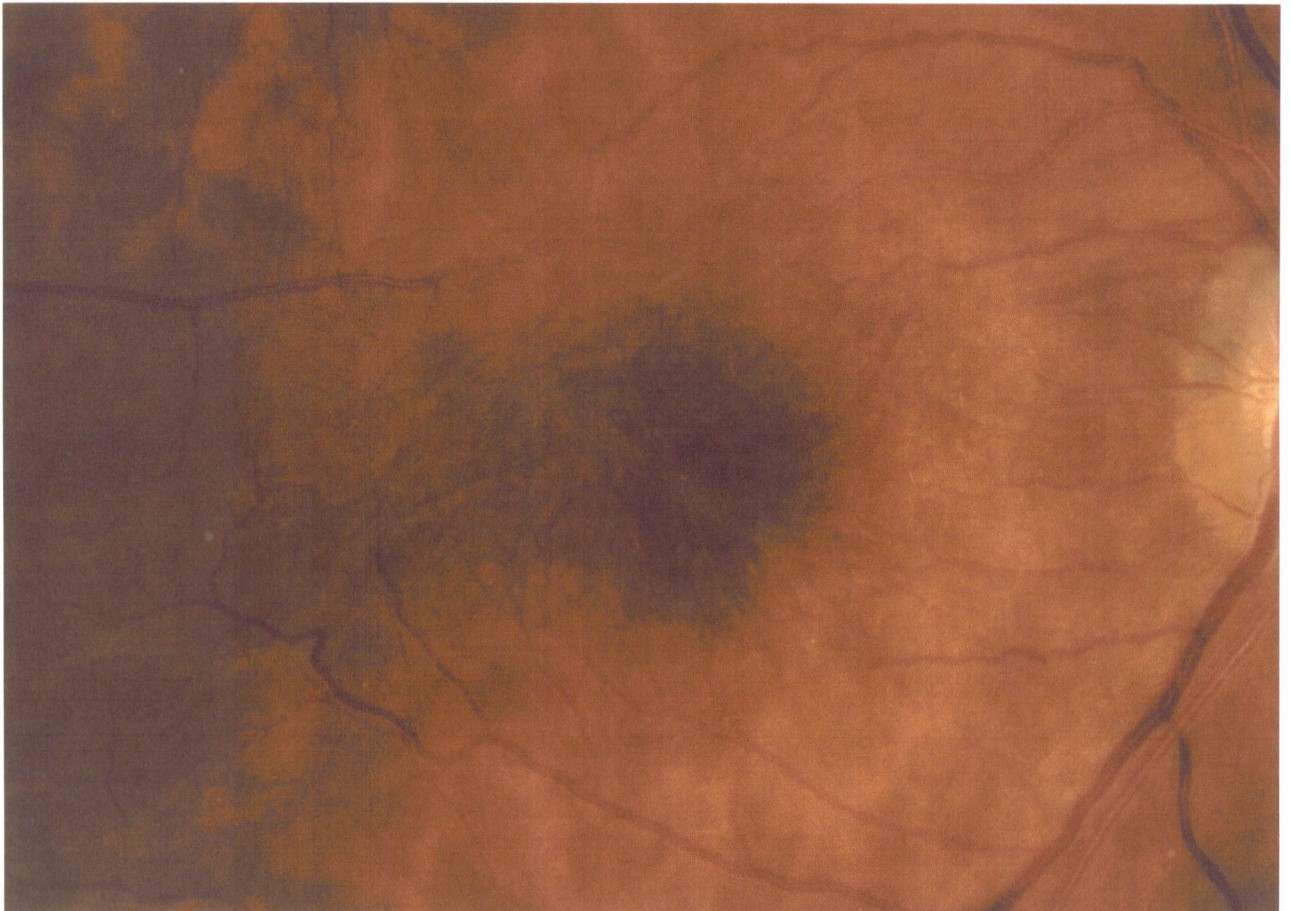

Macular hole: Vitreal macular traction separates the photoreceptors causing a red hole in the center of the fovea, reducing BCVA. A thin beam of slit lamp light will disappear over the fovea; this test is called the Watzke-Allen test. A macular hole is defined in stages 1 to 4 with a vitrectomy recommended for stages 2 and above. Don't worry about the definitions; just refer out macular holes when the BCVA is reduced to 20/40 or worse. A vitrectomy performed within 6 months of symptoms may have a better post-surgery BCVA than if a surgery is delayed longer than 2-3 years (**PPP idiopathic macular holes, 2013 pg. 7**). A tamponading air bubble has been frequently performed during the vitrectomy but recent evidence points towards not using an air bubble. There are mixed results from a vitrectomy with anatomical closure frequently accomplished, but improvement in BCVA less frequent, as is true for the treatment of an ERM. But since nothing else can help, a vitrectomy is frequently attempted. Virtually all of the patients will get a cataract after a vitrectomy but since most patients are at the cataract surgery "age" anyway, just let the patient know about it so it doesn't come as a surprise. Here is a photo of a macular hole (© AAO 2014):

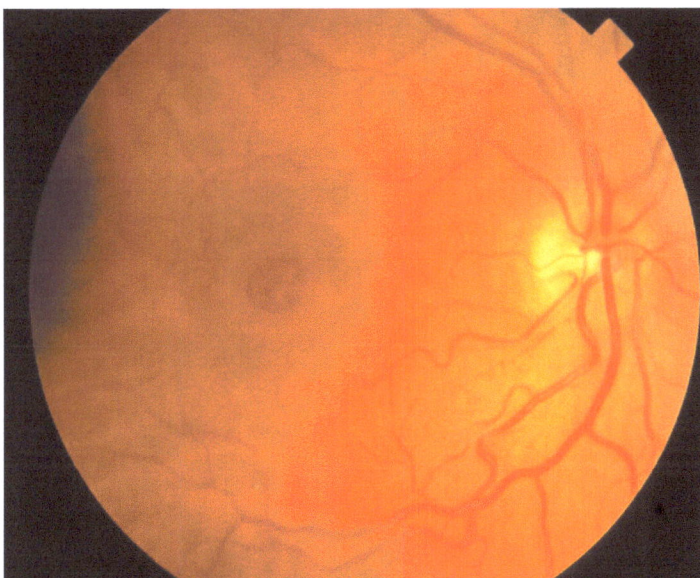

Here is one of my patients with a macular hole OS.

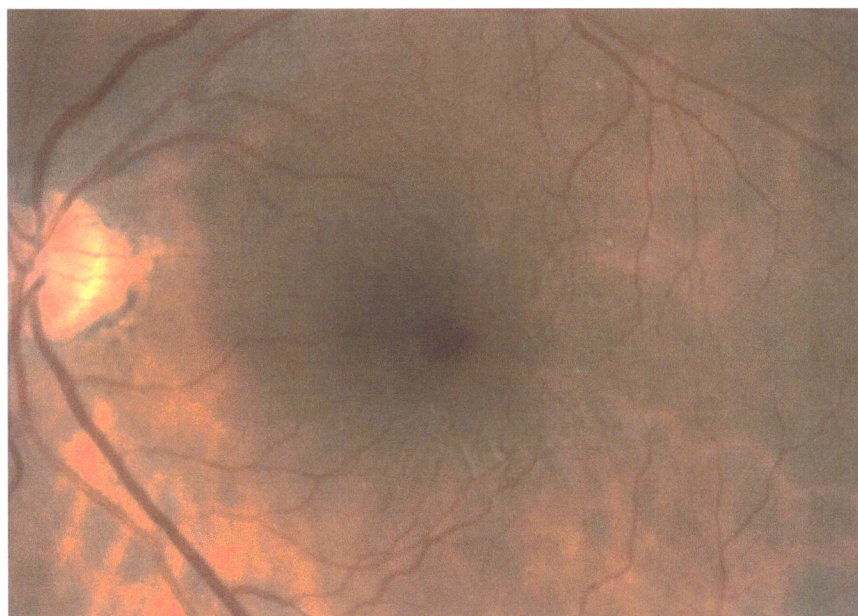

Here are some photos of a choroidal rupture with its circumferential scars around the optic nerve (© AAO 2014):

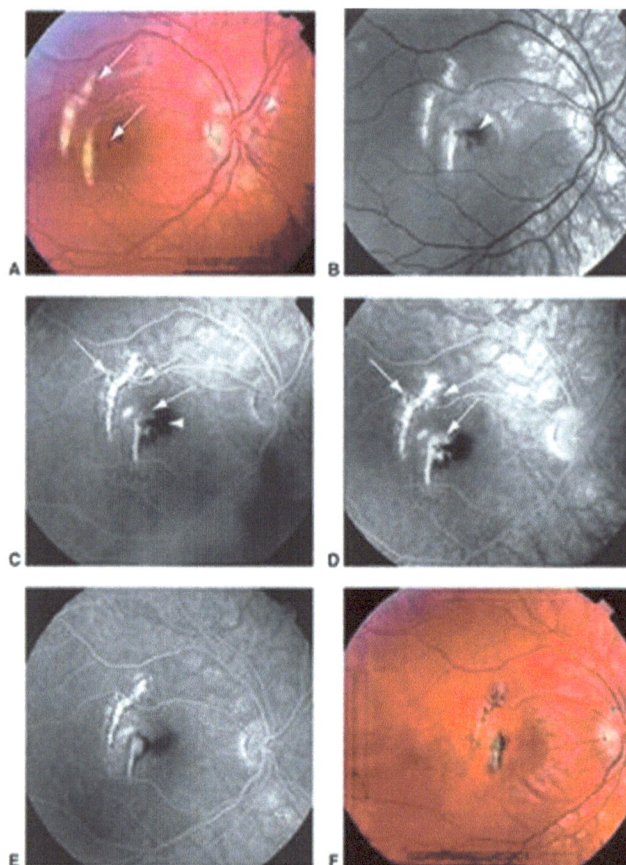

Color vision testing:

Insurance will not pay for the ishihara color vision test. They do typically pay for the Farnsworth panel D – 15 test, which is a better test anyway. The D - 15 test also has the advantage of showing acquired versus congenital color vision defects.[155] Look at the scoring card for the D 15 test. Congenital color defects tend to follow the protan/deutan error pattern line on the Farnsworth panel D-15 scoring card. Monocular color vision errors that do not follow the same error pattern on the scoring card, i.e. some errors are parallel with the deutan error line while other errors are parallel with the protan error line or are random, tend to be acquired color vision problems.[155] Tritan errors also tend to be acquired rather than congenital.[155] The old distinction that optic nerve diseases tend to have deutan/protan color vision defects and retinal diseases tend to be tritan color vision defects is not deemed to be reliable.[156] Instead, get the D-15 test and score each eye separately looking for the trends that I have noted above and that are shown in this figure. [157]

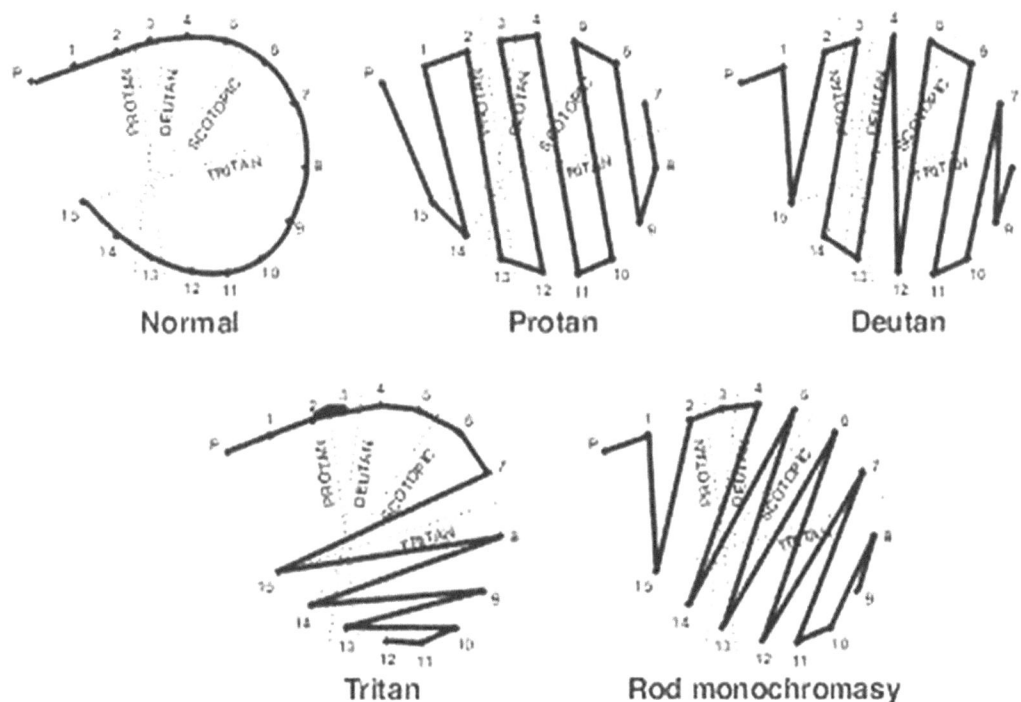

Figure 25. The Farnsworth Panel D-15 results from patients with various colour vision defects. The rod monochromatic results are idealised to illustrate the scotopic axis along 5-14. As a rule, rod monochromats give variable results with a tendency of crossing errors to fall along the 5-14 axis.

14. Lumps, bumps and lesions:

Growths outside of the globe fall into two main clinical categories, the conjunctive/cornea and the eyelids. The cornea is placed with the conjunctive since most corneal growths are extensions from the conjunctiva.[167] Your main job is to identify lesions that that can be followed and photographed each year or should be biopsied by an ophthalmologist. You don't have the luxury of having a tissue diagnosis on every lesion that you are looking at so, at a minimum, photograph all lesions each year.

Conjunctival lesions:

1. Non-pigmented lesions.

Conjunctival lesions with a tendency to grow into malignancies show certain features. These worrisome features include:[168]

- white surface (leukoplakia),
- a sessile (flat or *not* pedunctulated) tissue growth pattern
- inflammation or increased blood vessels feeding the lesion, representing potential growth
- thick or gelatinous tissue

All of these represent worrisome signs and these lesions should be biopsied. Sun exposure (actinic damage) can change a benign pingecula or pterygia into a squamous cell carcinoma, showing the features mentioned above. This transformation is much more common for people living along the equator than in northern latitudes. Squamous carcinomas of the conjunctiva and cornea used to be called CIN (conjunctival or corneal intraepithelial neoplasia) but are now called ocular surface squamous neoplasia (OSSN).[169]

The first photo is of a patient with CIN of the cornea, probably starting from the conjunctive and growing onto the cornea. The only case I've seen had a thickness to the growth that you cannot appreciate on this photo. The second photo is conjunctival CIN showing classic leukoplakia (both photos © AAO 2014):

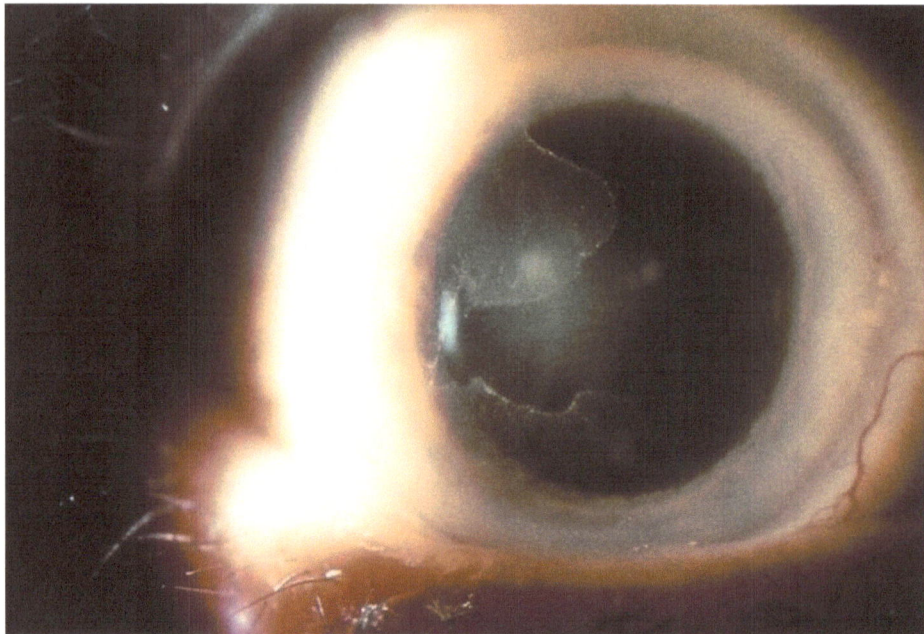

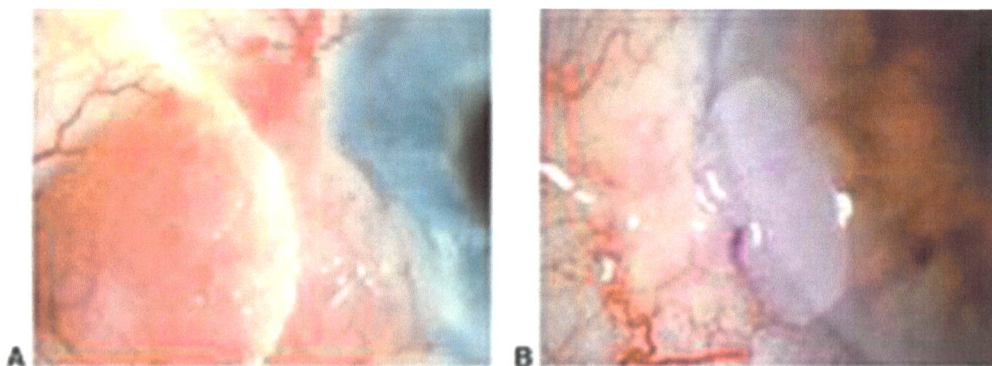

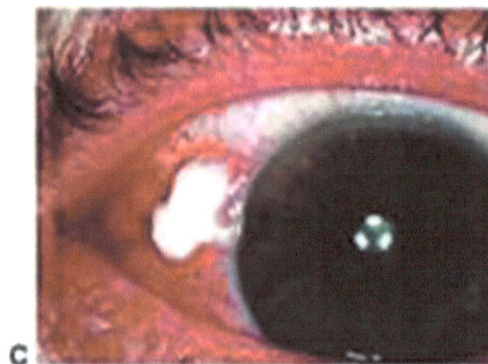

Lymphoid lesions tend to be thickened and gelatinous with a pink color (salmon patch). Benign lesions, papillomas, can be difficult to differentiate from malignant ones. Therefore, we biopsy all salmon colored, fleshy, thick lesions of the conjunctiva.[173] Two thirds of *conjunctival* lymphomas are localized and have not spread to the rest of the body (metastases). About two thirds of lymphomas in the pre-septal *skin* have metastized.[174]

The first photo is a benign papilloma while the next two are cancerous conjunctival CIN (all 3 © AAO 2014):

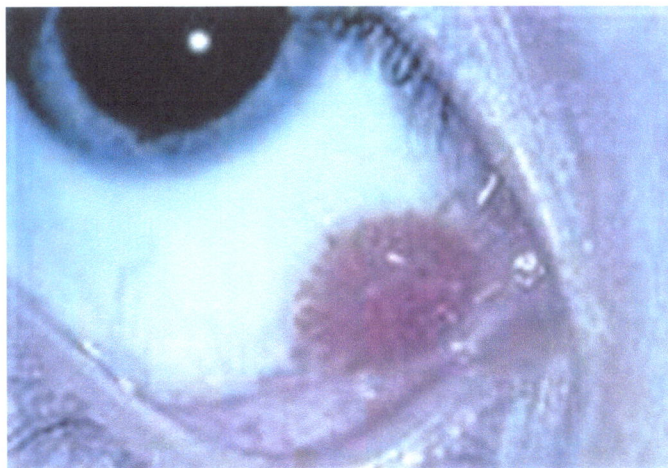

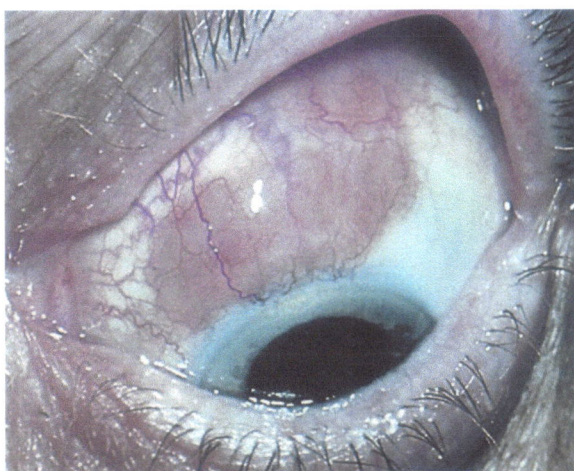

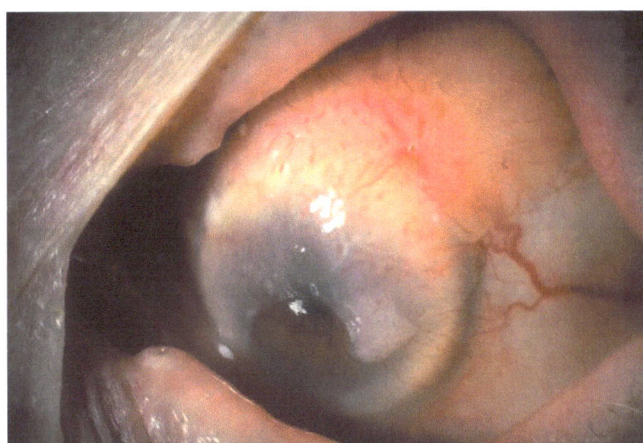

2. Pigmented conjunctival lesions:

Pigmented conjunctival lesions typically occur on the bulbar conjunctiva with conversion to a melanoma rare. Conversions to melanoma tend to occur in the inter-palpebral fissure secondary to sun exposure. Benign increased pigmentation tends to occur during puberty so photos may show benign increases in pigmentation during this time period. Watch for thickening of the pigmented tissue and irregular pigmentation to raise concern about biopsy. Primary acquired melanosis (**PAM**) is a pre-melanoma lesion that tends to be *unilateral*, begin in middle age, and occurs in light skinned people. *Bilateral* lesions tend to be benign, occur in darker skinned people and are called benign acquired melanosis (**BAM**). While PAM and BAM look the same clinically, a pigmented conjunctival lesion in whites should raise suspicion of malignancy. [171] PAM lesions greater than 3 clock hours in whites represent a greater than 20% chance of malignant transformation and should be biopsied. [171]

The first 2 photos are BAM. The third photo is PAM and the forth photo is of a conjunctival melanoma (all 4 photos © AAO 2014):

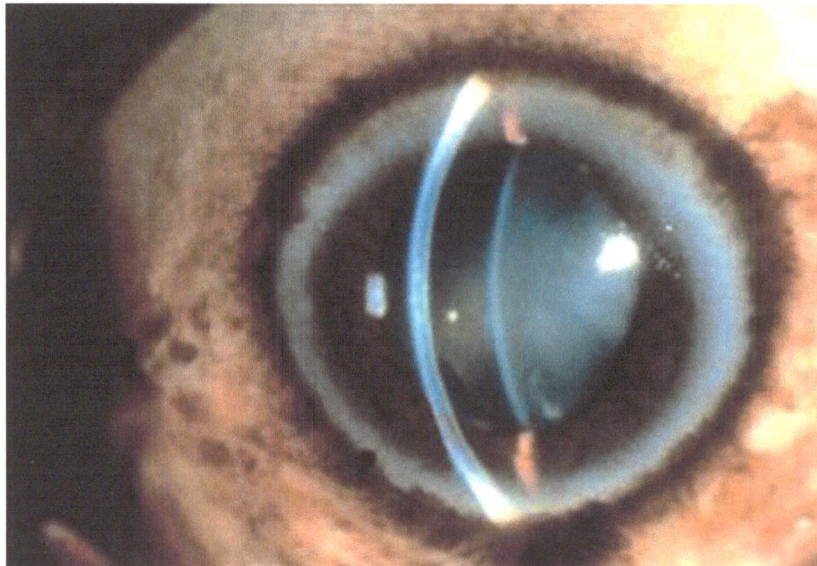

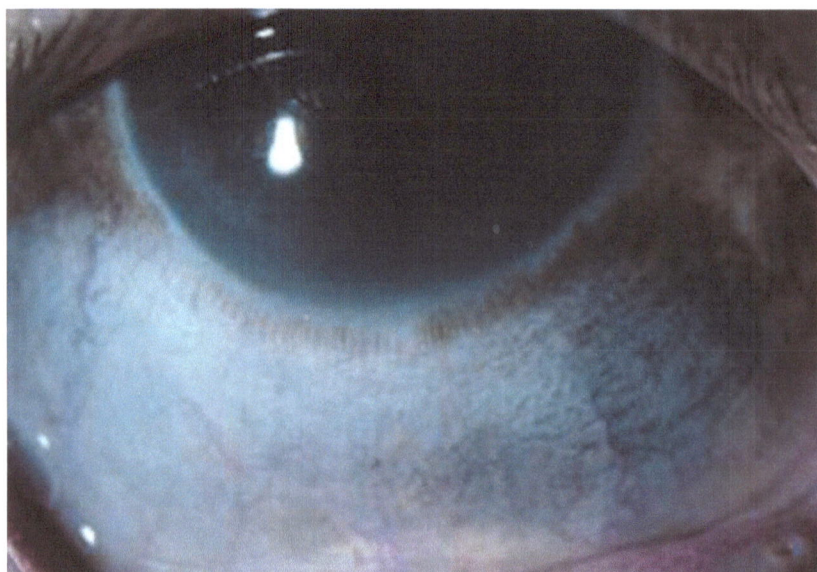

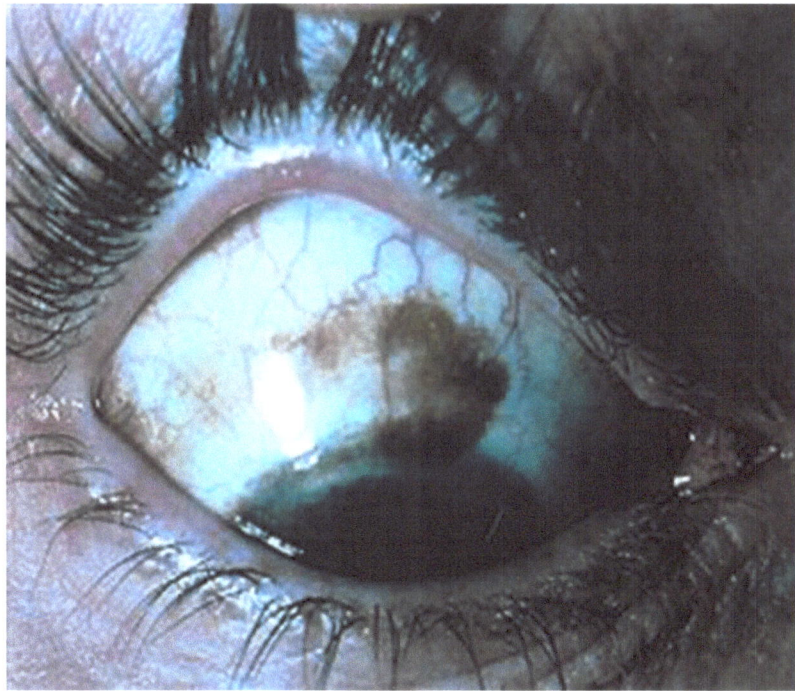

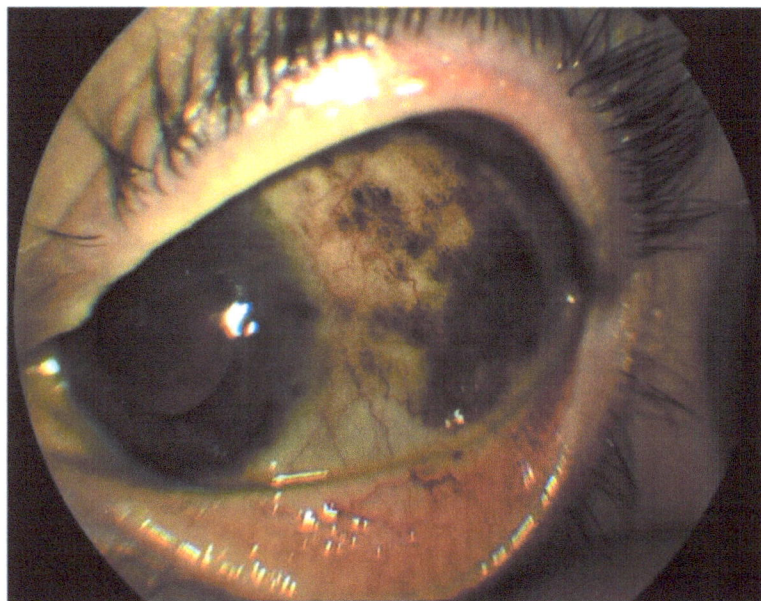

Two thirds of melanomas arise from PAM [172] and are thickened with blood vessels feeding them. Death rates for conjunctival melanomas are about 25%.[172]

So, the bottom line in pigmented conjunctival lesions is to ask the following questions, is the patient *white* (increasing the concern for malignant conversion) and are they in *middle or older age* (again, increasing the concern for malignancy).

Skin involvement of conjunctival pigmentation (nevus of Ota) is benign and occurs unilaterally in darkly pigmented people (Africans, Asians and Hispanics). Nevus of Ota lesions in whites have the potential to turn malignant, usually in the uveal tract and must be followed more closely.[170]

Eyelid lesions:

Eyelid growths are very common. Most of them are either cysts (fluid filled and diagnosed on SLE), yellow skin deposits called xanthelasma or viral infections forming papillomas. Papillomas have a typical appearance on exam and don't require excision or photos. The most common cancer of the eyelids is basal cell carcinoma, representing more than 90% of all eyelid cancers.[177] Here are photos of a basal cell carcinoma showing the ulcerated center and loss of eyelashes (all 3 © AAO 2014):

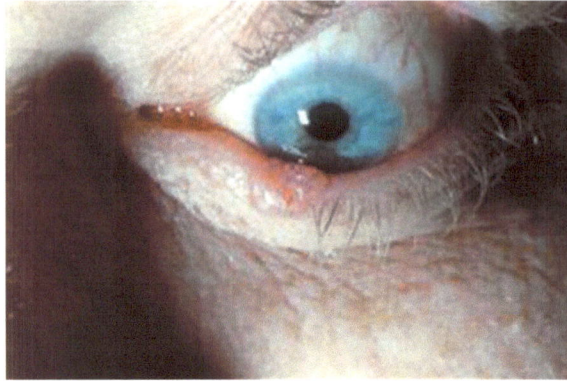

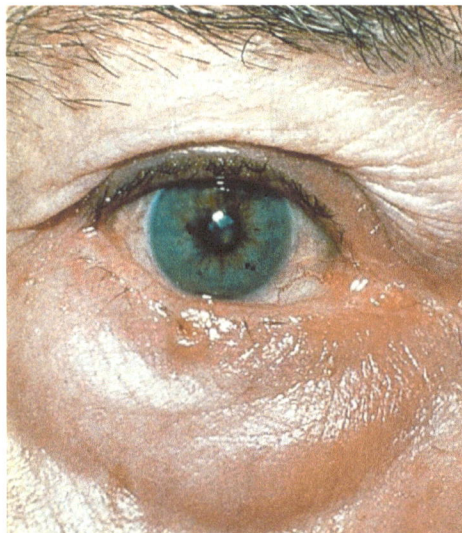

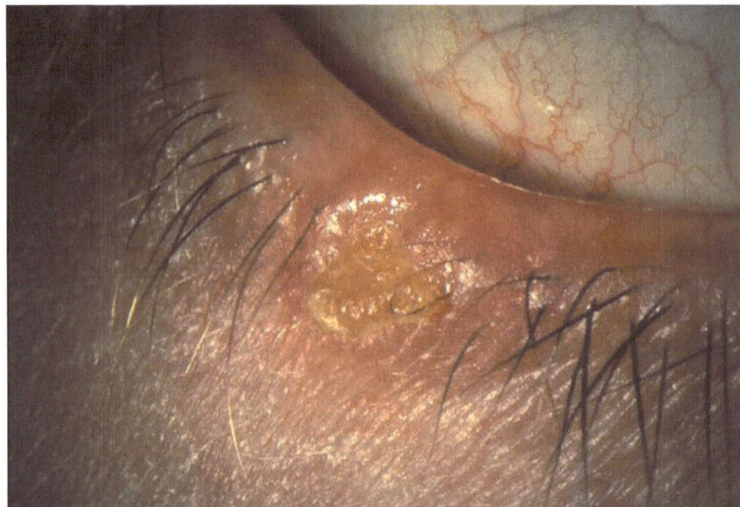

Basal cell carcinomas have a rolled up edge and an ulcer like center that is fairly typical. They need to be excised completely, ultimately using a Mohs surgical technique. Mohs is a specialized, fellowship requiring, technique used by dermatologists where the surgeon is trained to examine all of the margins of the tumor under a microscope while the patient is on the surgical table to insure that the entire tumor has been taken out. A Mohs surgeon may not be available in all areas. Here is a photo of a patient having undergone a Mohs surgery to remove a lesion (© AAO 2014):

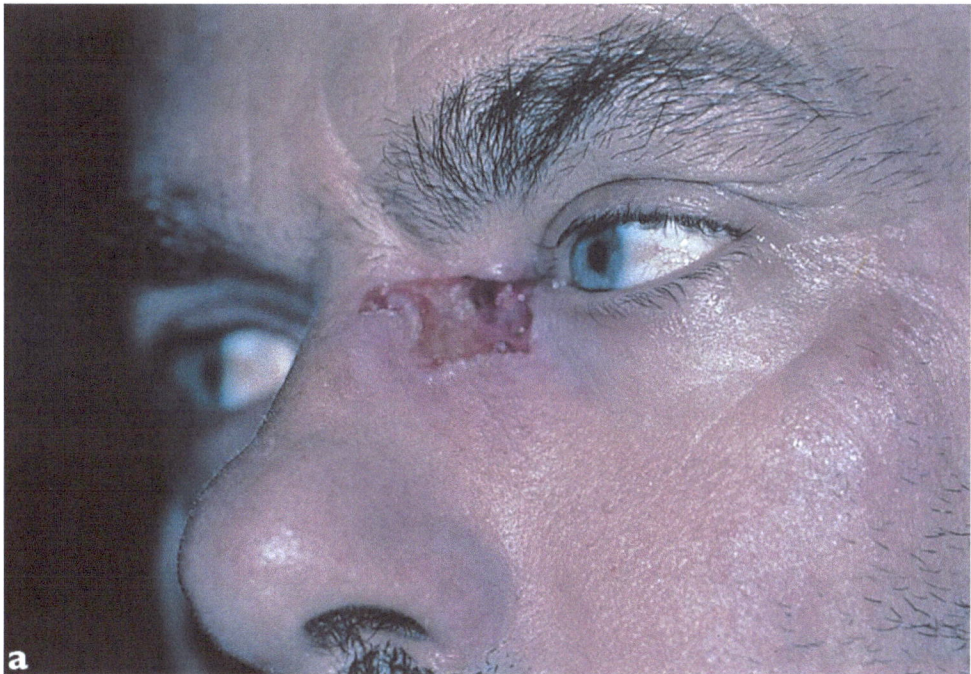

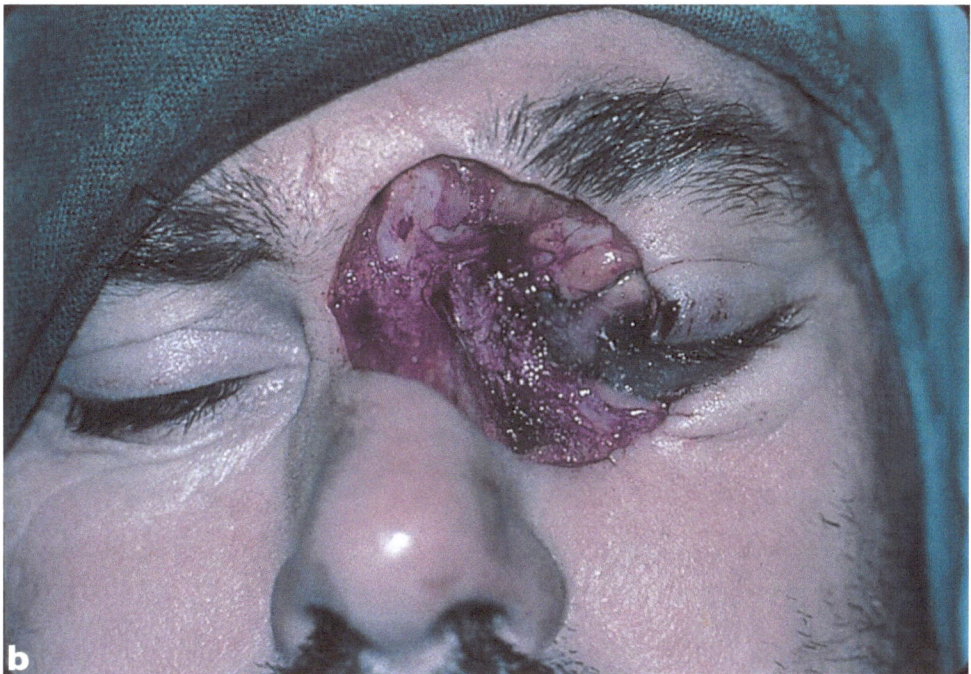

Clinically, loss of eyelashes and changing of the normal configuration of the eyelid (ectropion and entropion) are important clues for the need for referral and biopsy. Squamous cell carcinomas, sebaceous cell carcinoma, keratoacanthoma, and melanoma are all rare but possible growths that need to be biopsied. Here are photos of a squamous cell patient and then a keratoacanthoma patient (both © AAO 2014):

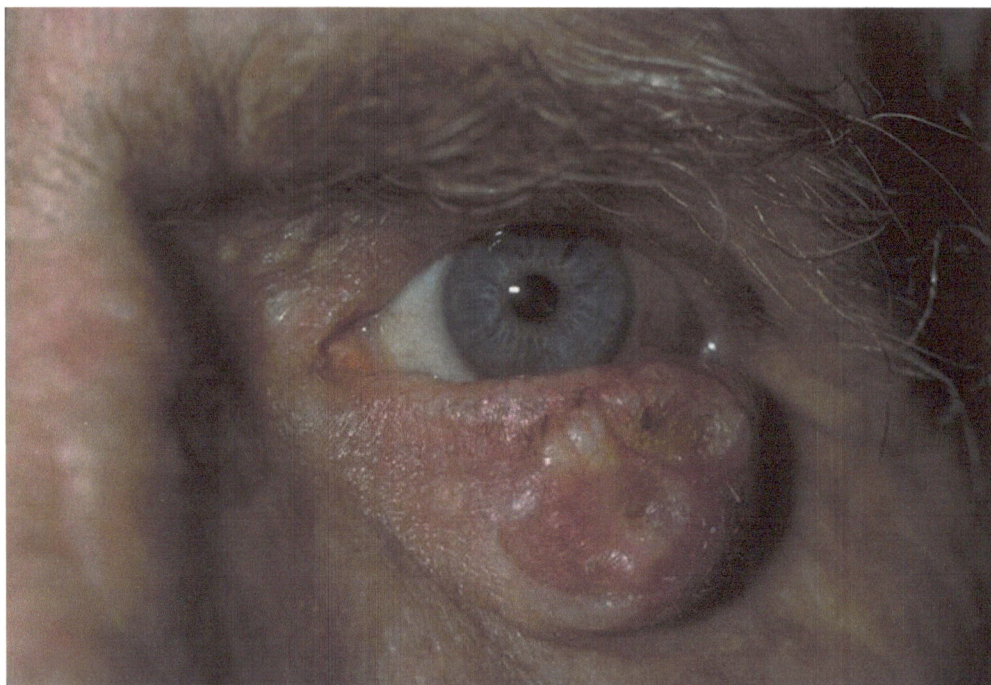

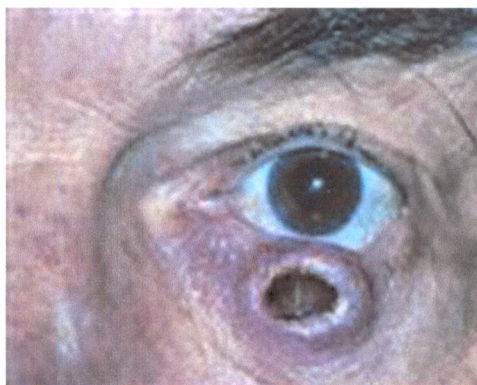

Here is a photo of benign molluscum contagiosum which looks like a basal cell but with the architecture of the eyelid intact (© AAO 2014):

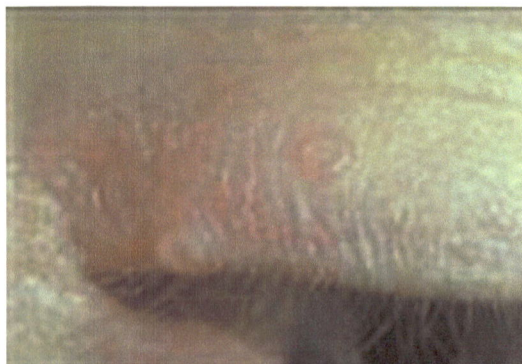

Here is a pre-operative and post-operative photo of a routine upper and lower eyelid blepharoplasty (© AAO 2014):

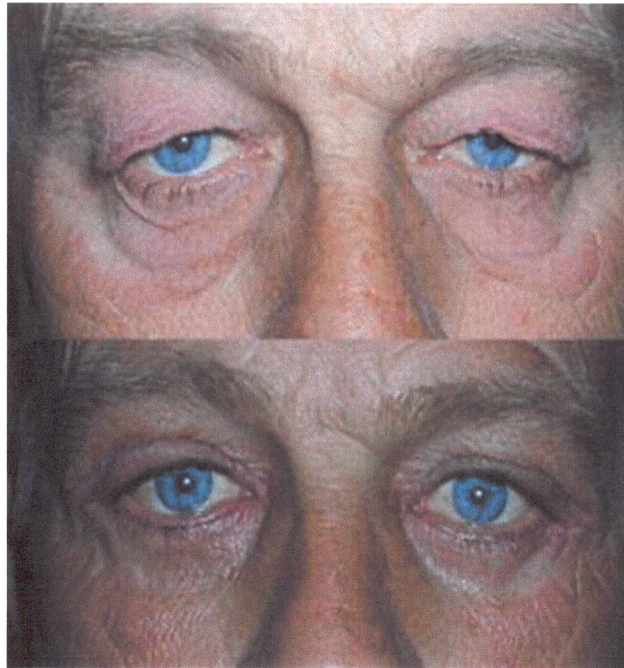

I've included a diagram (© AAO 2014) showing a Z-plasty correction for either tumor removal or scar revision. This is just to let you know what is available should the question arise from a patient. The two flaps created by the incision are switched.

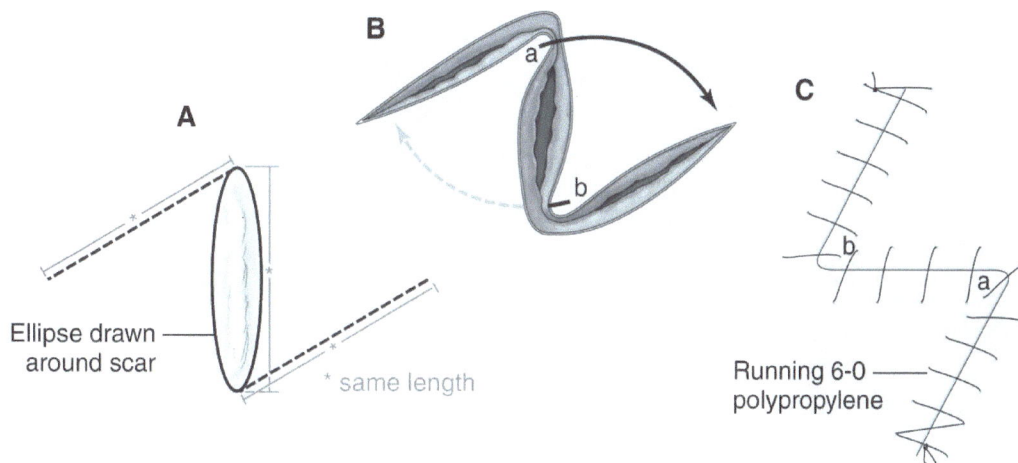

A

Ellipse drawn around scar

* same length

B

a

b

C

b

a

Running 6-0 polypropylene

EDITORIAL
IS CORNEAL THICKNESS AN INDEPENDENT RISK FACTOR FOR GLAUCOMA?

Felipe A. Medeiros, MD, PhD - *La Jolla, California*
Robert N. Weinreb, MD - *La Jolla, California*
Ophthalmology Volume 119, Number 3, March 2012

The Ocular Hypertension Treatment Study (OHTS) showed that central corneal thickness (CCT) was a significant predictor of which patients with ocular hypertension are at higher risk for converting to glaucoma.[1] Eyes with CCT of 555 μm or less had a 3-fold greater risk of developing glaucoma compared with eyes that had CCT of more than 588 μm. In a multivariable model including age, baseline intraocular pressure (IOP) (measured by Goldmann tonometer, Haag Streit, Koeniz, Switzerland), optic disc topography (cup:disc ratio), and visual field (pattern standard deviation), CCT retained its statistical significance as a predictor of glaucoma development, with a hazard ratio of 1.82 for each 40 μm thinner CCT.

The results of this report have been mistakenly interpreted by some investigators as demonstrating that CCT is an independent risk factor for the development of glaucoma. As Goldmann applination tonometry (GAT) measurements ultimately depend on CCT, it is impossible in the original model to completely disentangle the effects of both. For example, consider 2 patients with the same baseline GAT IOP of 24 mmHg, but with corneal thicknesses of 520 and 560 μm. The adjusted hazard ratio for CCT in the OHTS multivariable model would indicate that the risk for developing glaucoma for the one with the thin cornea would be 82% higher. However, it is impossible to determine from the original analysis whether the increased risk is caused by a true independent effect of corneal thickness per se, or simply due to the effect of CCT on GAT measurement error. In fact, using a correction formula proposed by Ehlers et al,[2] the patient with the thinner cornea would have "corrected" IOP close to the measured value of 24mmHg. In contrast, the "corrected" IOP would be 2.8 mmHg lower at approximately 21 mmHg for the patient with the thicker cornea. So, the increased risk could ultimately be caused by the fact that the first patient actually has a higher "true" IOP. However, some authors have suggested that the predictive effect of CCT is not fully accounted for by its induced GAT measurement error, but rather that there is a possible association between corneal thickness and structural measures possibly related to glaucoma risk, such as scleral or lamina cribosa thickness.

Whether corneal thickness is a true independent risk factor for glaucoma has remained an unanswered question. In this issue of *Ophthalmology*, Brandt et al[3] attempt to shed light on this issue. They evaluated whether the OHTS prediction model could be improved by correcting IOP for CCT using previously published formulas. The rationale of the authors was that if the influence of CCT on GAT fully explains the role of CCT as a predictive factor, than inclusion of CCT-corrected IOP values in the model would cause CCT to become nonsignificant. They show that models with CCT-corrected IOP do not perform better than the original model, as evaluated by *c*-statistics and calibration chi-squares. Additionally, CCT remains a statistically significant predictor in the multivariable model including CCT corrected IOP. Based on these results, the authors conclude that the influence of corneal thickness as a prognostic factor for primary open-angle glaucoma (POAG) is not entirely from its effect on IOP measurement error, but rather that CCT is a biomarker for structural and physical factors involved in the pathogenesis of glaucoma.

Although the results of Brandt et al[3] provide important clarification for the role of CCT as a risk factor for glaucoma development, caution should be exercised when concluding that they show that CCT is indeed a true biomarker or independent risk factor for glaucoma. A close analysis of the data actually suggests a decrease in the predictive ability of CCT when CCT-corrected IOP values were included in the model. The hazard ratios for CCT decreased from 1.84 in the original model to 1.38 in the model, which included IOP corrected by the Ehlers formula, for example. More importantly, it is likely that the correction formulas used by the authors did not fully capture the corneal-induced error on tonometric measurement. It has been shown that other factors besides corneal thickness may influence tonometric readings, such as corneal elasticity and viscoelasticity, and the formulas used by the authors do not fully take into account these factors.[4] The only way to fully evaluate the independent role of CCT as a prognostic factor would be to include in the predictive model IOP measurements obtained by a perfectly cornea-independent tonometer. As it does not require corneal applination, dynamic contour tonometry (DCT) measures have been proposed to be largely independent of corneal influence and agree closely with manometric readings.[5] Therefore, the inclusion of DCT measurements along with corneal thickness in a predictive model for glaucoma might provide a better assessment of the true independent value of CCT compared with the simple incorporation of CCT-corrected GAT values.

Brandt et al[3] concluded that the available formulas to correct IOP measurements do not improve the accuracy of prediction models for development of glaucoma. Although this conclusion is technically correct and expected, it should not be used to indicate that CCT provides an additional independent contribution to the prediction model besides its effect in correcting IOP. In fact, a close analysis of the data actually suggests just the opposite interpretation. The predictive abilities were similar between the original OHTS model including CCT and the models, which did not include CCT, but only CCT-corrected IOP. This could actually imply that CCT is relatively unimportant for the final predictive ability of the multivariable model, as long as one includes CCT-corrected IOP. For example, the model including IOP values corrected by the Ehlers formula (but excluding CCT) had a predictive ability almost identical to the original OHTS model. Such a result would hardly indicate a major true independent contribution of CCT as a prognostic factor for development of glaucoma.

In conclusion, the results of Brandt et al[3] suggest that the use of CCT correction formulas for GAT measurements is probably of little value in clinical practice. Instead of attempting to use these formulas, clinicians would probably find it advantageous to incorporate risk information as provided by validated predictive models for glaucoma development.[6,7] However, the conclusion that CCT is a true independent risk factor for glaucoma is not validated at this time and requires further investigations.

References

1. Gordon MO, Beiser JA, Brandt JD, et al. The Ocular Hypertension Treatment Study: baseline factors that predict the onset of primary open-angle glaucoma. Arch Ophthalmol 2002;120:714 –20; discussion 829–30.
2. Ehlers N, Bramsen T, Sperling S. Applination tonometry and central corneal thickness. Acta Ophthalmol (Copenh) 1975;53: 34–43.
3. Brandt JD, Gordon MO, Gao F, et al. Adjusting intraocular pressure for central corneal thickness does not improve prediction models for primary open-angle glaucoma. Ophthalmology 2012;119:437– 42.
4. Liu J, Roberts CJ. Influence of corneal biomechanical properties on intraocular pressure measurement: quantitative analysis. J Cataract Refract Surg 2005;31:146 –55.

5. Boehm AG, Weber A, Pillunat LE, et al. Dynamic contour tonometry in comparison to intracameral IOP measurements. Invest Ophthalmol Vis Sci 2008;49:2472–7.
6. Medeiros FA, Weinreb RN, Sample PA, et al. Validation of a predictive model to estimate the risk of conversion from ocular hypertension to glaucoma. Arch Ophthalmol 2005;123: 1351–60.
7. Gordon MO, Torri V, Miglior S, et al. Validated prediction model for the development of primary open-angle glaucoma in individuals with ocular hypertension. Ophthalmology 2007; 114:10 –9.

Adjusting Intraocular Pressure for Central Corneal Thickness Does Not Improve Prediction Models for Primary Open-Angle Glaucoma

James D. Brandt, MD,[1] Mae O. Gordon, PhD,[2,3] Feng Gao, PhD,[3] Julia A. Beiser, MS,[2] J. Phillip Miller, AB,[3] Michael A. Kass, MD,[2] for the Ocular Hypertension Treatment Study Group
Ophthalmology Volume 119, Number 3, March 2012

Purpose: To determine if the accuracy of the baseline prediction model for the development of primary open-angle glaucoma (POAG) in patients with ocular hypertension can be improved by correcting intraocular pressure (IOP) for central corneal thickness (CCT).

Design: Reanalysis of the baseline prediction model for the development of POAG from the Ocular Hypertension Treatment Study (OHTS) substituting IOP adjusted for CCT using 5 different correction formulae for unadjusted IOP.

Participants: A total of 1433 of 1636 participants randomized to OHTS who had complete baseline data for factors in the prediction model: age, IOP, CCT, vertical cup-to-disc ratio (VCDR), and pattern standard deviation (PSD).

Methods: Reanalysis of the prediction model for the risk of developing POAG using the same baseline variables (age, IOP, CCT, VCDR, and PSD) except that IOP was adjusted for CCT using correction formulae. A separate Cox proportional hazards model was run using IOP adjusted for CCT by each of the 5 formulae published to date. Models were run including and excluding CCT.

Main Outcome Measures: Predictive accuracy of each Cox proportional hazards model was assessed using the *c*-statistic and calibration chi-square.

Results: C-statistics for prediction models that used IOP adjusted for CCT by various formulas ranged from 0.75 to 0.77, no better than the original prediction model (0.77) that did not adjust IOP for CCT. Calibration chi-square was acceptable for all models. Baseline IOP, whether adjusted for CCT or not, was statistically significant in all models including those with CCT in the same model. The CCT was statistically significant in all models including those with IOP adjusted for CCT in the same model.

Conclusions: The calculation of individual risk for developing POAG in ocular hypertensive individuals is simpler and equally accurate using IOP and CCT as measured, rather than applying an adjustment formula to correct IOP for CCT.

Financial Disclosure(s): Proprietary or commercial disclosure may be found after the references.

Central corneal thickness (CCT) is among the strongest independent predictors for the development of primary open-angle glaucoma (POAG) in the Ocular Hypertension Treatment Study (OHTS)1 and the European Glaucoma Prevention Study (EGPS).2 The risk of developing POAG doubled for every 40 μm decrease in CCT from the overall mean of 573.3 μm in the OHTS and EGPS pooled sample.[3]Other independent predictive factors for the development of POAG in the OHTS and EGPS prediction models were baseline age, intraocular pressure (IOP), vertical cup-to-disc ratio (VCDR), and pattern standard deviation (PSD). Other studies have also reported similar associations between CCT and the incidence of POAG[4–6] or the prevalence of POAG.[7–9]

Although CCT and IOP have an independent effect on the risk of developing POAG, in fact, these 2 factors interact. When Goldmann applination tonometry (GAT) was introduced in the 1950s, the thickness of the cornea was recognized as a potential confounder to IOP measurement.[10] As optical and ultrasonic pachymeters became widely available, researchers realized that CCT varied greatly between individuals. Since then, a number of investigators have developed formulae to "adjust" IOP as measured by GAT for CCT.[11–15] These formulae have been based on cannulation studies of eyes during cataract surgery,[11,12,15] meta-analyses of published datasets,[14] or engineering models of the applanated cornea.[13] Why CCT is such a powerful predictor of POAG is unknown. Some clinicians think that the only reason CCT in the prediction model for the development of POAG is to correct IOP for mismeasurement. To address this issue, we correct IOP for CCT using published correction formulae described previously. If the influence of CCT on GAT fully explains CCT's role as a predictive factor, correcting IOP using these formulae should cause CCT to drop out of the predictive model. Neither the OHTS nor the EGPS prediction model for the development of POAG adjusted baseline IOP for corneal thickness by correction formulae. To determine whether doing so might improve the predictive ability of the model, we recalculated the predictive model for the development of POAG substituting the value of IOP adjusted for CCT for the unadjusted IOP. We compare the predictive accuracy of the original prediction model with unadjusted values of IOP to prediction models with values of IOP adjusted for CCT using correction formulae by Ehlers et al,[11] Whitacre et al,[12] Orssengo and Pye,[13] Doughty and Zaman,[14] and Kohlhaas et al.[15]

Materials and Methods

The OHTS is an unmasked randomized trial of the safety and efficacy of topical ocular hypotensive medication in preventing or delaying the development of POAG in individuals with ocular hypertension. The design and methods of OHTS have been described and can be found at http://ohts.wustl.edu (accessed March 1, 2011) and are briefly summarized in this article.[16,17] Eligibility criteria included age 40–80 years, a qualifying IOP μ24 mmHg and μ32 mmHg in 1 eye and μ21 mmHg and μ32 mmHg in the fellow eye. Both eyes had to meet eye-specific criteria, including gonioscopically open angles, normal and reliable visual fields, and normal optic discs. Individuals signed an informed consent approved by the institutional review board of each participating clinic.

Beginning in February of 1994, 1636 individuals were randomized to either observation or treatment with topical ocular hypotensive medication. All topical ocular hypotensive medications available commercially in the United States were available from the OHTS central pharmacy. Medication was selected at the clinician's discretion.

Follow-up Visits and Tests

Semiannual visits included an ocular and medical history, refraction, best-corrected visual acuity, full threshold Humphrey white on-white 30-2 visual field test, slit-lamp examination, IOP measurement, and direct ophthalmoscopy. In addition, annual visits included dilated fundus examination and stereoscopic optic disc photography. Intraocular pressure was measured by 2 certified study personnel, an operator and a recorder, using a calibrated GAT. The operator initially set the dial at 10 mmHg and then looked through the slit-lamp and adjusted the dial while the recorder read and recorded the results. This procedure was repeated on the same eye. If the 2 readings differed by μ2 mmHg, the average of the 2 readings served as the visit IOP. If the 2 readings differed by μ2 mmHg, a third reading was performed and the median of the 3 readings served as the visit IOP.

Central corneal thickness was measured at the clinical center by calibrated ultrasonic pachymeters (Pachette 500, DGH Technologies, Exton, PA). We began to collect CCT measurements in early 1999, approximately 2 years after randomization of the last participant. The protocol for measurement of CCT is described in a previously published article.[18]

Determination of Primary Open-Angle Glaucoma

Primary open-angle glaucoma was defined as the development of a reproducible visual field abnormality or reproducible, clinically significant optic disc deterioration attributed to POAG by the masked Endpoint Committee. Criteria for reproducible visual field abnormality were 3 consecutive reliable visual fields judged abnormal (corrected PSD of $P<5\%$ or a glaucoma hemifield test outside normal limits by STATPAC 2 criteria) by masked readers at the Visual Field Reading Center, University of California Davis, Sacramento, California.[19] Criteria for reproducible optic disc deterioration were 2 consecutive sets of optic disc photographs showing generalized or localized thinning of the optic disc neuroretinal rim compared with baseline stereoscopic optic disc photographs as determined by masked certified readers at the Optic Disc Reading Center, Bascom Palmer Eye Institute, Miami, Florida.[20] When either Reading Center determined the occurrence of a reproducible end point, the masked Endpoint Committee reviewed the participant's clinical and medical history to determine if the end point was due to POAG.

Statistical Analysis

The analysis dataset for this report consists of data collected prospectively in OHTS from the start of randomization in February 1994 to June 2002 when participants were managed according to their randomization assignment. This includes participants with complete data for factors in the prediction model for the development of POAG: baseline age, IOP, CCT, VCDR, and PSD.[3] A total of 1433 of the 1636 randomized participants had complete baseline data (717 observation participants and 716 medication participants). Participants completed a median follow-up of 7.0 years.

The analysis dataset included 102 incident cases of POAG in the observation group and 41 incident cases of POAG in the medication group. Cox proportional hazard models stratified by randomization group were run with the same predictors as the OHTS/EGPS prediction model: baseline age, IOP, CCT, VCDR, and PSD. For eye-specific variables, the average of right and left eyes was used. Separate Cox proportional hazards models stratified by randomization group were rerun with values for baseline IOP corrected for CCT using formulae published by Ehlers et al,[11] Whitacre et al,[12] Orssengo and Pye,[13] Doughty and Zaman,[14] and Kohlhaas et al.[15] Models with baseline IOP adjusted for CCT by various formulae were run with and without

CCT in the model to determine if CCT made an independent contribution to the risk of developing POAG. The Pearson correlation coefficient between unadjusted IOP and CCT in this sample is -0.03.[1] The Orssengo and Pye formula assumed a mean CCT for this sample of 1433 participants (580 _m) and a value for the radius of curvature of the central anterior surface of 7.80.[21] The formulae are given in Table 1 (available at http://aaojournal.org).

To compare the predictive accuracy of the prediction models, we calculated the *c*-statistic and calibration chi-square. The *c*-statistic ranges from 0.5 (chance) to 1.0 (perfect agreement) and indicates degree of agreement between the ordering of the estimated and observed probability of a dichotomous end point.[22] When CCT is included in the model, all formulae except the one by Orssengo and Pye[13] will have nearly identical *c*-statistics by definition because the formulae are linear adjustments to IOP that do not affect the relative ranking of unadjusted IOP to the risk of developing POAG. The calibration chi-square indicates whether the model over- or underestimates the actual number of POAG events for each model by decile of risk. A calibration chi-square of ≤20.00 indicates acceptable agreement between the predicted and observed event rates.[22] The *c*-statistic and calibration chi-square were calculated using the Design library from the Comprehensive R Archive Network (R Development Core Team, v2.3-0; available at: http://cran.r-project.org/web/packages/Design/index.html; accessed March 1, 2011).

Results

Figure 1 shows the distribution of the baseline IOP adjusted using the formulae from Ehlers et al,[11] Whitacre et al,[12] Orssengo and Pye,[13] Doughty and Zaman,[14] and Kohlhaas et al.[15] The correlations between adjusted IOPs calculated from these formulae were high (range, 0.87– 0.99). Table 2 (available at http://aaojournal.org) reports baseline age, VCDR, PSD, CCT, and IOP (unadjusted and adjusted for CCT) for participants who did and did not develop POAG. Table 3 reports the multivariate hazard ratio (HR) for the risk of developing POAG for baseline IOP, as adjusted for CCT by the aforementioned formulae. In all models, the HR for adjusted IOP was statistically significant with and without CCT in the model. There was little difference in the size of the HR in the models among the various IOP correction formulae. The HR with the lowest value for CCT adjusted IOP was 1.10 (95% confidence interval [CI], 1.04 –1.15) for the Orssengo and Pye[13] formula in a model that included CCT. The HR with the highest value for CCT adjusted IOP was 1.17 (95% CI, 1.12–1.23) for the Doughty and Zaman[14] formula and the Kohlhaas et al[15] formula in a model that did not include CCT (Table 3).

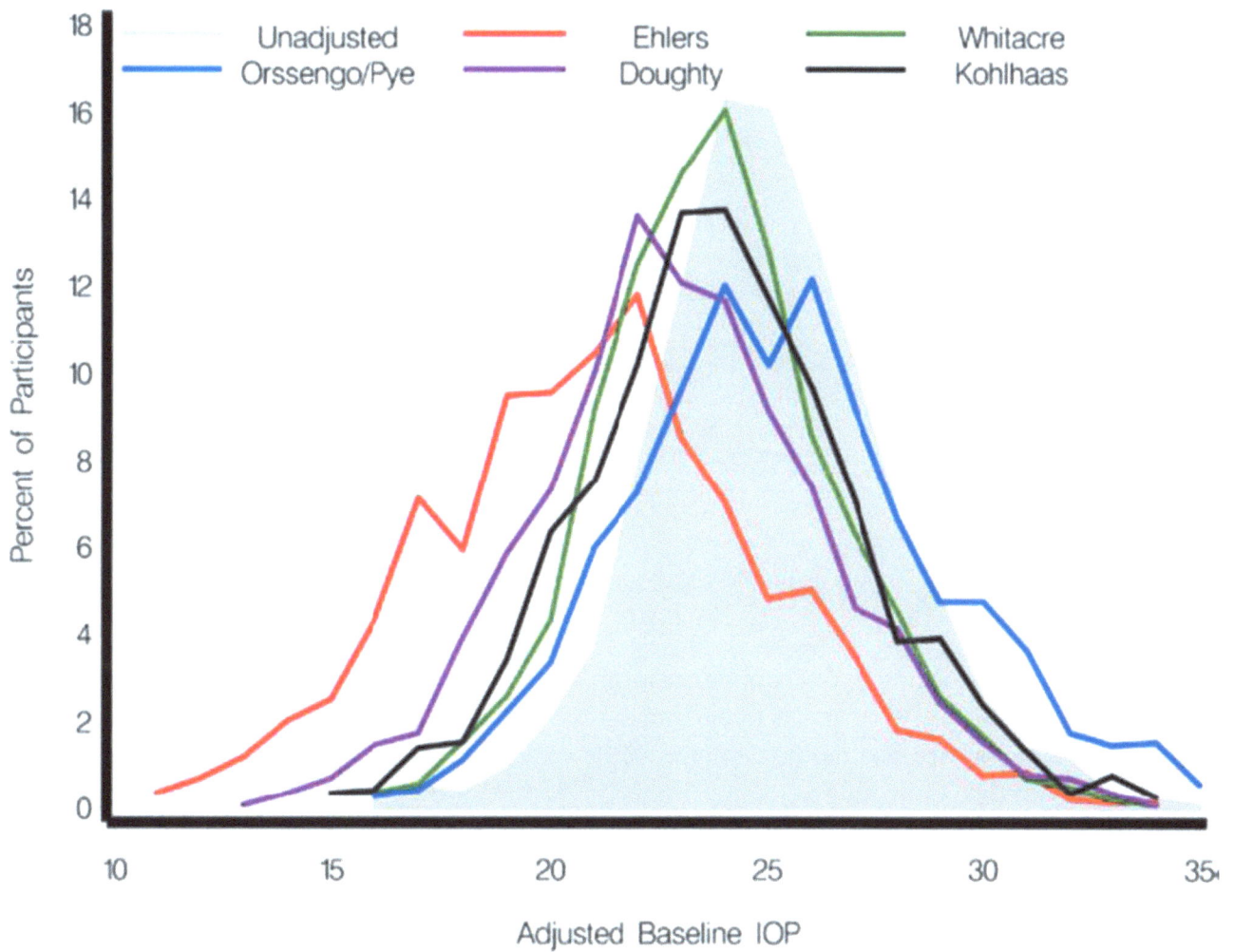

Figure 1. Distribution of baseline intraocular pressure (IOP) adjusted using the formulae from Ehlers et al,[11] Whitacre et al,[12] Orssengo and Pye,[13] Doughty and Zaman,[14] and Kohlhaas et al.[15]

Table 3. Comparison of Prediction Models for the Development of Primary Open-Angle Glaucoma When the Model Uses Unadjusted Intraocular Pressure Versus Intraocular Pressure Adjusted by Formula for Central Corneal Thickness:

OHTS prediction model*	Baseline IOP (mmHg) HR (95% CI)	CCT (40 µm decrease) HR (95% CI)	C-Statistic	Calibration Chi-Square Square
Baseline age, unadjusted IOP, CCT, PSD, and VCDR				
Model with CCT (as above)	1.11 (1.05–1.17)	1.84 (1.53–2.21)	0.774	7.86
Model without CCT	1.12 (1.06–1.19)	n/a	0.747	3.57
Ehlers et al[11] adjusted IOP				
Model with CCT	1.11 (1.05–1.17)	1.38 (1.07–1.77)	0.774	7.86
Model without CCT	1.16 (1.12–1.21)	n/a	0.770	4.60
Whitacre et al[12] adjusted IOP				
Model with CCT	1.11 (1.05–1.17)	1.69 (1.40–2.05)	0.774	7.86
Model without CCT	1.16 (1.10–1.22)	n/a	0.754	2.59
Orssengo and Pye[13] adjusted IOP				
Model with CCT	1.10 (1.04–1.15)	1.42 (1.12–1.80)	0.775	5.32
Model without CCT	1.15 (1.11–1.19)	n/a	0.770	3.83
Doughty and Zaman[14] adjusted IOP				
Model with CCT	1.11 (1.05–1.17)	1.50 (1.20–1.87)	0.774	7.86
Model without CCT	1.17 (1.12–1.23)	n/a	0.766	2.14
Kohlhaas et al[15] adjusted IOP				
Model with CCT	1.11 (1.05–1.17)	1.55 (1.25–1.91)	0.774	7.86
Model without CCT	1.17 (1.12–1.23)	n/a	0.763	2.43

CCT = central corneal thickness; CI = confidence interval; HR = hazard ratio; IOP = intraocular pressure; n/a = not applicable; OHTS = ocular hypertension treatment study; POAG = primary open-angle glaucoma; PSD = pattern standard deviation; VCDR = vertical cup-to-disc ratio. All models are run with and without CCT in the model.

Hazard ratios and 95% CIs may differ slightly from previous reports because of slight differences in dataset freeze date. This report includes data from February 28, 1994 to June 1, 2002.

The CCT was a statistically significant independent predictor for the development of POAG in all prediction models that included baseline age, VCDR, PSD, CCT, and IOP (unadjusted and adjusted for CCT). In models that included both adjusted IOP and CCT, the HR for CCT ranged from 1.38 (95% CI, 1.07–1.77) to 1.69 (95% CI, 1.40–2.05) depending on the correction formula (Table 3). The Pearson correlation coefficient between CCT and CCT adjusted IOP ranged from -0.53 (Kohlhaas et al[15]) to -0.71 (Ehlers et al[11]). C-statistics for predictive accuracy of models using baseline IOP corrected for CCT by formula ranged from 0.754 to 0.775overall, no better than the original prediction model for the development of glaucoma (0.774) that used unadjusted IOP (Table 3).There was virtually no difference in the c-statistic among models using various correction formulae for CCT. The maximum difference between the highest and lowest c-statistic among all models in Table 3 was 0.027.

Calibration chi-squares of all prediction models in Table 3 were < 20.00, indicating an acceptable fit. The calibration chi-square for the original prediction model with unadjusted IOP was 7.86 and ranged from 2.43 to 7.86 for models using CCT-adjusted IOP. Models that included both CCT-adjusted IOP and CCT performed slightly worse (higher calibration chi-square of 5.32 to 7.86) partly because of the high correlation between CCT-adjusted IOP and CCT.

Discussion

The predictive accuracy of the OHTS/EGPS prediction model for the development of POAG was not improved by correcting IOP for CCT using formulae published by Ehlers et al,[11] Whitacre et al,[12] Orssengo and Pye,[13] Doughty and Zaman,[14] and Kohlhaas et al.[15] C-statistics for these prediction models using corrected IOP ranged from 0.75 to 0.77, no better than the original OHTS prediction model (0.77) that did not adjust IOP for CCT. Calibration chi-squares did not differ meaningfully among these models (2.14 –7.86, < 20 is acceptable) and was no better than the original prediction model (7.86). This finding should not be a surprise. Tonometry is influenced by the material properties of the cornea, of which CCT is but one component. Current correction formulae for IOP use only CCT[11,12,14,15] (or CCT +corneal curvature)[13] to "adjust" IOP estimates. One engineering model of applination suggests that Young's modulus, a measure of material "stiffness" known to vary widely between individuals, has a stronger impact on GAT error than does CCT.[23] Available formulae do not seem to adequately correct the measurement of IOP for the biomechanical properties of the cornea on GAT.

There are several reasons that might explain why CCT is a strong predictor for the development of POAG. Central corneal thickness can be measured with high reliability in one sitting.[18] Because CCT is relatively stable over the lifetime of an adult, a single measurement of CCT is adequate in most patients. By comparison, IOP reflects transient factors that may or may not be relevant to the risk of developing POAG. The test–retest agreement between multiple IOPs at a given OHTS visit is high, but the test–retest agreement between 6-month visits is low to moderate.[24] The OHTS may not have captured information that is important to ascertaining the relationship of IOP to the risk of developing POAG. In OHTS, IOP was measured 2 to 3 times per visit during normal office hours. No diurnal measurements were taken. Our study highlights the fact that all tonometry techniques provide only an *estimate* of "true" IOP, a physiologic parameter that can vary greatly within the individual. Goldmann tonometry is widely considered a reference standard in the conduct of clinical and regulatory trials, but even when IOP is adjusted for CCT using specialized formulae, the adjusted IOP suffers from the inherent variability of IOP and inaccuracy of IOP measurement.

In the OHTS, IOP measured by GAT was used to determine participant eligibility, to guide treatment decisions, and to construct the predictive model for the development of POAG. Had the OHTS been carried out with a perfectly accurate, cornea-independent tonometer (something that does not exist), it is entirely possible that IOP might have been a more powerful predictor for the development of POAG and that CCT might have been a less powerful predictor. However, it is worth noting that in the Early Manifest Glaucoma Trial (EMGT), IOP was not used to determine eligibility or treatment decisions, and thus the influence of CCT on GAT measurements had no opportunity to affect the incidence rate of glaucoma progression. In the EMGT, CCT was found to be an independent predictive factor for progression of POAG.[25] In the population-based, longitudinal Barbados Eye Studies, CCT measured at 9 years from baseline was an independent risk factor for incident glaucoma.[6] In the population-based Los Angeles Latino Eye Study (LALES), the prevalence of glaucoma was higher among individuals with thin CCTs than among individuals with normal or thick CCTs across all levels of IOP.[9] The LALES investigators explored whether adjusting each IOP individually for CCT using the Doughty and Zaman algorithm[14] changed this relationship and found almost no change in the association between thin CCT and higher prevalence of glaucoma. LALES investigators concluded " ...there is an independent risk related to CCT itself."[9] In combination with the present study, the findings of the EMGT, Barbados Eye Studies, and LALES suggest that the influence of CCT on glaucoma risk is caused by more than just tonometry artifact.

Available formulae to correct IOP measurements for CCT do not improve the accuracy of the original prediction model for the development of POAG. Nor did we find that any formula outperformed the other formulae as judged by the *c*-statistic and calibration chi-square. We caution that the accuracy of the original OHTS prediction model is estimated from data averaged over a large sample of ocular hypertensive individuals and that its accuracy in predicting outcome for a single ocular hypertensive individual cannot be reliably estimated.

In conclusion, the 5-year risk of developing POAG for an individual with ocular hypertension can be simply calculated from age, IOP, CCT, VCDR, and PSD using the risk calculator available at http://ohts.wustl.edu/risk (accessed March 1, 2011), which can be downloaded free of charge.[3] The results of our analyses suggest that the influence of corneal thickness as a prognostic factor for the development of POAG is not entirely through its effect on IOP measurement, but that CCT is a biomarker for structural or physical factors involved in the pathogenesis of POAG.

References

1. Gordon MO, Beiser JA, Brandt JD, et al, Ocular Hypertension Treatment Study Group. The Ocular Hypertension Treatment Study: baseline factors that predict the onset of primary open angle glaucoma. Arch Ophthalmol 2002;120:714 –20; discussion 829–30.
2. European Glaucoma Prevention Study (EGPS) Group. Predictive factors for open-angle glaucoma among patients with ocular hypertension in the European Glaucoma Prevention Study. Ophthalmology 2007;114:3–9.
3. Ocular Hypertension Treatment Study Group, European Glaucoma Prevention Study Group. Validated prediction model for the development of primary open-angle glaucoma in individuals with ocular hypertension. Ophthalmology 2007;114: 10–9.
4. Medeiros FA, Sample PA, Zangwill LM, et al. Corneal thickness as a risk factor for visual field loss in patients with preperimetric glaucomatous optic neuropathy. Am J Ophthalmol 2003;136:805–13.
5. Zeppieri M, Brusini P, Miglior S. Corneal thickness and functional damage in patients with ocular hypertension. Eur J Ophthalmol 2005;15:196 –201.
6. Leske MC, Wu SY, Hennis A, et al, BESs Study Group. Risk factors for incident open-angle glaucoma: the Barbados Eye Studies. Ophthalmology 2008;115:85–93.
7. Wolfs RC, Klaver CC, Vingerling JR, et al. Distribution of central corneal thickness and its association with intraocular pressure: the Rotterdam Study. Am J Ophthalmol 1997;123: 767–72.
8. Nemesure B, Wu SY, Hennis A, Leske MC, Barbados Eye Study Group. Corneal thickness and intraocular pressure in the Barbados Eye Studies. Arch Ophthalmol 2003;121:240–4.

9. Francis BA, Varma R, Chopra V, et al, Los Angeles Latino Eye Study Group. Intraocular pressure, central corneal thickness, and prevalence of open-angle glaucoma: the Los Angeles Latino Eye Study. Am J Ophthalmol 2008;146:741– 6.

10. Goldmann H, Schmidt T. Applination tonometry [in German]. Opthalmologica 1957;134:221– 42.

11. Ehlers N, Bramsen T, Sperling S. Applination tonometry and central corneal thickness. Acta Ophthalmol (Copenh) 1975; 53:34–43.

12. Whitacre MM, Stein RA, Hassanein K. The effect of corneal thickness on applination tonometry. Am J Ophthalmol 1993; 115:592– 6.

13. Orssengo GJ, Pye DC. Determination of the true intraocular pressure and modulus of elasticity of the human cornea in vivo. Bull Math Biol 1999;61:551–72.

14. Doughty MJ, Zaman ML. Human corneal thickness and its impact on intraocular pressure measures: a review and met analysis approach. Surv Ophthalmol 2000;44:367– 408.

15. Kohlhaas M, Boehm AG, Spoerl E, et al. Effect of central corneal thickness, corneal curvature, and axial length on applanation tonometry. Arch Ophthalmol 2006;124:471– 6.

16. Gordon MO, Kass MA, Ocular Hypertension Treatment Study Group. The Ocular Hypertension Treatment Study: design and baseline description of the participants. Arch Ophthalmol 1999;117:573– 83.

17. Kass MA, Heuer DK, Higginbotham EJ, et al, Ocular Hypertension Treatment Study Group. The Ocular Hypertension Treatment Study: a randomized trial determines that topical ocular hypotensive medication delays or prevents the onset of primary open-angle glaucoma. Arch Ophthalmol 2002;120: 701–13; discussion 829–30.

18. Brandt JD, Beiser JA, Kass MA, Gordon MO, Ocular Hypertension Treatment Study (OHTS) Group Central corneal thickness in the Ocular Hypertension Treatment Study (OHTS). Ophthalmology 2001;108:1779–88.

19. Johnson CA, Keltner JL, Cello KE, et al, Ocular Hypertension Study Group. Baseline visual field characteristics in the Ocular Hypertension Treatment Study. Ophthalmology 2002;109: 432–7.

20. Feuer WJ, Parrish RK II, Schiffman JC, et al, Ocular Hypertension Treatment Study Group. The Ocular Hypertension Treatment Study: reproducibility of cup/disk ratio measurements over time at an optic disc reading center. Am J Ophthalmol 2002;133:19 –28.

21. Emsley HH. The Schematic Eye-Unaided and Aided. In: Visual Optics. Volume 1: Optics of Vision. 5th ed. London: Butterworths; 1973:346.

22. D'Agostino RB Sr, Grundy S, Sullivan LM, Wilson P, CHD Risk Prediction Group. Validation of the Framingham coronary heart disease prediction scores: results of a multiple ethnic groups investigation. JAMA 2001;286:180 –7.

23. Liu J, Roberts CJ. Influence of corneal biomechanical properties on intraocular pressure measurement: quantitative analysis. J Cataract Refract Surg 2005;31:146 –55.

24. Bhorade AM, Gordon MO, Wilson B, et al, Ocular Hypertension Treatment Study Group. Variability of intraocular pressure measurements in observation participants in the Ocular Hypertension Treatment Study. Ophthalmology 2009;116: 717–24.

25. Leske MC, Heijl A, Hyman L, et al, EMGT Group. Predictors of long-term progression in the Early Manifest Glaucoma Trial. Ophthalmology 2007;114:1965–72.

Footnotes and Financial Disclosures

Originally received: December 6, 2010.
Final revision: March 8, 2011.
Accepted: March 9, 2011.
Available online: June 25, 2011. Manuscript no. 2010-1526.

1. Department of Ophthalmology and Vision Science, University of California, Davis.
2. Department of Ophthalmology and Visual Sciences, Washington University School of Medicine, St. Louis, Missouri.
3. Division of Biostatistics, Washington University School of Medicine, St.Louis, Missouri.

Presented at: The Association for Research in Vision and Ophthalmology, May 4, 2010, Fort Lauderdale, Florida.

Group members of the Ocular Hypertension Treatment Study Group appear online (https://vrcc.wustl.edu/clinics.html).

Financial Disclosure(s):

The author(s) have made the following disclosure(s): James D. Brandt: Alcon Laboratories (consultant, lecturer); Allergan (consultant); Endo Optiks (consultant); Glaukos (consultant); Merck (lecturer); Pfizer (consultant, employee).

Supported by grants from the National Eye Institute and the National Center on Minority Health and Health Disparities, National Institute of Health, Bethesda, Maryland (EY09341, EY09307, EY091369, and Core Grant EY02687); Merck Research Laboratories, White House Station, New Jersey; Pfizer Inc., New York, New York; and an unrestricted grant from Research to Prevent Blindness, New York, New York. Correspondence: Mae O. Gordon, PhD, Washington University School of Medicine, Department of Ophthalmology and Visual Sciences, 660 S. Euclid, Box 8203, St. Louis, MO 63110. E-mail: mae@vrcc.wustl.edu.

Graves' disease patients can be _____
    euthyroid
    hypothyroid
    hyperthyroid

You order what lab tests when you suspect proptosis?
    TSH
    T4
    T3
    TSI

Significant proptosis is defined as _____
    greater than 22mm in all patients
    greater than 24 mm in all patients
    greater than 22mm in whites and 24 mm in blacks

Which of the following disease have to be ruled out in every ptosis patient?
    Horner's (anisocoria, miosis on ptosis side)
    Myasthenia Gravis (ptosis worse in PM and with effort)
    3rd CN palsy (binocular diplopia)
    superior orbital growths causing a ptosis
    CPEO (bilateral limitation of EOMs, bilateral, never diplopia, normal pupils)

You should lift the eyebrow in a ptosis examination to _____
    confuse the patient
    eliminate the effect of brow lifting on the underlying ptosis
    satisfy insurance requirements

Bell's palsy is defined as a _____
    peripheral 7th nerve palsy
    central 7th nerve palsy
    no nerve palsy

It is important to do a Hertel exophthalmometer test on every ptosis patient in order to:
    make sure that a proptotic eye is not making the non-proptotic eye look like it has a ptosis
    make sure that a proptotic eye is not making the proptotic eye look like it has a ptosis
    to charge the insurance company for the Hertel test

Which of the following is the most common cause of a ptosis?
    aponeurosis dehiscence
    neurogenic (3rd CN palsy, Horner's, Myasthenia Gravis)
    myogenic (congenital)
    orbital masses
    pseudo ptosis (proptotic eye making the other eye look ptotic)

Pupil involved 3rd nerve palsies present as:
    eye down and out
    ptosis on the same side
    anisocoria with mydriasis on the same side

The second step to do once you suspect a CN palsy is to:
    check for other palsies
    check your own pulse
    check the IOP

The most important sign in an isolated 3rd cranial nerve (CN) palsy is:
    measuring for an APD
    measuring the VFCF
    looking for pupil involvement

CN palsies present as:
    non-commitant deviations (different deviations in different gaze positions
    commitant deviations (same deviations in different gaze positions)
    no set pattern

A 4th nerve palsy presents with the eye _____
    down and out
    up and in
    looking lateral

A 6th nerve palsy presents with the eye _____
    looking medial
    looking lateral
    up and in

A 5th nerve palsy presents as _____
    a droopy face
    a numb face
    a herpes simplex infection eruption

A 7th nerve palsy presents as _____
    a droopy face
    a numb face
    a herpes zoster eruption

Prevalence studies:
    measure new cases over time
    measure all existing cases at a single point in time
    measure existing new cases at a single point in time

Incidence studies:
    measure new cases over a given time period
    measure existing cases at a single point in time

Corrected corneal thickness (CCT):
    isn't a settled issue
    is a mis-measurement of true IOP
    is a risk factor of true IOP
    all of the above

Risk factors for OAG include:
    - a high IOP, corrected or uncorrected for CCT
    - older age, especially above 65 yo
    - family history
    - an enlarged cup to disk ratio (C/D)
    - disk hemorrhages
    - nerve fiber layer defects (NFLD)
    - asymmetry in IOP, C/D ratio between the two eyes
    - a large diurnal curve
    - race might be
    - ALL OF THE ABOVE

Primary angle closure glaucoma is defined as:
    the presence of iris nodules
    the presence of iris bombe
    the presence of transillumination defects

Secondary angle closure glaucoma is defined as:
    the presence of iris bombe
    the presence of something pushing or drawing the iris into the TM
    it doesn't exist

A laser peripheral iridectomy (LPI) is performed for:
    PACG
    POAG
    SOAG
    I don't know what an LPI is

Examples of SACG include:
     phacomorphic
     neo-Vascular glaucoma
     plateau iris configuration
     iridocorneal endothelial syndrome (ICE)
     malignant glaucoma
     all of the above

Topamax has been reported to:
     cause the ciliary body to rotate backward into the retina
     cause the ciliary body to rotate anteriorly into the vitreous

Patients at risk of acute ACG are:
     age > 50 yo
     hyperopes
     have early cataract enlargement of the lens
     all of the above

The main concern after a CRVO has occurred is:
     - the development of NVG 90 days later
     - the development of a CRVO in the other eye due to undiagnosed or untreated glaucoma
     - a second ischemic CRVO after an initial non-ischemic CRVO in the same eye
     - all of the above

The first differentiation of the types of glaucoma is based on what test?
     - pachymetry
     - HVF
     - gonioscopy
     - HRT

Plateau iris is defined as angle closure in the presence of a:
     large C/D ratio
     patent LPI
     normal HVF
     Plateau iris is treated by:
     calling for help
     an iridoplasty
     performing a second LPI in the presence of a patent one

The iridocorneal endothelial syndrome is:
     common
     very rare
     has a normal iris and corneal appearance

Primary in primary angle closure glaucoma means:
    the mechanism is known
    the mechanism is not known
    iris bombe is present
    iris bombe is not present

The treatment of PACG is:
    SLT
    glaucoma drops
    an LPI
    an iridoplasty

Secondary in SACG means:
    the mechanism is known
    the mechanism is not known
    iris bombe is present

A CRVO is concerning for what type of glaucoma?
    NVG
    NVE
    NVD

The reason why a CRVO in one eye should be treated for glaucoma OU is:
    because Dr. Sedgewick said so
    untreated glaucoma is a significant risk factor for CRVO
    the risk of a second CRVO episode in the same eye or other eye is very small
    the vision in a CRVO is typically intact

In order of increasing incidence, NVI and NVG occur in:
    CRVO > HRVO > BRVO
    HRVO > CRVO > BRVO
    BRVO > HRVO > CRVO

Ocular ischemic syndrome differs from diabetic retinopathy:
    OIS has more hemorrhaging in the macula
    OIS has more hemorrhaging in the periphery
    Diabetic retinopathy has carotid bruits

The following are to be thought of in unilaterally high IOP:
    Possner-Schlossman
    NVG
    PDG
    ghost cell glaucoma
    ACG
    steroid use in one eye
    Sturge-Weber
    all of the above

The following present unpredictable risk factors in the diagnosis of glaucoma:
    diurnal variation
    visual field damage lags behind optic nerve damage
    low tension glaucoma can be difficult to detect
    the rate of ON damage increases as damage advances

What percentage of DH were missed by the glaucoma specialists in the OHTS?
    20%
    50%
    80%
    86%

Pachymetry should be done:
    at every visit since corneal thickness changes over time
    at the last visit
    at the initial visit

Optic disk photos should be done:
    yearly
    once every 5 years
    never
    only when medicare pays for them

The correct endpoint in Goldmann tonometry is:
    when the inner mires completely overlap
    when the inner mires barely touch
    when the inner mires overlap as much as they underlap

The addition of the VFI is:
    not important since it adds nothing
    is an easy to understand test for patients and doctors
    is too expensive to do

The 3 major tests looking for early glaucoma damage:
    - include the HRT/OCT/GDX
    - are equivalent in ability to detect early glaucoma damage
    - should be done at every visit you do the VF
    - all of the above

Gonioscopy:
    - can be replaced with an ultrasound of the anterior angle
    - should be done routinely at every yearly visit
    - is critical to perform to detect an open angle as well as the degree of angle pigmentation

Lowering of the IOP has been conclusively shown to reduce the chances of developing glaucoma damage in:

normal tension glaucoma
low tension glaucoma
ocular hypertension
all of the above

A choroidal nevus is present in:

0.1% of the American population
1% of the American population
10% of the American population

The main concern about a choroidal nevus is:

- a conversion to a sub retinal neovascular membrane
- a melanoma
- nothing ever happens to a choroidal nevus

Shields found that the following were risk factors for metastasis of the nevus:

(relative risk)
- 8.8 tumor thickness <1.0 to 1.1 to 3.0 mm
- 3.2 documented growth of the tumor
- 2.9 margin of the nevus touching the optic disk
- 1.9 blurred vision as a presenting symptom
- all of the above

The correct summary of the Shields article is:

- virtually all tumors les than1.0 mm in thickness are benign
- virtually all tumors thicker than 3.0 mm are melanomas
- lesions 1.1 to 2.9 mm thick are benign with the risk increasing with thickness
- all of the above

What percentage of uveal melanomas are found in Caucasians?

10%
50%
80%
89%

The small (1.0 to 3.5 mm thick) COMS study showed the following risk factors for the conversion to melanoma:

thickness > 2.0 mm
subretinal fluid
orange pigment (lipofuscin, not yellow drusen)
tumor edge within 3 mm from disk
all of the above

CSME is defined as:
- retinal edema within 500 um of the fovea
- hard exudates associated with macular edema within 500 um of the fovea
- macula edema 1 disk diameter (DD) in size within 1 DD of the fovea
- all of the above

"High risk" NPDR is defined as:
- having IRMA in 1 quadrant
- venous beading in 2 quadrants
- diffuse intra retinal bleeding in 4 quadrants
- all of the above. This is the 4-2-1 rule

Proliferative diabetic retinopathy is defined as:
really bad macular swelling
swelling at the optic disk
new blood vessel growth

Type 1 diabetes is defined as:
a resistance to insulin at the cell level
a lack of insulin production
an over production of insulin

Type 2 diabetes is defined as:
a resistance to insulin at the cell level
a lack of insulin production
an over production of insulin

Which of the following is a worse form of diabetic retinopathy?
blot and dot hemorrhages
cotton wool spots
IRMA

The ETDRS showed that with 1 of the high risk 4-2-1 rule categories the risk of developing high risk PDR is:
15%
45%
100%

The ETDRS showed that having 2 of the high risk 4-2-1 categories carries what risk of developing high risk PDR?
15%
45%
100%

The DCCT study showed that tight control of blood sugars in type 1 diabetics reduced the chances of:
  kidney damage
  nerve damage
  retinal damage
  all of the above

The differential diagnosis of diabetic retinopathy includes:
  CRVO
  OIS
  HTN retinopathy
  hyper coagulation states
  all of the above

Uveitis is defined as an inflammation of the:
  uveal tract
  iris
  ciliary body
  choroid
  all of the above

Which of the following statements is true?
  - a significant uveitis starting 3-5 days after intraocular surgery is endophthalmitis until proven otherwise
  - a uveitis starting 1 day after surgery is TASS
  - a persistent uveitis 6 months after intraocular surgery could be endophthalmitis
  - all of the above

My mnemonic for remembering the etiologies of an iritis is:
  JBGHUTFL
  PAIR for HLA b27 diseases
  both of the above

My mnemonic for remembering posterior uveitis is:
  S's and T's
  A's and B's
  Y's and Z's

Circulating auto-antibodies to the thyroid gland are thought to be the cause of Grave's disease:
  true
  false

The antibodies to the thyroid gland in TED can cause inflammation in the orbital tissue as well:
  true
  false

TED is usually a self-limiting disease:
    true
    false

Smoking can increase the length of a TED episode from 1 to 2.5 years:
    true
    false

TED patients can be:
    hypothyroid
    euthyroid
    hyperthyroid
    all of the above

The problem in thyroid eye disease is:
    an antibody affecting the thyroid only
    an antibody affecting the orbital tissues only
    an antibody affecting the whole body

On average, a TED episode lasts:
    1 year in non-smokers
    2-3 years in smokers
    can vary a lot
    all of the above

TED patients can be:
    hyperthyroid (90%)
    euthyroid (6%)
    hypothyroid (4%)
    all of the above

On an orbital CT, TED affects the tendons of the EOMs:
    true
    false

On an orbital CT, pseudo tumor affects the EOM tendons:
    true
    false

TED is suspected when:
    the sclera is seen below or above the limbus
    a deep, dull orbital pain is present
    the upper lid lags behind the limbus in down gaze
    all of the above

A normal reading for the Hertel is:
    22mm for whites
    24 mm for blacks
    a difference of 2 mm or less
    all of the above

The biggest ocular fear in TED is:
    exophthalmos
    proptosis
    optic nerve impingement

The following lab tests should be ordered in all patients failing a Hertel:
    TSH
    TSI
    T4, T3
    HVF 24-2
    D 15 color vision testing
    all of the above

Patients with normal labs on the initial w/u should get an orbital CT, thin cuts, axial and coronal:
    true
    false

Patients with a FBS should have their upper lid everted to r/o FBs trapped there:
    true
    true

A 30 gauge diabetic needle at a tangential angle is best to remove a corneal FB:
    true
    true

Examine the iris and lens in all corneal FB patients to r/o penetration of the globe:
    true
    true

Abelson's triad includes:
    itchy eyes are allergic conjunctivitis
    sticky, red eyes are bacterial conjunctivitis
    watery eyes that burn or are sandy are dry eyes
    watery eyes with SEIs in non SCL wearers is viral
    all of the above

Insurance will pay for:
    the D 15 color test
    the Ishihara color test
    the lantern color test

Worrisome signs of carcinoma/ cornea of the conjunctive include:
    leukoplakia
    gelatinous thickening of the lesion
    sessile formation
    injected BVs supplying the lesion
    all of the above

Sun exposure seems to be the cause of most carcinomas of the conjunctiva:
    true
    false

The classic sign of a conjunctival lymphoid lesion is:
    a salmon colored patch
    an injected, red patch
    a patch with leukoplakia
    a papillomatous pattern

You only biopsy a lesion that you are sure is a cancer:
    true
    false

Primary acquired melanosis (PAM) of the conjunctiva occurs:
    in light complexion peoples
    unilateral
    middle age
    all of the above

Benign acquired melanosis occurs:
    in darker skinned people
    is bilateral
    are benign as the name implies
    all of the above

PAM is a pre-malignant lesion:
    true
    false

BAM is a pre-malignant lesion:
    true
    false

PAM of 3 clock hours in size presents what chance of malignancy?
    20%
    50%
    80%
    0%

What percentage of melanomas arise from PAM?
    0%
    20%
    50%
    67%

Skin involvement of conjunctival pigmented lesions is called:
    a big nevus
    nevus of Toledo
    nevus of ota

Nevus of Ota in pigmented peoples is:
    benign
    premalignant
    malignant

Nevus of Ota:
    is benign in Caucasions
    can be malignant in Caucasions
    is malignant in African Americans

Basal cell carcinomas:
    are the most common eyelid carcinoma
    have ulcerated centers
    typically destroy eyelash hair
    have a rolled up edge
    could use the Mohs surgical technique
    all of the above

Medicare may or may not pay for an upper lid bleph:
    true
    false

Medicare will not pay for a lower lid bleph:
    true
    false

Choroidal ruptures tend to occur circumferentially around the ON:
    true
    false

Most corneal carcinomas arise from the limbus or conjunctiva:
    true
    false

About what percentage of lymphomas in the eyelid skin have mediatized?

0%

20%

50%

67%

<u>Cornea</u>

Steroids for Corneal Ulcer Trial (SCUT):

A prospective, randomized, multicenter Steroids for Corneal Ulcer Trial (SCUT) showed no benefit in BCVA (P = 0.39) nor in size of the scar (P = 0.69) after starting PF 48 hours after using moxifloxacin in treating corneal ulcers (n = 399 patients followed for 12 months)[145] Since PF is contraindicated for fungal, acanthamoeba, norcardia, and HSV infections and you don't really know which organism is causing the ulcer, don't use steroids in corneal ulcer treatment.

<u>Glaucoma</u>

Los Angles Latino Eye Study (LALES):

The graph below shows the LALES *prevalence* risk of POAG at baseline exam IOP, 5,970 people screened with no prior history of glaucoma, follow up = none (prevalence data, you should know this by now), n = 167 patients diagnosed with OAG at the time of screening, using uncorrected IOP[55]

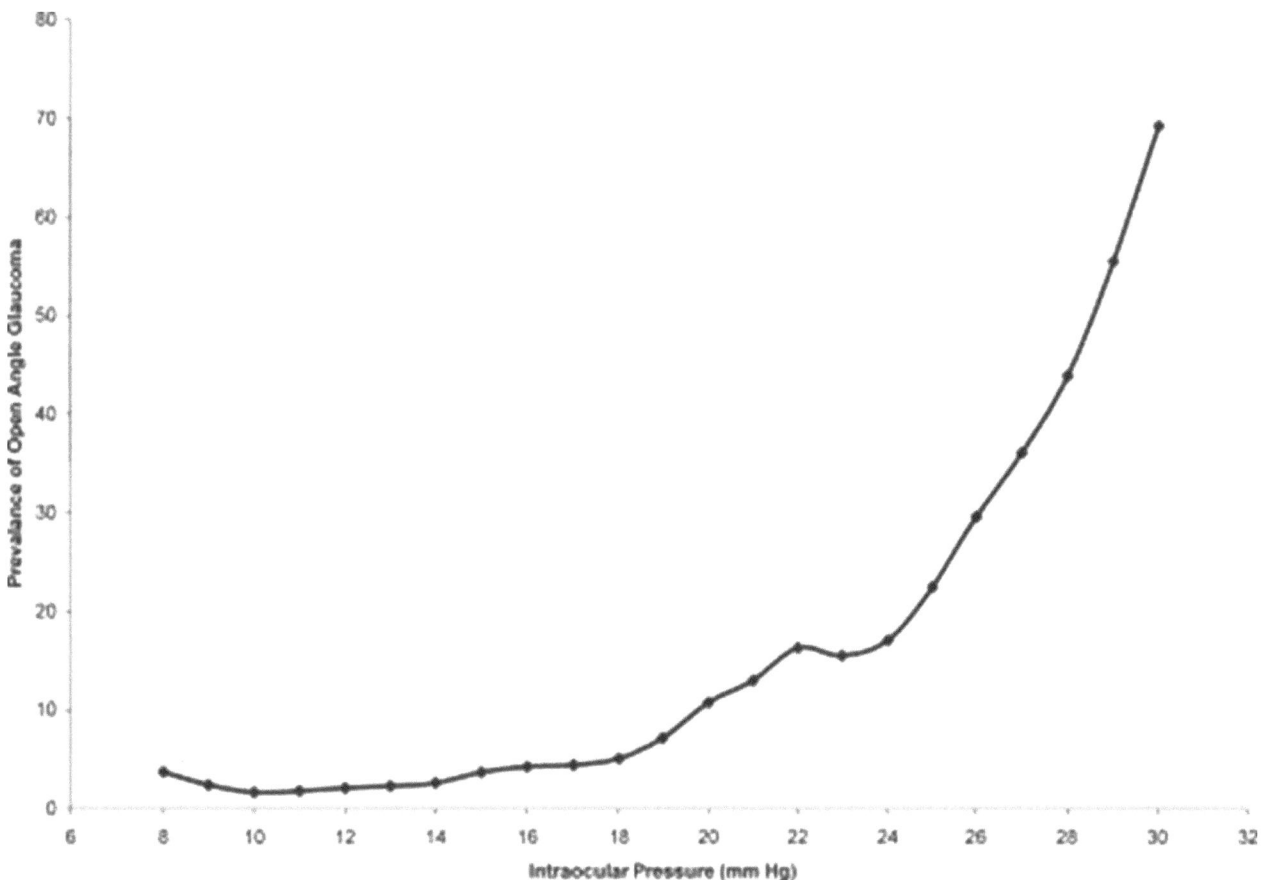

Below is the graph from LALES measuring the *prevalence* risk of glaucoma versus CCT broken into 3 strata, no follow up (prevalence data), 5,970 screened, n = 167 patients had undiagnosed glaucoma at the baseline exam[55]

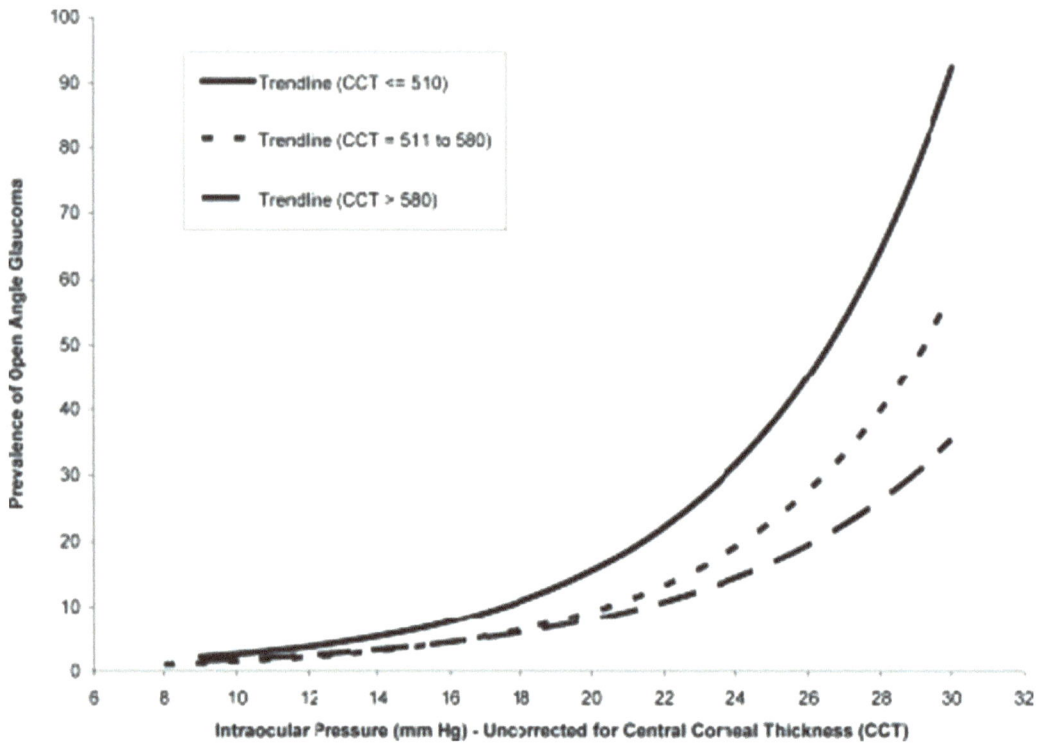

The graph below shows the LALES *incidence* risk of POAG in 3,772 people followed for 4 years (87 developed glaucoma) using baseline uncorrected IOP with conversion to glaucoma after adjusting for covariate factors.[57] The graph after the next one is the same data after factoring in CCT. [57]

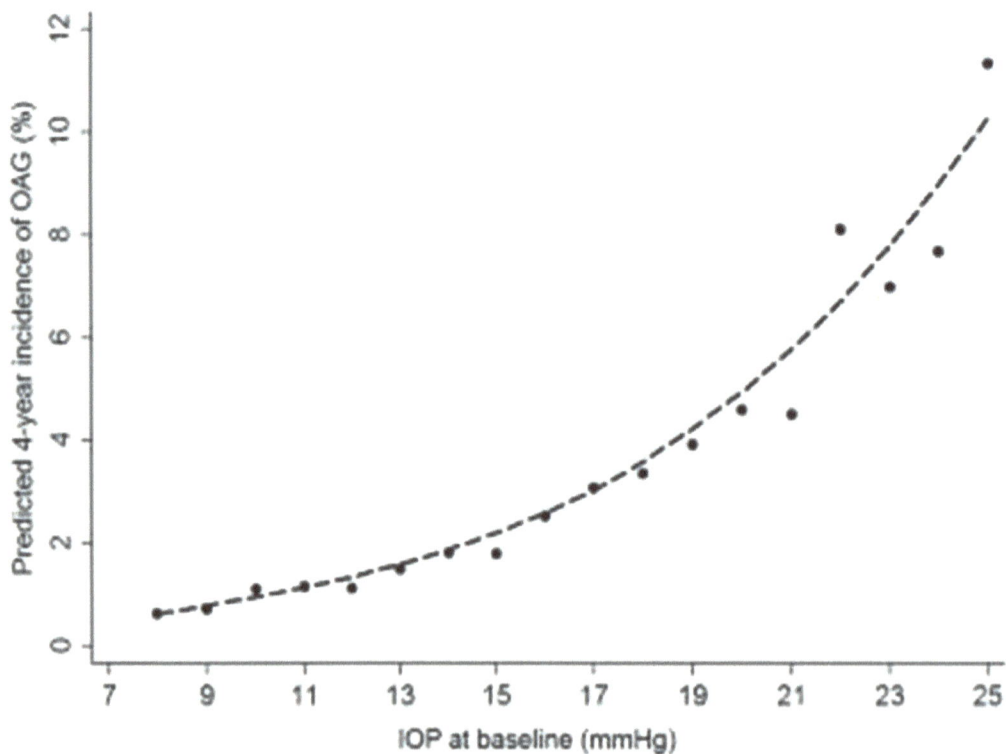

The LALES *incidence* of OAG once CCT is factored into the IOP:

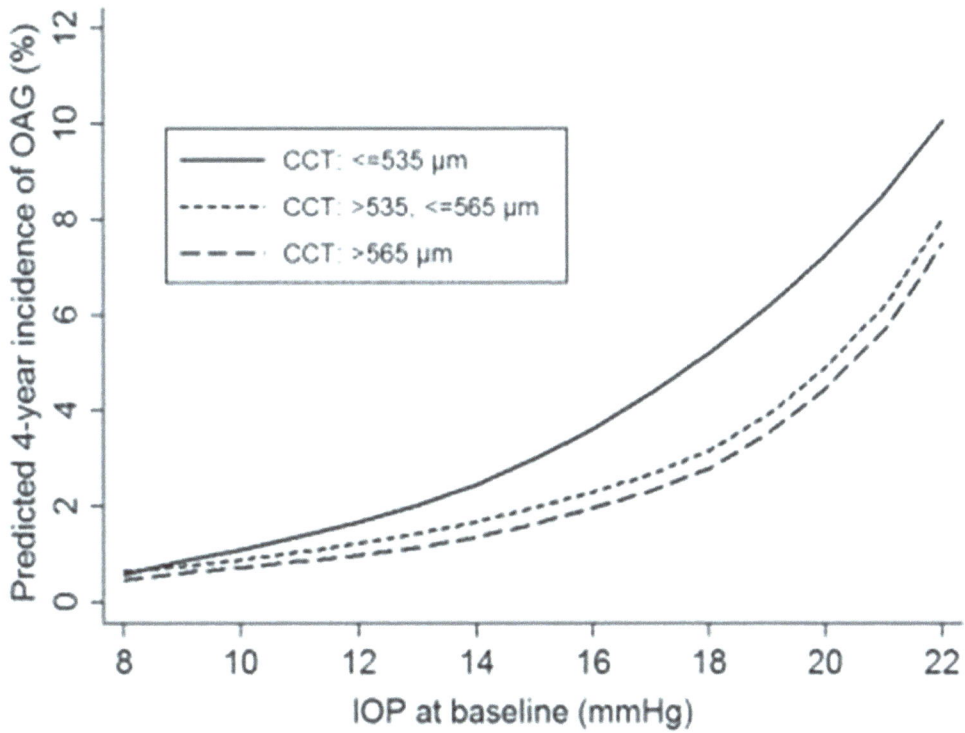

Yazd Study of Glaucoma:

Prevalence of glaucoma of all types, Yazd Iran, 1,990 patients screened, n = 87 [66] with glaucoma

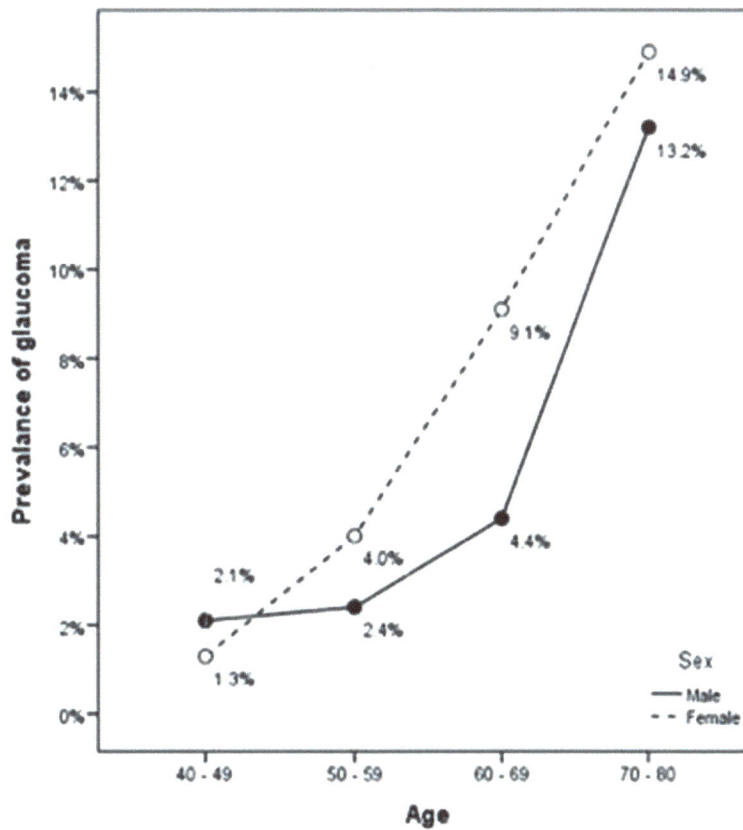

Barbados Study of Glaucoma:

Incidence risk of developing POAG over 9 years of follow-up at the given baseline uncorrected IOP. 3,222 people screened, n = 125 patients developing OAG[56]

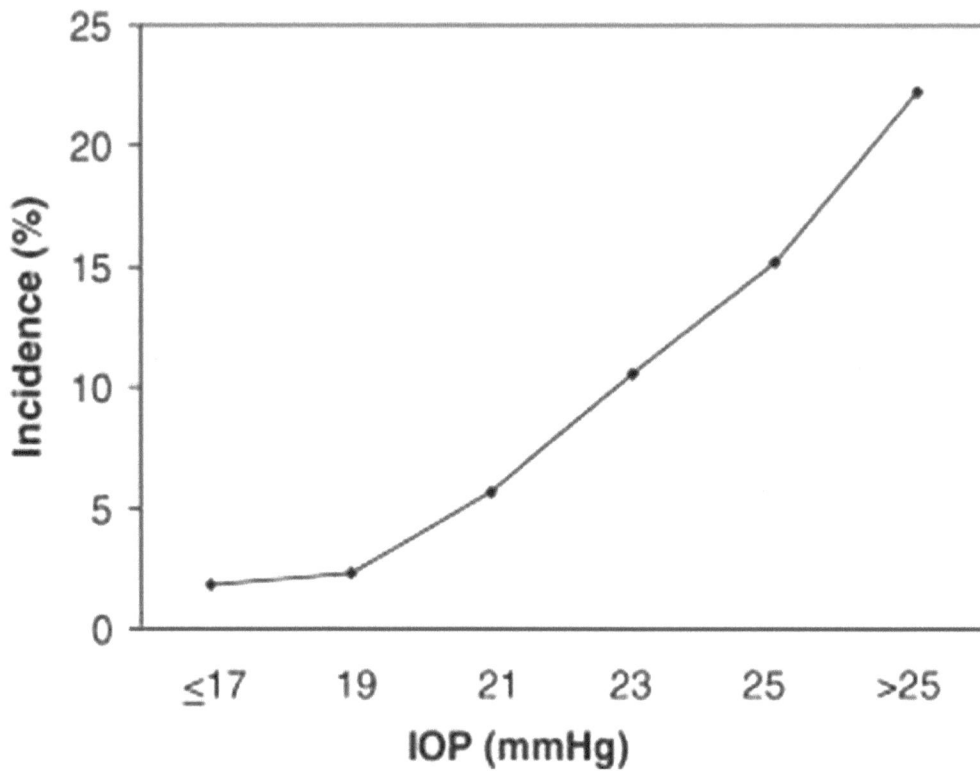

Incidence of Glaucoma from 4 separate studies according to age [67]

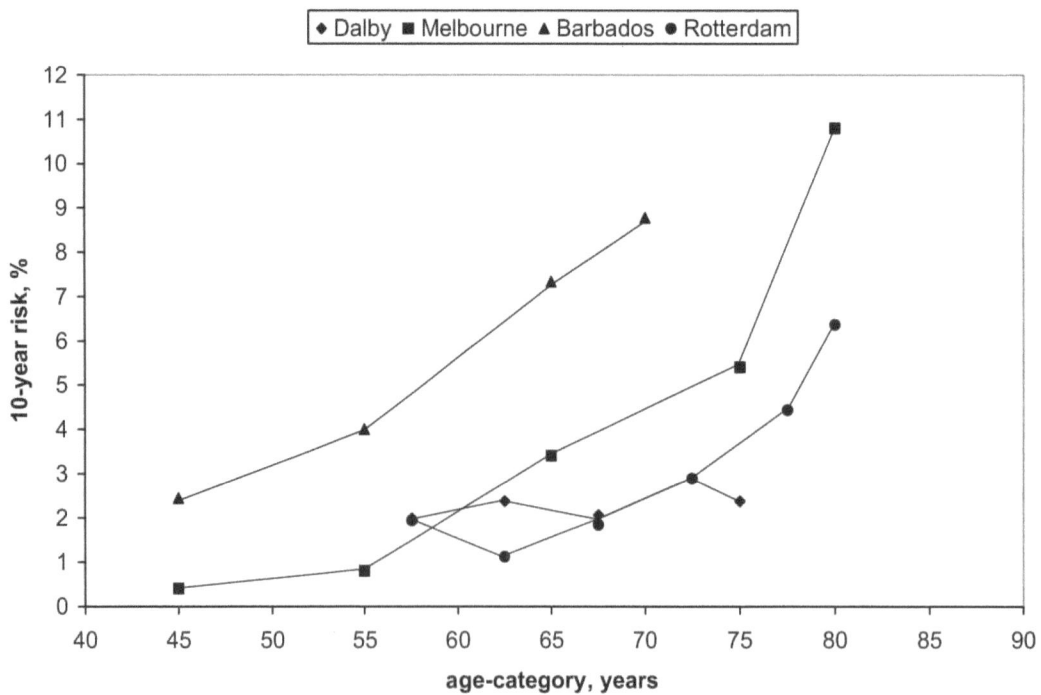

The Ocular Hypertensive Treatment Study (OHTS), a prospective, multicenter, randomized study that randomized a total of 1,408 patients with ocular hypertension to two groups, half received OAG treatment and half did not. Patients completed 60 months of follow up. Treatment reduced the risk of developing OAG by over 50% (9.5% vs 4.4%, P < 0.0001).[65] This completed phase 1 of the OHTS. Due to the obvious benefits of treatment, both groups were then given treatment for OAG. They were followed for an additional time for a total of 13 years, to determine if the non-treatment group had a penalty for delaying treatment. They did not (P < 0.77). [107] This was phase 2 of the OHTS. In the graph below, you can see that the two groups, represented by the two lines, diverge before the gray area, which represents phase 1. The time between phase 1 and phase 2 is represented by the grey column. Phase two, the area after the grey line, shows that the two groups had their risk of developing OAG parallel, meaning that delaying treatment in the non-treatment group did not induce a penalty on them.[107]

86% (604/703) of the abnormalities on the initial and deemed reliable HVF tests in the Ocular Hypertensive Treatment Study (OHTS) were not confirmed abnormal upon retesting.[18]

279 eyes that converted to OAG in the OHTS were compared to 279 eyes that did not convert in the same study.[194] No significant difference was found in the alpha and beta zones around the optic nerve. Currently, I don't place much importance on the alpha and beta peripapillary areas in early OAG detection and treatment.

The OHTS found, in multivariate analysis (taking into account demographic data and other known risk factors that could skew the results), a 3.7 fold increase (P < 0.001) in the risk of developing POAG in patients who had a disk hemorrhage (n=3,236 eyes screened, 168 eyes developed glaucoma with a mean follow up 31 months since the discovery of the hemorrhage, 128 eyes had disk hemorrhages. Most of the eyes, 87%, that had a disk hemorrhage did not develop glaucoma).[74] A 3.7 fold increase in risk is a big risk factor. The patients in the OHTS were dilated once a year. Again, all of the DH in the study mentioned earlier resolved in less than 1 year. The researchers suggest that this relative risk could have been even higher with more frequent DFEs, since more DH may have been discovered.

Interestingly, they found that **84% of the hemorrhages were missed on live exams (by glaucoma specialists!) and were only discovered during a photo review of the disks**. This shows the importance of having a mental checklist that you go through when examining the optic nerve, noting the C/D ratio and looking for NFLD and disk hemorrhages in all of the patients.

In the Ocular Hypertensive Treatment Study using uncorrected for CCT IOP readings, 90% of the subjects had IOP > 22, which makes sense as it is the ocular hypertensive study after all. Whereas after using the Ehlers correction calculation for CCT, **only 48% had an IOP > 22.[58] Almost half of the ocular hypertensives in the OHTS were not hypertensives after IOP readings were corrected for CCT**!

The OHTS researchers found that the African-American participants had corneas 23.5 um thinner than non-Africans and, combined with the larger optic nerve sheaths and therefor C/D found in Africans, eliminated race as a significant risk factor for the development of glaucoma in the multivariate analysis.[70] In support of this, the basic clinical science course glaucoma volume says: "in OHTS, African-American patients were 59% more likely than Caucasions to develop glaucoma in a univariate analysis, but this relationship was not present after corneal thickness and baseline vertical cup-to-disk ratio were factored into the multivariate analysis (African-American patients had lower corneal thickness and larger vertical cup-to-disk ratios on average)"[87]

The 13 year reduction in onset of glaucoma in the prospective, randomized OHTS study. Phase 1: n= 1,408, follow up= 60 months.[65] Phase 2: n = 1,159, follow up = 13 years for both phase 1 and 2 combined. [107]

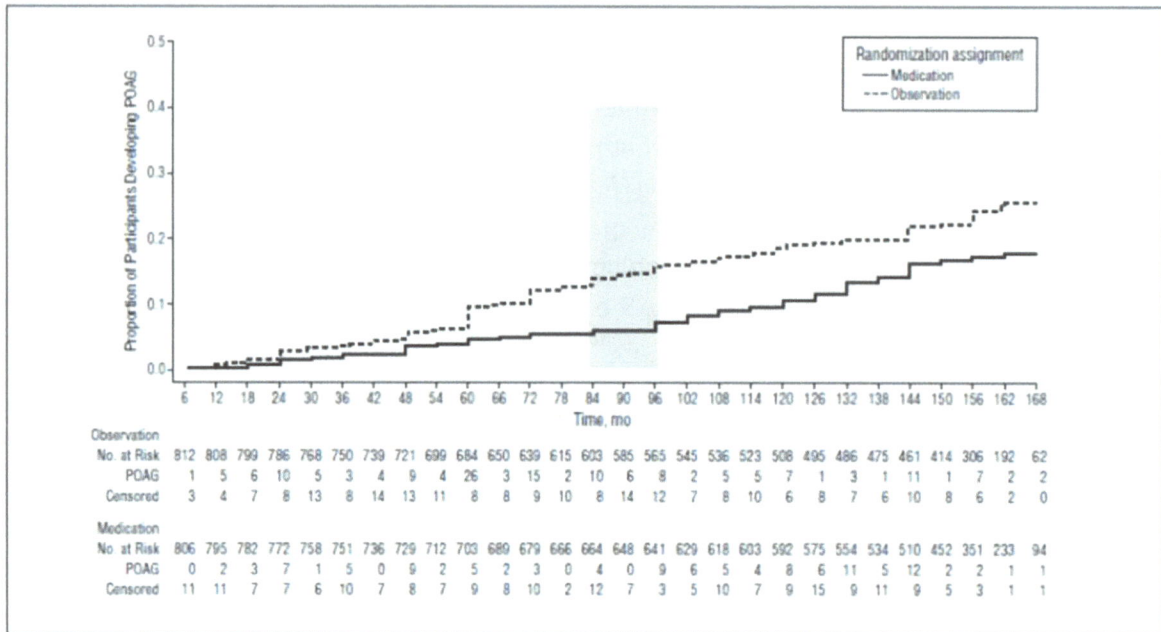

Figure 4. Survival plot of the cumulative probability of developing primary open-angle glaucoma (POAG) over the entire course of the study (February 1994 to March 2009) by randomization group. The number of participants at risk are those who have not developed POAG at the beginning of each 6-month period. Participants who did not develop POAG and withdrew before the end of the study are censored from their last completed visit. Participants who did not develop POAG and died are censored at their date of death. The shaded column indicates initiation of medication in the original observation group.

The graph below[74] looks at the OHTS data in another way. It asks the question, "did treatment show a reduced risk of developing disk hemorrhages?" The graph shows the strong trend of more frequent DHs in the *untreated* group. But while not statistically significant (p = 0.13) over a follow up of 31 months since the discovery of the DH, it leads one to believe that with a longer follow up, the difference may have become statistically significant (my words). Most of the patients who developed a disk hemorrhage did not develop POAG (87%) over the mean follow up of 31 months. So it may be that while it takes a while to develop OAG after a DH occurs, a DH should certainly increase your concern about a patient eventually developing glaucoma.

Budenz et al · Optic Disc Hemorrhages in the OHTS

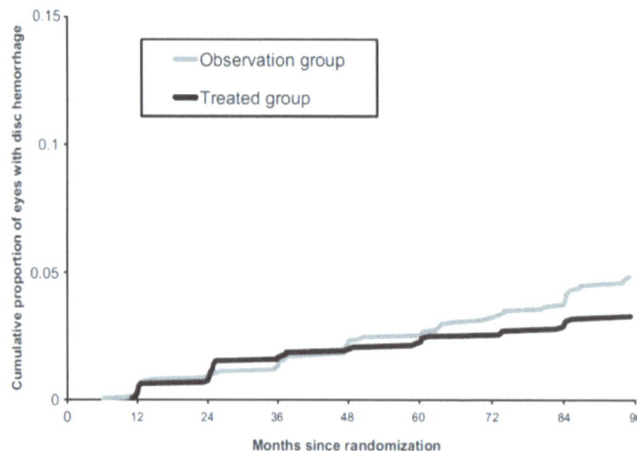

Figure 2. Graph showing cumulative incidence of optic disc hemorrhages in annual photographs of treated and observation groups.

Early Manifest Glaucoma Trial (EMGT):

The EMGT was a prospective, multicenter, randomized study of 44,243 patients screened for glaucoma. 255 had newly diagnosed OAG via loss on HVF. Half were randomized to "treatment" with ALT and Betaxolol or to "no treatment." The results showed a 50% reduction in progression of OAG with treatment, progression was reduced from 76% to 59% (P < 0.0001. n=227 completed the median follow up of 8 years).[106, 197] Notice that the treated group continued to have VF loss despite treatment and that this study did not measure or randomize CCT initially.

EMGT researchers found the same number of patients with DH on any of the live clinical exams as those found on photo review (140 versus 141). They did miss a significant number of DH on live clinical exams when of all the exams were compared for photo review (P< 0.0001). [198] The authors attributed this difference to the EMGT forms requiring the ophthalmologists to actively check off whether a DH was present or not. The OHTS did not have this on their exam sheet and this probably accounts for the marked difference between the two studies in the rate of DH found on exam. This supports my recommendation to have a mental list that you go through when examining a disk in a glaucoma patient. They also concluded that, while the presence of DH shortened the time to OAG progression, there was no evidence to support the idea that patients with a DH are being undertreated or are in need of a more intensive IOP lowering treatment.[198] The same number of patients in both the treated group and the non-treated group developed DH (P = 0.93).[198] Here is a graph showing the frequency of DH found on live exam versus photo review as well as the number of eyes that had multiple number of DHs. [198]

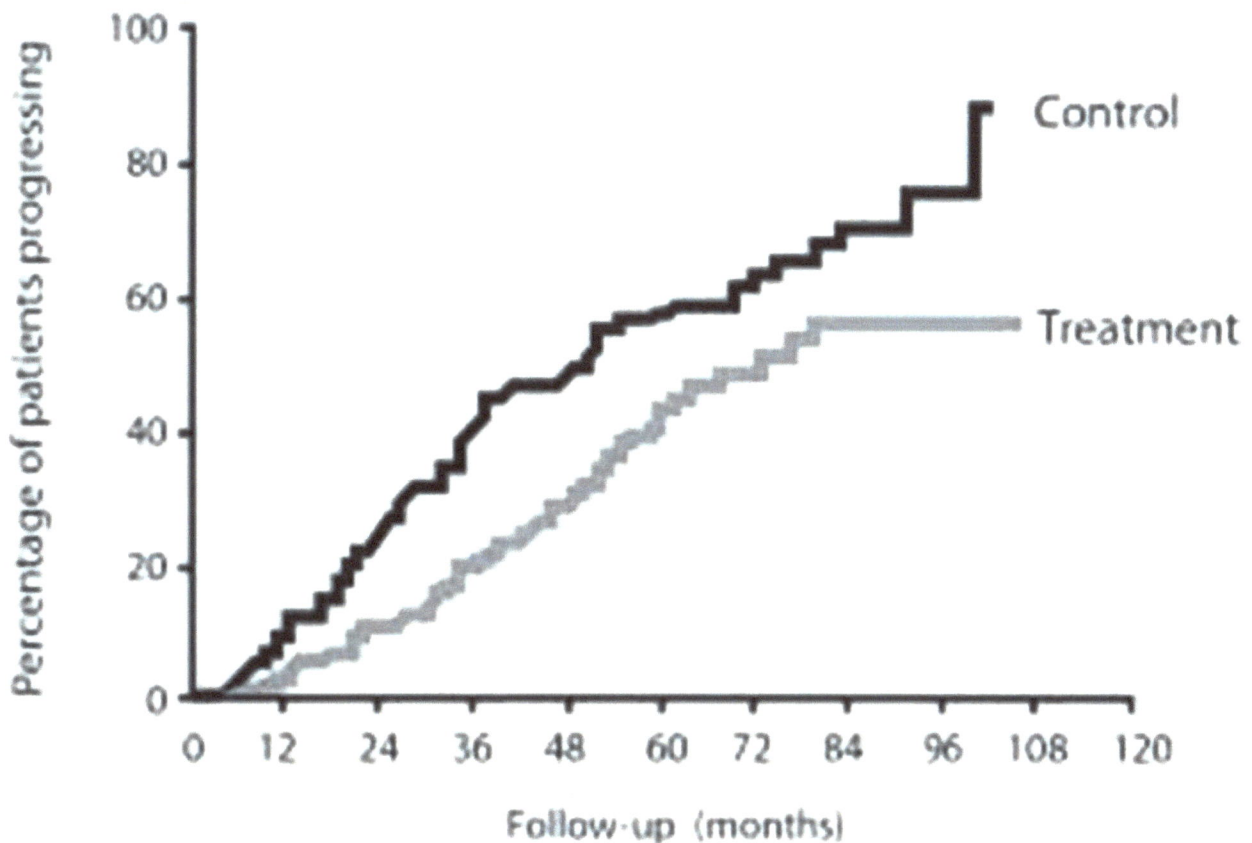

This graph is also from the Early Manifest Glaucoma Trial (EMGT). Notice that the prevalence of OAG increases with age, most eyes had mild glaucoma at the time of diagnosis in all age groups, and that the extent of VF loss is similar in all age groups from age 60 and older.[165]

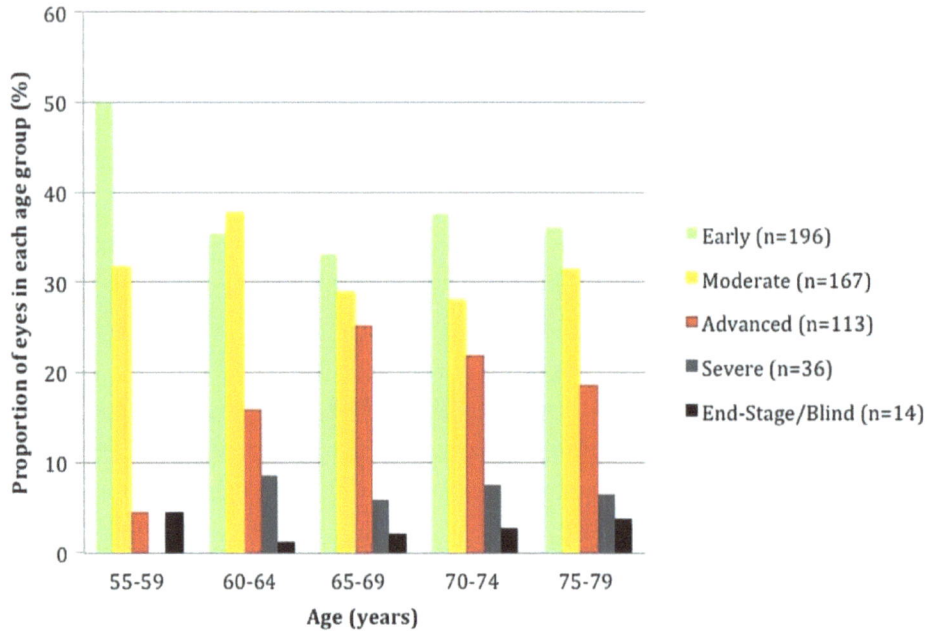

In his review of the untreated arm of the EMGT, Heijl identified 118 patients with untreated open angle glaucoma (OAG) followed for at least 6 years.[109] 46 patients had high tension (uncorrected for CCT) OAG patients, 57 had NTG and 15 had pseudoexfoliation. Normal tension glaucoma patients had a yearly dB MD loss rate of -0.36 dB/year, ocular hypertension patients had a yearly loss of -1.31 dB/year and pseudoexfoliation patients had a yearly loss of -3.13 dB/year.[109] So, glaucoma progression rates of VF loss vary a lot from one type of OAG to another.

Time to progression of glaucoma in patients with existing glaucoma for 3 different types of OAG (n = 118) in the "no treatment" of the EMGT cohort (3 graphs above); high tension glaucoma (HTG), normal tension (NTG) and pseudo exfoliation (PEXG) in the EMGT[109]

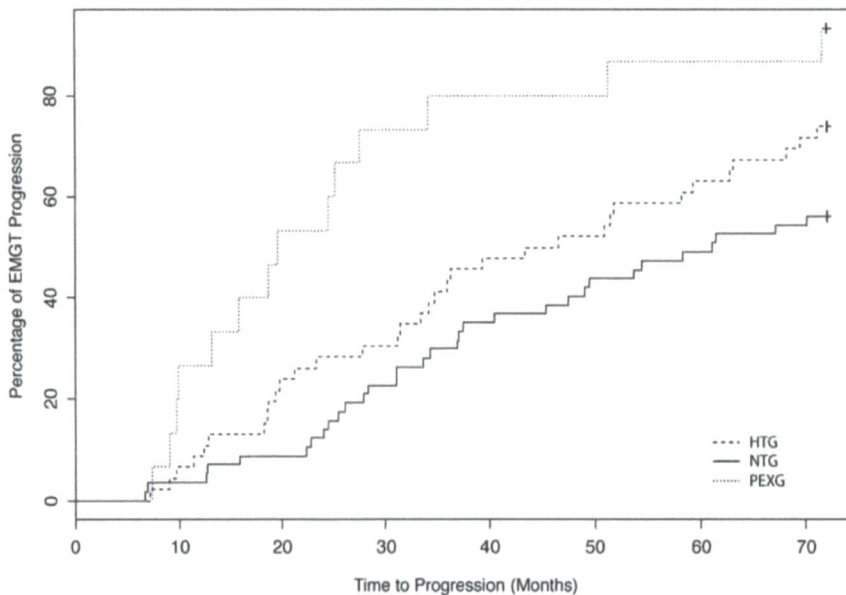

This graph shows the importance of preserving remaining rim tissue as damage becomes more severe:

- In early disease, dB function changes relatively slowly compared to structure

- In later disease, dB function changes relatively faster compared to structure

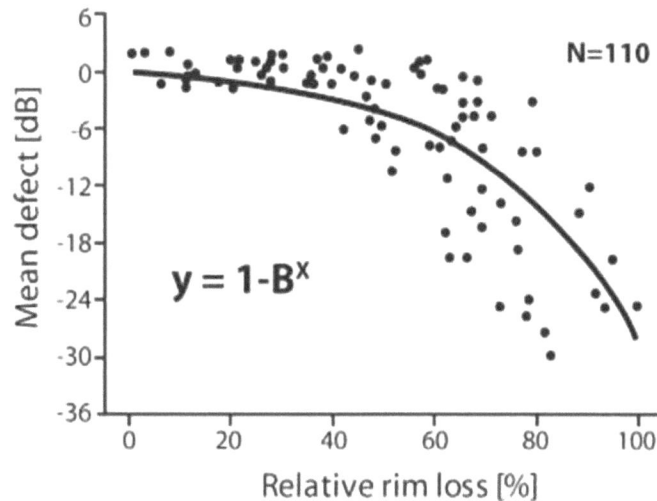

$$y = 1 - B^x$$

Graph: Mean defect [dB] (y-axis, from 6 to -36) vs Relative rim loss [%] (x-axis, 0 to 100). N=110.

The Olmstead study showed that the most important risk factor for the development of blindness was the presence of VF loss of optic nerve damage at the time of initial diagnosis of glaucoma[95] (60,666 patient's charts reviewed, n = 295 patients with OAG, mean follow-up of 15 years). This is probably related to patients being on the rapidly descending slope at the time of diagnosis (my words).

Normal Tension Glaucoma Study (NTGS):

A prospective, randomized, multicenter study that 140 eyes randomized to treatment or no treatment and followed patients for an average of 5.6 years. The "proportion surviving" is defined as patients that maintained no further damage (i.e. did not progress in glaucoma damage) after a number of covariates were factored out (P < 0.0001). The NTGS found reducing IOP by 30% with drops reduced the 5 year chances of developing glaucoma from 35% to 12% (P < 0.0001, n = 140 eyes, follow up was 7-8 years, again half received treatment, half did not). [96] So, lowering IOP works in normal tension glaucoma just as it does in high IOP glaucoma.

"In any individual patient, there is no characteristic abnormality of the optic disk or VF that distinguishes NTG from POAG with higher IOPs" .[54]

Prospective, randomized, multicenter study, the Normal Tension Glaucoma Study (NTGS) showing a reduced risk of further damage from NTG in patients already diagnosed with NTG (n=140 eyes)[108]

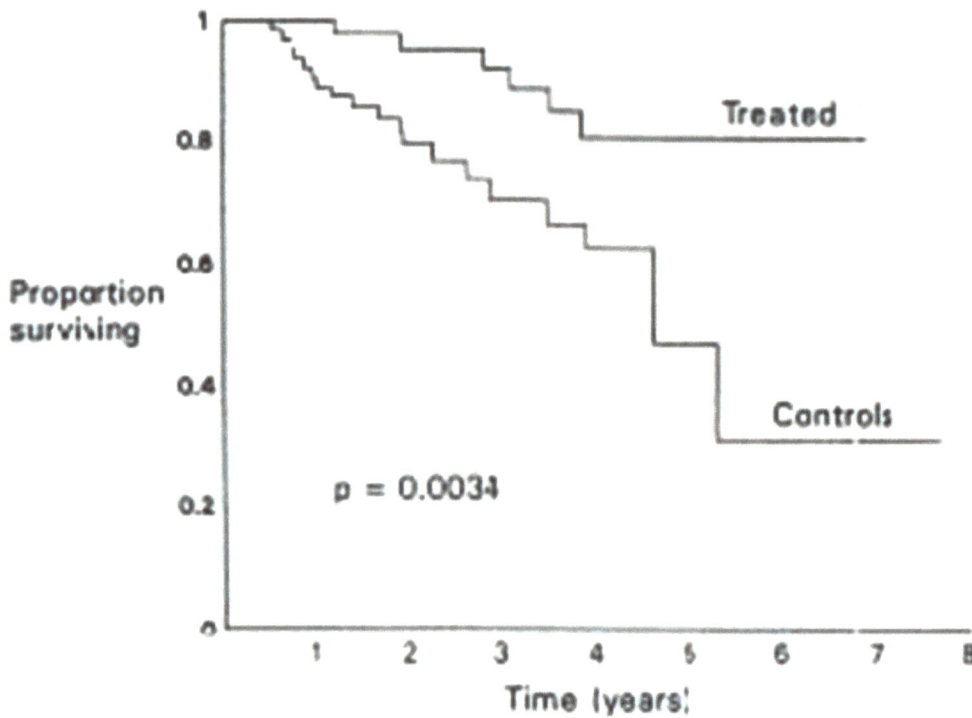

European Glaucoma Prevention Study:

The European Glaucoma Prevention Study[80] followed 1,077 people for a median of 59 months and showed a multivariate increased risk of 1.4X for the development of POAG in patients with a C/D asymmetry of > 0.4.

The European Glaucoma Prevention Study[99] graph below shows a scatter plot of CCT for 854 patients versus the age of the patients. It shows a very gradual thinning of the cornea as the age of the patient's increases but not an obvious trend. This is not the best design for a study to answer this question, but it's the best I've got.

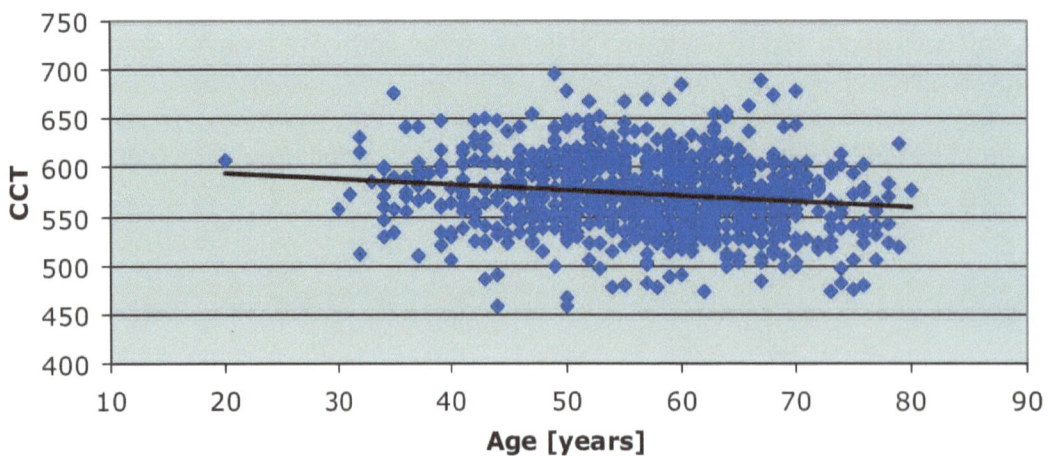

Blue Mountains Eye Study:

The Blue Mountains Eye Study in Australia examined 3,126 people, or 6,252 eyes, and found a statistically significant difference in developing OAG between patients with C/Ds < 0.3 and > 0.3. The > 0.3 group had (*prevalence* data) significantly more OAG than the other group (P< 0.0001).[81] The authors do state though that "the highest positive predictive factor for any asymmetric variable used as a test for OAG was only 17%, and this occurred at a cup-to-disk ratio asymmetry of 0.3 or more, which was found in only 10% of patients with OAG and 1% of patients without OAG."[81] So, they don't recommend diagnosing POAG on a C/D ratio asymmetry of > 0.3 but it certainly is another red flag. I use a difference of 0.3 or higher as an increased risk factor for POAG in my clinic.

Singapore Glaucoma Study:

Here is the *prevalence* of glaucoma for various C/D ratios from the Singapore study (1,232 people screened, n = 61 eyes having glaucoma):[73]

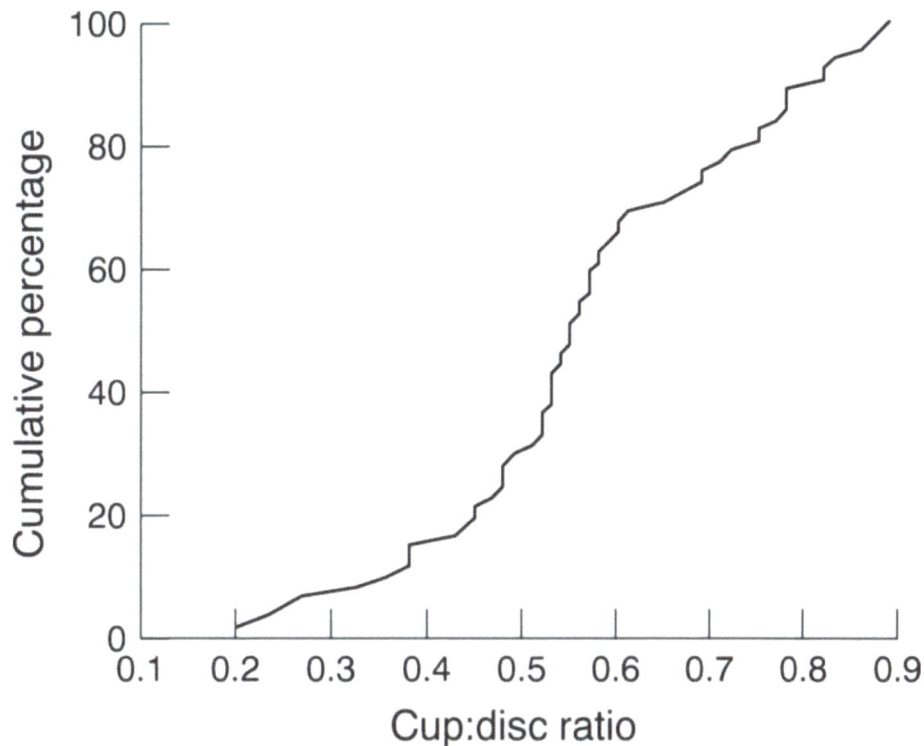

Alice Williams Glaucoma Study of IOP Asymmetry:

The following graph is from a *prevalence* study of patients initially diagnosed with glaucoma by Alice Williams[77] (A total of 652 patients were examined, half with newly diagnosed glaucoma and half without glaucoma) showing the chance of having glaucoma with varying pre-treatment IOP differences between the two eyes of the same patient. A difference of > 6mm confers a 57% chance of having glaucoma and was determined to be a "great risk" for the development of glaucoma.[77] This was a prevalence study, IOP was measured at only one point in time for each patient, so the time frame for OAG development was not able to be calculated.

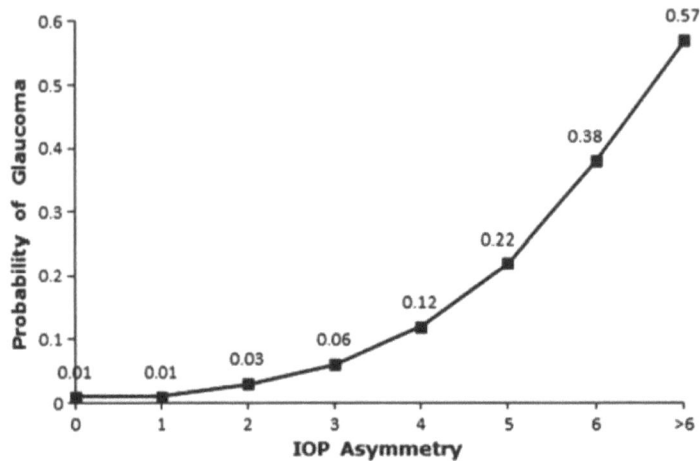

FIGURE 3. Probability of having glaucoma as a function of intraocular pressure (IOP) asymmetry between the fellow eyes. Probability of having glaucoma was calculated with a logistic regression analysis, assuming an overall worldwide prevalence of 2%.

Tube versus Trab Study:

Treatment of glaucoma in the Tube vs. Trabeculectomy Study (n=157, half had trab, half had a tube),[114] a prospective, randomized, multicenter study, 30% of the tube patients and 47% of the trab patients had a failure to control IOP after 5 years of follow-up (P= 0.002): Further, 50% of both groups lost two or more lines of BCVA once "other" causes of reduced vision related to the surgery (such as endophthalmitis and corneal edema) are factored in. And, this doesn't take into account the significant lifelong risk of endophthalmitis that exists after both of these surgeries. This is why we reserve glaucoma surgery as a method of last resort to lower IOP in the US. In Britain, with its socialized medicine, it is cheaper to do trabs as a first line therapy so they do a lot more glaucoma surgery. Recent new technologies such as Minimally Invasive Glaucoma Surgery (MIGS) promise to reduce complication rates but so far have not lowered IOP as well as trabs and tubes do.

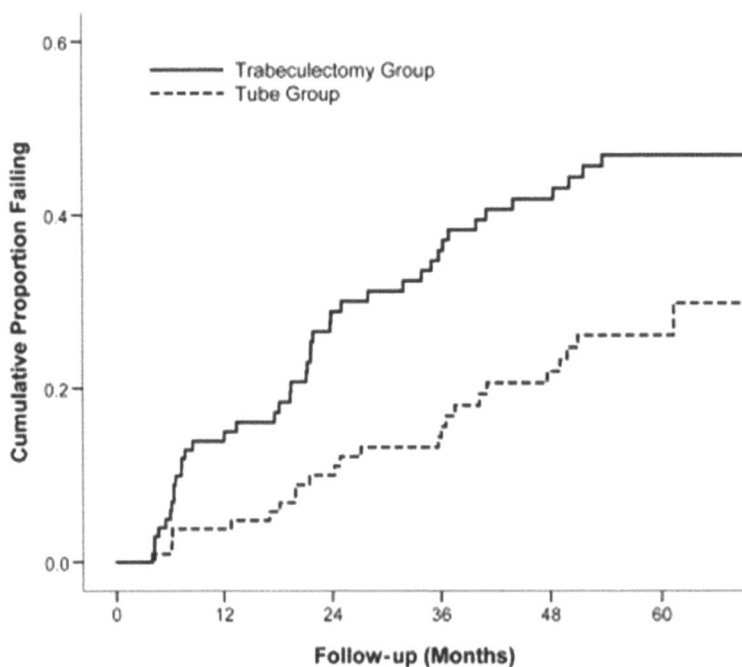

Advanced Glaucoma Intervention Study (AGIS):

This graph, from the prospective, randomized, multicenter Advanced Glaucoma Intervention Study (AGIS), plots the risk of worsening OAG over time between two groups of patients with uncontrolled OAG (N = 501 eyes of 401 patients total for both groups).[90] One group received treatment in the order of ALT-trab-trab and the other group received treatment in the order of trab-ALT-trab. For both groups, those patients with > 3mmHg showed a significant worsening of HVF (as noted as a higher AGIS score) over a mean follow up of about 7.5 years (p = 0.0006). It shows that IOP variation in OAG patients is statistically significantly correlated with worsening of OAG, despite treatment. The only significant difference (at < 0.1%) between the two groups after 8 years of follow-up on multivariate analysis was the slope of the rate of change, for reasons not identified.[163]

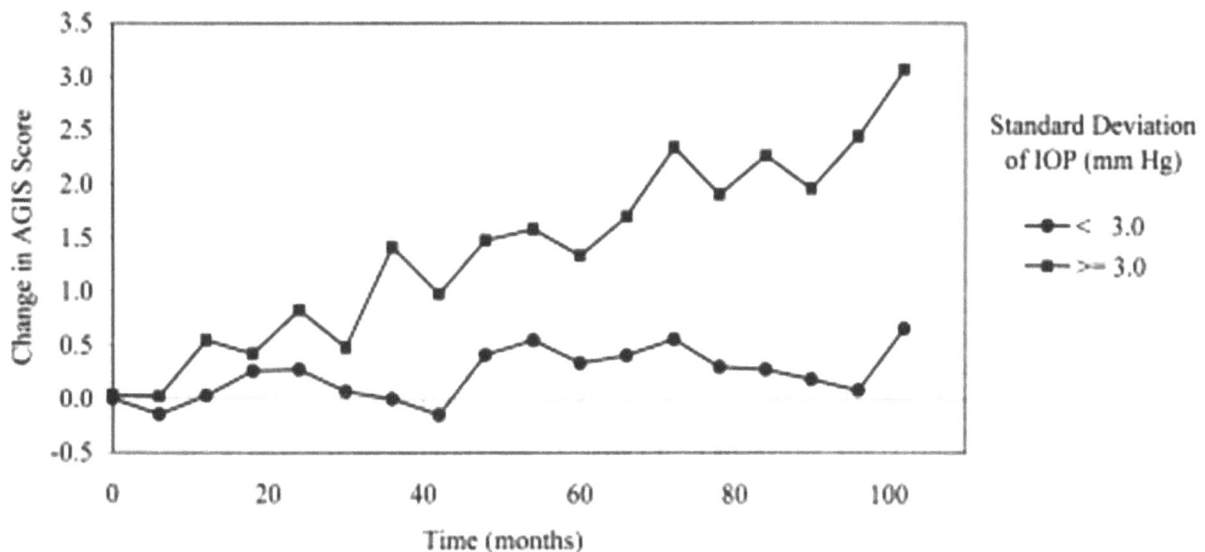

The Collaborative Initial Glaucoma Treatment Study:

The Collaborative Initial Glaucoma Treatment Study (n = 604 who finished the study, mean follow up was 7.2 years)[182] was designed to compare two different treatments for newly diagnosed OAG. Which two treatments are irrelevant for our discussion here. The study showed that just as many patients improved their VF result as those that worsened over a 5 year follow up period (p > .20). In those patients that had Improvements in their VF, improvements stopped after 5 years of treatment while those with worsening VF results continued to worsen (p < 0.0053). Here is a scatter plot of all 604 patients broken out by the VF mean deviation (MD) at the baseline exam versus the 5 year VF MD change over the 5 year period showing that no matter how significant the damage was at baseline, before diagnosis and treatment (as evidenced by a more negative MD value on the plot), there were patients above and below the substantial gain and loss lines. i.e., the effect of VF improvement and VF worsening existed for all ranges of baseline MD values over the 5 year period. [182] (There are asterisks in the substantial gain and loss area for all measurements of baseline dB). I don't have a p value comparing worse baseline MD with better baseline MD over the 5 years of follow up but the results are obvious on the graph. Up to a 3 db of change on the VF MD was not considered significant and only those VF results with higher and lower than a 3 db change over a 5 year period were considered "substantial". You can see that the bulk of patients neither improved nor worsened but rather stayed steady over the 5 year follow up period of this graph.

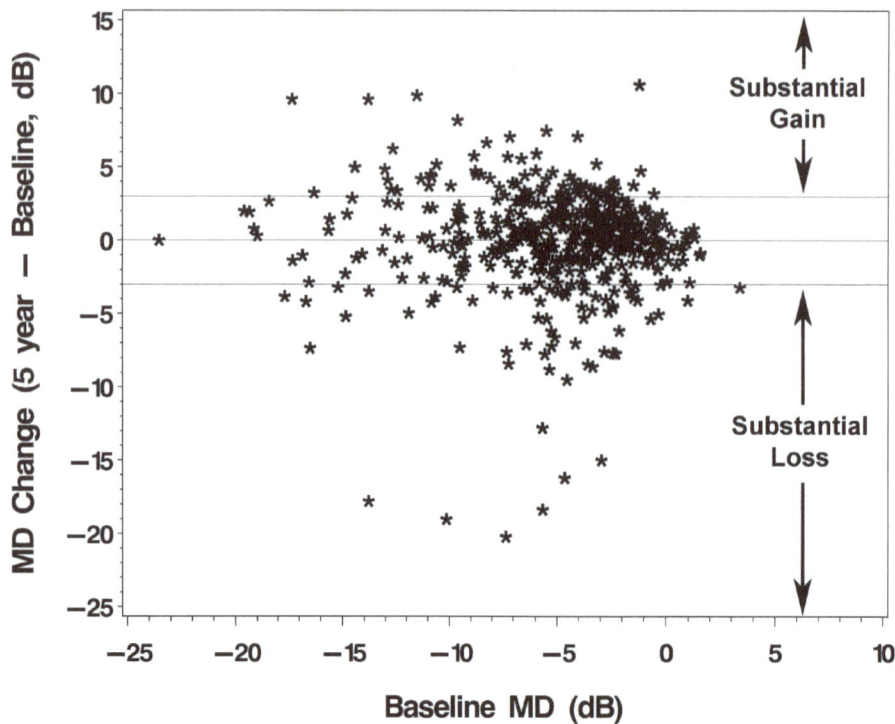

Glaucoma Laser Treatment Trial:

Selective Laser Trabeculoplasty (SLT): This is an in-office laser treatment that opens up the drainage meshwork so that the aqueous drains out of the TM better. SLT has been shown in the Glaucoma Laser Treatment Trial (GLT)[113], a prospective, randomized, multicenter study (n=203, median follow up 7 years) to be even more effective than drops in the initial treatment of OAG. If it does not work, drops are as effective after failed SLT as before SLT, so we have little to lose by trying SLT first. The success of SLT seems to be related to how much pigment is in the drainage meshwork, with more pigment increasing the chance for a response. Overall, 70% of patients will respond to SLT. The effect of SLT lasts for an average of 3-5 years and slowly wears off. SLT does reduce the need for drops during those 3-5 years and for many patients this makes non-compliance less of an issue. I treat 3600 initially. I have had some patients with IOP's above 40 respond to SLT with IOP going down to 15 within 1 week after SLT. This is an exceptional response, however, and is not typical. As you can tell, I am a fan of SLT!

<u>Retina</u>

Small Collaborative Ocular Melanoma Study (COMS):

The Small Collaborative Ocular Melanoma Study (COMS), a retrospective study of 240 patients with lesions 1.0 mm to 3.5 mm thick followed for an average of 3.3 years, showed that the risk of a lesion <2 mm in thickness showing <u>growth</u> over a mean follow up of 3.3 years was 2% (not 0%), and the risk of a lesion > 2 mm in thickness showing <u>growth</u> over a mean follow up of 3.3 years was 14%.[116] Documented growth is a real concern for conversion and deserves a referral to an oncology center.

The following graph shows the risk of growth versus the thickness (height) of the lesion being followed. It shows the higher risk for lesions > 2.5 mm thick in the small COMS.[116]

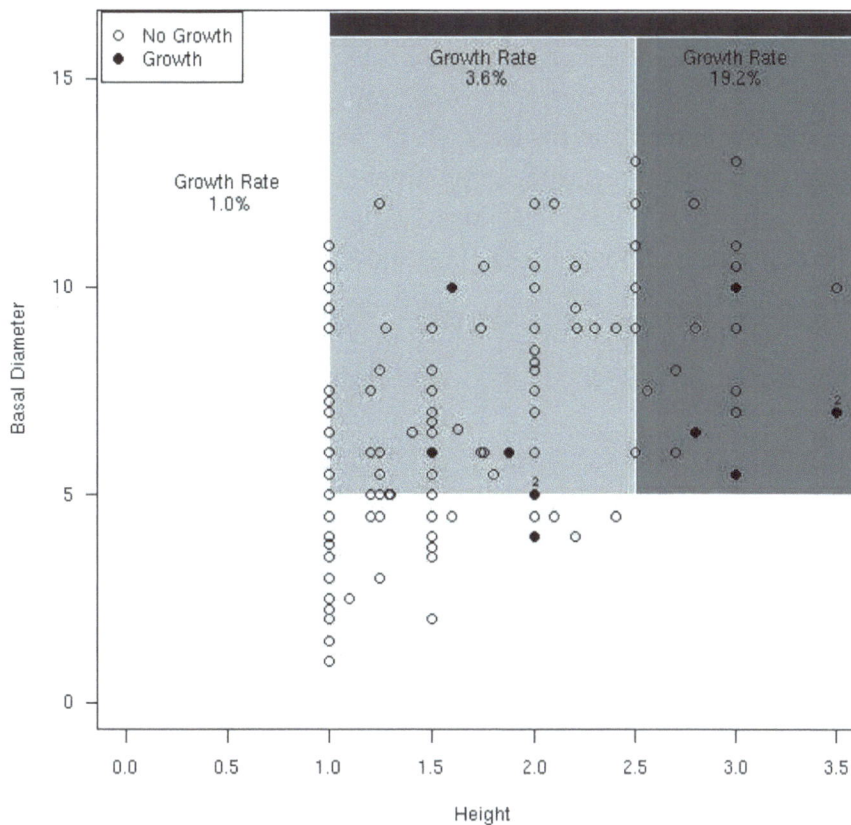

CRVO Study:

A retrospective study examined 492,488 patients for the onset of CRVO and BRVO, (mean follow up was about 2 years).[184,185] On a multivariate analysis, the risk factors for CRVO were, in descending order:

> hypercoagulable state
> hypertension
> POAG
> ARMD
> stroke but not diabetes [184,185,85]

In the same patient population, the risk factors for BRVO were in descending order:

> hypertension
> stroke, but not POAG or diabetes

Early Treatment Diabetic Retinopathy Research Group (ETDRS):

The ETDRS categorizes "high risk" NPDR with the 4-2-1 rule:[128]

> - diffuse intra-retinal bleeding (> 20 in total) in the 4 quadrants of the retina
> - venous beading in any 2 of the quadrants
> - IRMAs in any 1 quadrant

The ETDRS showed that having 1 of these 4-2-1 categories carries a 15% chance of developing "high risk" PDR, with high risk requiring PRP within 1 year. Having 2 categories carries a 45% risk of needing PRP within 1 year. Therefore, patients having 1 or more of the 4-2-1 categories should be considered for early PRP128 despite not meeting the requirements of the DRS. This is partially due to the risk of losing the patient to follow up and developing complications without treatment. Also, patients today are getting treatments earlier due to the development of anti-VEGF injections, OCT and micro-pulse lasers, which were all developed since the DRS/ETDRS studies were done. Also, OCT testing has supplanted IVFA in the diagnosis of diabetic retinopathy since an IVFA carries a small but significant risk of anaphylactic death from the injection of the dye into the veins. So, don't be too surprised if your patient does not get an IVFA before an FML or vitreal injection or seems to be getting anti-VEGF or laser treatments before the ETDRS/DRS would normally call for them.

The Diabetes Control and Complications Trial Research Group (DCCT):

The DCCT study[133] showed that tight control of a type 1 diabetic's blood sugar significantly reduced the complications from diabetes for the kidney, nerve, and eyes. The incidence of retinopathy increased slightly during the first 18 months and then decreased by 74% at a mean follow up of 6.5 years. There were also more severe hypoglycemic episodes associated with tight control as well but these complications were considered worth the overwhelming benefit to the eyes, kidneys and nerves. A similar result was shown in type 2 diabetics in a large European study.[134]

RISE and RIDE Studies:

Anti VEGF injections are changing the way CSME and PDR are being treated. In a study combining the RIDE and RISE data, two randomized, prospective, multi-center clinical studies, 759 patients with CSME and no PDR were randomized to monthly anti-VEGF injections of two different concentrations for 24 months versus a third group receiving saline ("sham") injections. At 24 months of treatment, the saline group was also injected with anti-VEGF medicine due to the obvious benefit of the treatment. 707 patients completed the 36 months of follow up. The risk of developing PDR is graphed below. © Elsevier. Anti-VEGF injections reduced the risk of development of PDR significantly (P < 0.0001). The risk of developing PDR in the saline group (sham) was reduced to the same rate as the treatment group after anti-VEGF treatments were started at the 25 month point (the vertical dotted line). Note how 36% of the sham group developed new onset PDR during this study and that at least 15% of the treated group still developed PDR at the end of 36 months of anti-VEGF injections. [193]

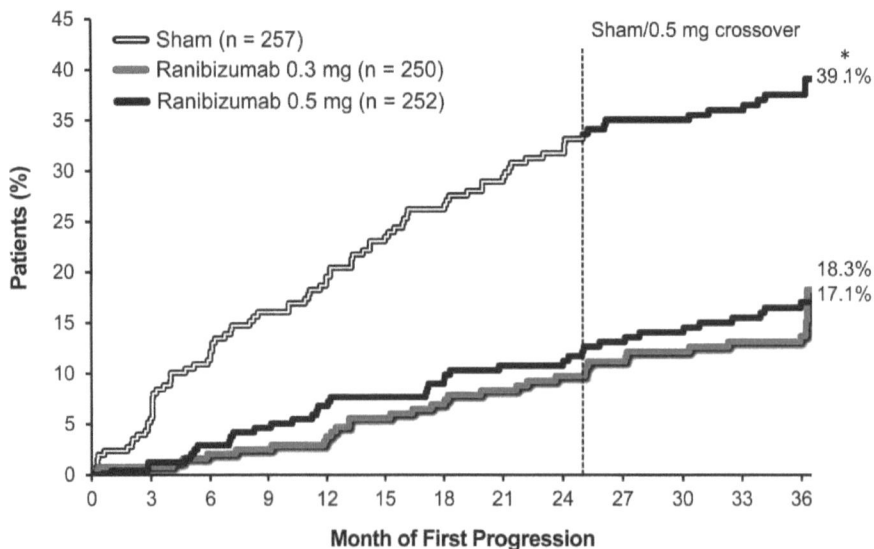

ANCHOR and MARINA:

Two prospective, randomized, blind, multicenter studies confirmed the efficacy of anti-VEGF treatments. The ANCHOR[200] study looked at classic wet ARMD (n = 420 who completed the 2 year follow up, p < 0.0001 better BCVA with treatment) and the MARINA study [201] looked at occult wet ARMD lesions (713 completed the 2 year follow up, p < 0.0001 better BCVA with treatment). [The difference between classic and occult wet ARMD have little meaning in the anti-VEGF period.

Due to the cost of the drug and clinic time involved in monthly injections (forever), various alternatives have been looked at for the frequency of long term anti-VEGF injections. A study of ABCHOR, MARINA and HORIZON patients looked at maintenance of vision improvement after 7 years of follow up (the SEVEN-UP study). 203 The frequency of anti-VEGF injections was dropped from monthly in the "initial study" period to an "as needed" basis. The SEVEN-UP study showed not only a loss of the BCVA gain seen during the initial 2 year Q monthly injections (seen at the 2 year mark in the graph below) but a worsening of BCVA from baseline after 7.3 years of follow up. There doesn't seem to be any way around it, Avastin or Leucentis anti-VEGF injections need to be given Q month or every 2 months forever. The new longer acting agents, Eylea are proving to extend this dosing regimen substantially.

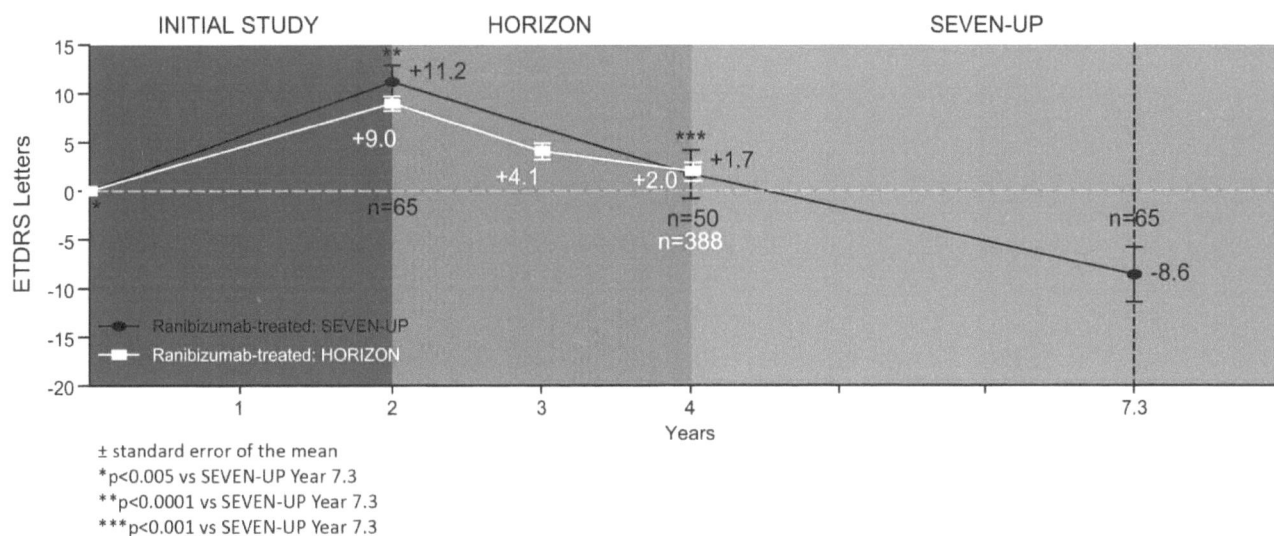

± standard error of the mean
*p<0.005 vs SEVEN-UP Year 7.3
**p<0.0001 vs SEVEN-UP Year 7.3
***p<0.001 vs SEVEN-UP Year 7.3

AREDS 2 and AREDS 1:

Your main job is to identify the degree of ARMD, place patients on Age Related Eye Disease 2 Study vitamins (AREDS 2), a prospective, randomized, multicenter study (n = 3,694, average follow up 5 years)[151], encourage cessation of smoking, start home Amsler grid use when appropriate, take a photo and refer out all the patients who might be getting close to developing wet ARMD. The AREDS 2 vitamins are in addition to a MVI taken every day. Lutein and Zeaxanthin were not included in the AREDS 1 study, a prospective, randomized, multicenter study (n = 3,597, average follow up 6.3 years)[149] as they were not commercially available at that time.[149] The zinc in the AREDS 1 formula increased hospital admissions for prostate issues and the vitamin A has been linked to lung cancer in smokers and ex-smokers. Because of these risks, the AREDS 2 study was done to see if substituting lutein and Zeaxanthin for vitamin A and reducing the zinc content would keep the efficacy in the ARMD formation. It did.[151] Again, substituting Lutein and Zeaxanthin and/or omega 3 long chain fatty acids (DHA and EPA) for vitamin A (beta carotene) did not reduce the effectiveness of AREDS 1 vitamins in reducing the chance of developing advanced ARMD, as shown in the AREDS 2 study.[151] I have intermediate patients RTC Q 6 months for DFE. The risk of progressing from intermediate to advanced ARMD in the AREDS 1 study was 18% over 5 years.[152]

Mini-TRUST:

The MIVI-TRUST study, a prospective, multi-center, double blind study with n = 652 eyes injected with Ocriplasmin or saline, showed that injecting a truncated form of the human serum protease, Plasmin significantly resolved VTR (P < 0.001) as well as closure of macular holes (P < 0.001). 195 The FDA has approved Ocriplasmin when 2 conditions are met. Symptoms consistent with vitreal-retinal traction (metamorphopsia, blurry vision, relative blind spot, etc.) and an OCT that shows VRT. A macular hole without VRT would not qualify for Ocriplasmin injections.

REFERENCES

1. Kline LB, Foroozan R, *Neuro-Ophthalmology Review Manual seventh edition*. New Jersey; SLACK, 2013, page 129.
2. Girkin CA, *Evaluation of Pupillary Light Responses as an Objective Measure of Visual Function*. Ophthalmology Clinics of North America 2003; 16:143-153. Reprinted from reference # 2, figure 3, 4, and 5 page 146, copyright 2003 with permission from Elsevier.
3. Kline LB, Foroozan R, *Neuro-Ophthalmology Review Manual seventh edition*. New Jersey; SLACK, 2013, page 125.
4. Kline LB, Foroozan R, *Neuro-Ophthalmology Review Manual seventh edition*. New Jersey; SLACK, 2013, page 135.
5. Kline LB, Foroozan R, *Neuro-Ophthalmology Review Manual seventh edition*. New Jersey; SLACK, 2013, page 126.
6. Kline LB, Foroozan R, *Neuro-Ophthalmology Review Manual seventh edition*. New Jersey; SLACK, 2013, page 131.
7. Martinez-Thompson JM, Diehl NN, et al. *Incidence, Types and Lifetime Risk of Adult-Onset Strabismus*. Ophthalmology 2014; 121: 877-882. Reprinted from this reference #7, figure 1 page 879, and figure 2 page 880, copyright 2014 with permission from Elsevier.
8. Kline LB, Foroozan R, *Neuro-Ophthalmology Review Manual seventh edition*. New Jersey; SLACK, 2013, page 69.
9. Levin LA, *Relevance of the Site of Injury of Glaucoma to Neuroprotective Strategies*. Survey of Ophthalmology 2001; 45: S243-S249. Reprinted from this reference #9, figure 6 page S246, copyright 2001 with permission from Elsevier.
10. Jonas JB, Muller-Bergh JA, et al. *Histomorphometry of the Human Optic Nerve*. Invest Ophthalmology Vis Sci 1990; 31: 736-744.
11. Kline LB, Foroozan R, *Neuro-Ophthalmology Review Manual seventh edition*. New Jersey; SLACK, 2013, page 10.
12. Kline LB, Foroozan R, *Neuro-Ophthalmology Review Manual seventh edition*. New Jersey; SLACK, 2013, page 9.
13. Am Academy of Ophthalmology, *Basic Clinical Science Course Glaucoma*. California; AAO, 2010, pages 69-71.
14. Heijl A, Patella VM, *Essential Perimetry Third Edition*. California, Zeiss, 2002, page70.
15. Heijl A, Patella VM, *Essential Perimetry Third Edition*. California, Zeiss, 2002, page 5.
16. Kline LB, Foroozan R, *Neuro-Ophthalmology Review Manual seventh edition*. New Jersey; SLACK, 2013, page 8-9.
17. Meira-Freitas D, Tatham A, et al. *Predicting Progression of Glaucoma from Rates of Frequency Doubling Technology Perimetry Change*. Ophthalmology 2014; 121: 498-507.
18. Ketlner JL, Johnson CA, et al. *Confirmation of Visual Field Abnormalities in the Ocular Hypertensive Treatment Study*. Arch Ophthalmology 2000; 118: 1187-1194.
19. Heijl A, Patella VM, *Essential Perimetry Third Edition*. California, Zeiss, 2002, page 85.
20. Heijl A, Patella VM, *Essential Perimetry Third Edition*. California, Zeiss, 2002, page 76.
21. Heijl A, Patella VM, *Essential Perimetry Third Edition*. California, Zeiss, 2002, page 11.
22. Heijl A, Patella VM, *Essential Perimetry Third Edition*. California, Zeiss, 2002, page 87.

23. Am Academy of Ophthalmology, *Basic Clinical Science Course Glaucoma*. California; AAO, 2010, page 61.

24. Spry PGD, Johnson CA. *Assessing Visual Function in Clinical Practice*. Ophthalmology Clinics of N. America 2000; 13: 361-382. Reprinted from this reference #24, figure 4 page 364, copyright 2000 with permission from Elsevier.

25. Garway-Heath DF, Poinoosawmy D, et al. *Mapping the Visual Field to the Optic Disc in Normal Tension Glaucoma Eyes*. Ophthalmology 2000; 107: 1809-1815. Reprinted from this reference #25, figure 2 page 1810 and figure #7 page 1813, copyright 2000, with permission from Elsevier.

26. Heijl A, Patella VM, Essen*tial Perimetry Third Edition*. California, Zeiss, 2002, page 17- 18.

27. Heijl A, Patella VM, *Essential Perimetry Third Edition*. California, Zeiss, 2002, page 28-29 and 66.

28. Heijl A, Patella VM, *Essential Perimetry Third Edition*. California, Zeiss, 2002, page 7.

29. Heijl A, Patella VM, *Essential Perimetry Third Edition*. California, Zeiss, 2002, page 13.

30. Heijl A, Patella VM, *Essential Perimetry Third Edition*. California, Zeiss, 2002, page 47.

31. Katz J, Sommer A, et al. *Comparison of analytic Algorithms for Detecting Glaucomatous Visual Field Loss*. Arch Ophthalmology 1991; 109: 1684-1689.

32. Heijl A, Patella VM, *Essential Perimetry Third Edition*. California, Zeiss, 2002, page 10.

33. Heijl A, Patella VM, *Essential Perimetry Third Edition*. California, Zeiss, 2002, page 54.

34. Heijl A, Patella VM, *Essential Perimetry Third Edition*. California, Zeiss, 2002, page 58.

35. Am Academy of Ophthalmology, *Basic Clinical Science Course Glaucoma*. California; AAO, 2010, page 47.

36. Am Academy of Ophthalmology, *Basic Clinical Science Course Glaucoma*. California; AAO, 2010, page 44.

37. Am Academy of Ophthalmology, *Basic Clinical Science Course Glaucoma*. California; AAO, 2010, page 45.

38. Am Academy of Ophthalmology, *Basic Clinical Science Course Glaucoma*. California; AAO, 2010, page 109.

39. Scott A, Kotecha A, et al. *YAG Laser Peripheral Iridotomy for the Prevention of Pigmentary Dispersion Glaucoma*. Ophthalmology 2011; 118: 468-473.

40. Siddiqui Y, Ten Hulzen RD, et al. *What is the risk of developing Pigmentary Glaucoma from Pigment Dispersion Syndrome?*. Am J Ophthalmology 2003; 135: 794-799.

41. Am Academy of Ophthalmology, *Basic Clinical Science Course Glaucoma*. California; AAO, 2010, page 87.

42. Am Academy of Ophthalmology, *Basic Clinical Science Course Glaucoma*. California; AAO, 2010, page 103.

43. Gesternblith AT, Rabinowitz MP. *The Wills Eye Manual, 6th edition*. Pennsylvania; Lippincott, 2012, page 221.

44. Gesternblith AT, Rabinowitz MP. *The Wills Eye Manual, 6th edition*. Pennsylvania; Lippincott, 2012, page 217.

45. Am Academy of Ophthalmology, *Basic Clinical Science Course Glaucoma*. California; AAO, 2010, page 10.

46. Gesternblith AT, Rabinowitz MP. *The Wills Eye Manual, 6th edition*. Pennsylvania; Lippincott, 2012, page 216.

47. Gesternblith AT, Rabinowitz MP. *The Wills Eye Manual, 6th edition*. Pennsylvania; Lippincott, 2012, page 215.

48. Am Academy of Ophthalmology, *Basic Clinical Science Course Glaucoma*. California; AAO, 2010, page 127.

49. Gesternblith AT, Rabinowitz MP. *The Wills Eye Manual, 6th edition*. Pennsylvania; Lippincott, 2012, page 233.

50. Gesternblith AT, Rabinowitz MP. *The Wills Eye Manual, 6th edition*. Pennsylvania; Lippincott, 2012, page 236.

51. Kitazawa Y, Shirato S, et al. *Optic Disk Hemorrhage in Low Tension Glaucoma*. Ophthalmology 1986; 93: 853-857.

52. Am Academy of Ophthalmology, *Basic Clinical Science Course Glaucoma*. California; AAO, 2010, page 110.

53. Am Academy of Ophthalmology, *Basic Clinical Science Course Glaucoma*. California; AAO, 2010, page 23.

54. Am Academy of Ophthalmology, *Basic Clinical Science Course Glaucoma*. California; AAO, 2010, page 79.

55. Francis BA, Varma R, et al. *Intraocular Pressure, Central Corneal Thickness, and Prevalence of Open Angle Glaucoma: The Los Angles Latino Eye Study*, Am J Ophthalmology 2008; 146: 741-746. Reprinted from this reference #55, figure 2 page 743, and figure 1 page 743, copyright 2008 with permission from Elsevier.

56. Nemesure B, Honkanen R, et al. *Incident Open-Angle Glaucoma and Intraocular Pressure*. Ophthalmology 2007; 114: 1810-1815. Reprinted from this reference #56, figure 2 page 1813, copyright 2007 with permission from Elsevier.

57. Jiang X, Varma R, et al. *Baseline Risk Factors that Predict the Development of Open-Angle Glaucoma in a Population*. Ophthalmology 2012; 119: 2245-2253. Reprinted from this reference #57, figure 1, page 2249, copyright 2012 with permission from Elsevier.

58. Brandt JD, Beiser JA, et al. *Central Corneal Thickness in the Ocular Hypertension Treatment Study (OHTS)*. Ophthalmology 2001; 108: 1779-1788.

59. Gesternblith AT, Rabinowitz MP. *The Wills Eye Manual, 6th edition*. Pennsylvania; Lippincott, 2012, page 211.

60. Collaborative Normal Tension Glaucoma Study Group. *Comparison of Glaucomatous Progression between Untreated Patients with Normal Tension Glaucoma and Patients with Therapeutically Reduced Intraocular Pressures*. Am J Ophthalmology 1998; 126: 487-497.

61. Jiang X, Varma R, et al. *Baseline Risk Factors that Predict the Development of Open-Angle Glaucoma in a Population*. Ophthalmology 2012; 119: 2245-2253.

62. Gordon MO, Beiser JA, et al. *The Ocular Hypertension Treatment Study: Baseline Factors that Predict the Onset of Primary Open-Angle Glaucoma*. Arch Ophthalmology 2002; 120: 714-720, table3.

63. Am Academy of Ophthalmology, *Basic Clinical Science Course Glaucoma*. California; AAO, 2010, page 25.

64. Medeiros FA, Weinreb RN. *Is Corneal Thickness an Independent Risk Factor for Glaucoma?* Ophthalmology 2012; 119: 435-436.

65. Kass MA, Heuer DK, et al. *The Ocular Hypertension Treatment Study*. Arch Ophthalmology 2002; 120: 701-713.

66. Pakravan M, Yazdani S, et al. *A Population Survey of the Prevalence and Types of Glaucoma in Central Iran*. Ophthalmology 2013; 120: 1977-1984. Reprinted from this reference #66, figure 2 page 1981, copyright 2013 with permission from Elsevier.

67. Czudowska MA, Ramdas WD, et al. *Incidence of Glaucomatous Visual Field Loss: A Ten Year Follow-up from the Rotterdam Study*. Ophthalmology 2010; 117: 1705-1712. Reprinted from this reference #67, figure 2 page 1712, copyright 2010 with permission from Elsevier.

68. Schoff EO, Hattenhauer MG, et al. *Estimated Incidence of Open-angle Glaucoma in Olmsted County, Minnesota*. Ophthalmology 2001; 108: 882-886. Reprinted from this reference #68, figure 1 page 884, copyright 2001 with permission from Elsevier.

69. Am Academy of Ophthalmology, *Basic Clinical Science Course Glaucoma*. California; AAO, 2010, page 84.

70. Gordon MO, Beiser J, et al. *The Ocular Hypertensive Treatment Study*. Arch Ophthalmology 2002; 120: 714-720.

71. Leske MC, Wu SY, et al. *Risk Factors for Incident Open-Angle Glaucoma*. Ophthalmology 2008; 115: 85-93.

72. Am Academy of Ophthalmology, *Basic Clinical Science Course Glaucoma*. California; AAO, 2010, page 46.

73. Foster PJ, Buhrmann R, et al. *The Definition and Classification of Glaucoma in Prevalence Surveys*. British J Ophthalmology 2002; 86: 238-242. Reprinted from this reference #73, figure 1 page 239, copyright 2002 with permission from BMJ Publishing Group Ltd..

74. Budenz DL, Anderson DR, et al. *Detection and Prognostic Significance of Optic Disc Hemorrhages during the Ocular Hypertension Treatment Study*. Ophthalmology 2006; 113: 2137-2143. Reprinted from this reference #74, figure 2 page 2139, copyright 2006 with permission from Elsevier.

75. Gesternblith AT, Rabinowitz MP. *The Wills Eye Manual, 6th edition*. Pennsylvania; Lippincott, 2012, page 208.

76. Am Academy of Ophthalmology, *Basic Clinical Science Course Glaucoma*. California; AAO, 2010, page 48.

77. Williams AL, Gatla S, et al. *The Value of Intraocular Pressure Asymmetry in diagnosing Glaucoma*. J Glaucoma 2013; 22: 215-218. Reprinted this reference #77, figure 3 page 217, copyright 2013 with permission from Lippincott Williams and Wilkens/Wolters Kluwer Health. Promotional and commercial use of the material in print, digital or mobile device format is prohibited without the permission of the publisher Lippincott Williams and Wilkins. Please contact journalpermissions@lww.com for further information.

78. Levine RA, Demirel S, et al. *Asymmetries and Visual Field Summaries as Predictors of Glaucoma in the Ocular Hypertension Treatment Study*. Invest Ophthalmology Vis Sci 2006; 47: 3896-3903.

79. Gesternblith AT, Rabinowitz MP. *The Wills Eye Manual, 6th edition*. Pennsylvania; Lippincott, 2012, page 204.

80. European Glaucoma Prevention Group. *Predictive Factors for Open-Angle Glaucoma among Patients with Ocular Hypertension in the European Glaucoma Prevention Study*. Ophthalmology 2007; 114: 3-9.

81. Ong LS, Mitchell P, et al. *Asymmetry in Optic Disc Parameters: The Blue Mountains Eye Study*. Invest Ophthalmology Vis Sci 1999; 40: 849-857.

82. Am Academy of Ophthalmology, *Basic Clinical Science Course Glaucoma*. California; AAO, 2010, page 8.

83. Am Academy of Ophthalmology, *Basic Clinical Science Course Glaucoma*. California; AAO, 2010, page 78.

84. OHTS and EGPS groups. *Validated Prediction Model for the Development of Primary Open-Angle Glaucoma in Individuals with Ocular Hypertension*. Ophthalmology 2007; 114: 10-19.

85. Am Academy of Ophthalmology, *Basic Clinical Science Course Retina*. California; AAO, 2013, page 129.

86. Am Academy of Ophthalmology, *Basic Clinical Science Course Retina*. California; AAO, 2013, page 128.

87. Am Academy of Ophthalmology, *Basic Clinical Science Course Glaucoma*. California; AAO, 2010, page 76 and 77.

88. Am Academy of Ophthalmology, *Basic Clinical Science Course Glaucoma*. California; AAO, 2010, page 22.

89. Liu JHK, Sit AJ, et al. *Variation of 24-hour Intraocular Pressure in Healthy Individuals: Right Eye Versus Left Eye*. Ophthalmology 2005; 112: 1670-1675. Reprinted from this reference #89, figure 1 and 2 page 1671, copyright 2005 with permission from Elsevier.

90. Nouri-Mahdavi K, Hoffman D, et al. *Predictive Factors for Glaucomatous Visual Field Progression in the Advanced Glaucoma Intervention Study*. Ophthalmology 2004; 111: 1627-1635. Reprinted from this reference #90, figure 1, page 1635, copyright 2004 with permission from Elsevier.

91. Medeiros FA, Weinrib RN, et al. *Long-Term Intraocular Pressure Fluctuations and Risk of Conversion from Ocular Hypertension to Glaucoma*. Ophthalmology 2008; 115: 934-940. Reprinted from this reference #91, figure 3, page 938, copyright 2008 with permission from Elsevier.

92. Quigley HA, Dunkelberger GR, et al. *Retinal Ganglion Cell Atrophy Correlated with Automated Perimetry in Human Eyes with Glaucoma*. Am J Ophthalmology 1989; 107: 453-464.

93. Quigley HA, Addicks EM, et al. *Optic Nerve Damage in Human Glaucoma*. Arch Ophthalmology 1982; 100: 135-146.

94. Caprioli J, Garway-Heath DF, et al. *A Critical Reevaluation of Current Glaucoma Management*. Ophthalmology 2007; 114: S1-S41. Reprinted from this reference #94, figure 8 page S8, copyright 2007 with permission from Elsevier.

95. Hattenhauer MG, Johnson DH, et al. *The Probability of Blindness from Open-Angle Glaucoma*. Ophthalmology 1998; 105: 2099-2104.

96. Collaborative Normal Tension Glaucoma Study Group. *Comparison of Glaucomatous Progression between Untreated Patients with Normal-Tension Glaucoma and Patients with Therapeutically Reduced Intraocular Pressure*. Am J Ophthalmology 1998; 126: 487-497.

97. Am Academy of Ophthalmology, *Basic Clinical Science Course Glaucoma*. California; AAO, 2010, page 80.

98. Keltner JL, Johnson CA, et al. *Confirmation of Visual Field Abnormalities in the Ocular Hypertension Treatment Study*. Arch Ophthalmology 2000; 118: 1187-1194.

99. European Glaucoma Prevention Study Group. *Central Corneal Thickness in the European Glaucoma Prevention Study*. Ophthalmology 2007; 114: 454-459. Reprinted from this reference #99, figure 3 page 457, copyright 2007 with permission from Elsevier.

100. Brandt JD, Gordon MO, et al. *Adjusting Intraocular Pressure for Central Corneal Thickness does not Improve Prediction Models for Primary Open-Angle Glaucoma*. Ophthalmology 2012; 119: 437-442.

101. Gesternblith AT, Rabinowitz MP. *The Wills Eye Manual, 6th edition*. Pennsylvania; Lippincott, 2012, page 443.

102. Am Academy of Ophthalmology, *Basic Clinical Science Course Glaucoma*. California; AAO, 2010, page 38.

103. Am Academy of Ophthalmology, *Basic Clinical Science Course Glaucoma*. California; AAO, 2010, page 118.

104. Am Academy of Ophthalmology, *Basic Clinical Science Course Glaucoma*. California; AAO, 2010, page 30 and 34.

105. Lin SC, Singh K, et al. *Optic Nerve Head and Retinal Nerve Fiber Layer Analysis*. Ophthalmology 2007; 114: 1937-1949.

106. Heijl A, Leske MC, et al. *Reduction of Intraocular Pressure and Glaucoma Progression, Results from the Early Manifest Glaucoma Trial*. Arch Ophthalmology 2002; 120: 1268-1279. Reprinted from this

reference #106, figure 2, page 1272, copyright © 2002 American Medical Association. All rights reserved.

107. Kass MA, Gordon MO, et al. *Delaying Treatment of Ocular Hypertension, the Ocular Hypertension Treatment Study.* Arch Ophthalmology 2010; 128: 276-287. Reprinted from this reference #107, figure 4 page 281, copyright 2010 with permission from Elsevier.

108. Collaborative Normal Tension Glaucoma Study Group. *Comparison of Glaucomatous Progression between Untreated Patients with Normal-Tension Glaucoma and Patients with Therapeutically Reduced Intraocular Pressure.* Am J Ophthalmology 1998; 126: 487-497. Reprinted from this reference #108, figure 4, page 492, copyright © 1998 American Medical Association. All rights reserved.

109. Heijl A, Bengtsson B, et al. *Natural History of Open-Angle Glaucoma.* Ophthalmology 2009; 116: 2271-2276. Reprinted from this reference #109, figure 2 page 2274, copyright 2009 with permission from Elsevier.

110. Am Academy of Ophthalmology, *Basic Clinical Science Course Glaucoma.* California; AAO, 2010, page 160.

111. Gesternblith AT, Rabinowitz MP. *The Wills Eye Manual, 6th edition.* Pennsylvania; Lippincott, 2012, page 209.

112. Am Academy of Ophthalmology, *Basic Clinical Science Course Glaucoma.* California; AAO 2010, page 169.

113. Glaucoma Laser Trial Research Group. *The Glaucoma Laser Trial (GLT) and Glaucoma Laser Follow-up Study, 7.* Am J Ophthalmology 1995; 120: 718-731.

114. Gedde SJ, Schiffman JC, et al. *Treatment Outcomes in the Tube Versus Trabeculectomy (TVT) Study after Five Years of Follow up.* Am J Ophthalmology 2012; 153: 789-803. Reprinted from this reference #114, figure 3 page 795, copyright 2012 with permission from Elsevier.

115. Am Academy of Ophthalmology, *Basic Clinical Science Course Ophthalmic Pathology and Intraocular Tumors.* California, AAO 2014, page 266.

116. Singh AD, Mokashi AA, et al. *Small Choroidal Melanocytic Lesions, Features Predictive of Growth.* Ophthalmology 2006; 113: 1032-1039. Reprinted from this reference #116, figure 3 page 1036, copyright 2006 with permission from Elsevier.

117. Shields CL, Shields JA, et al. *Risk Factors for Growth and Metastasis of Small Choroidal Melanocytic Lesions.* Ophthalmology 1995; 102: 1351-1361.

118. Singh AD, Turell ME, et al. *Uveal Melanoma: Trends in Incidence, Treatment, and Survival.* Ophthalmology 2011; 1881-1885.

119. Gesternblith AT, Rabinowitz MP. *The Wills Eye Manual, 6th edition.* Pennsylvania; Lippincott, 2012, page 354.

120. Am Academy of Ophthalmology, *Basic Clinical Science Course Retina.* California; AAO, 2014, page 255 and 282.

121. Marmor MF, Kellner U, et al. *Revised Recommendations on Screening for Chloroquine and Hydroxychloroquine Retinopathy.* Ophthalmology 2011; 118: 415-422.

122. Am Academy of Ophthalmology, *Basic Clinical Science Course Retina.* California; AAO, 2013, page 267.

123. Gottsch JD, Schein OD. *Taking TASS to task.* Ophthalmology 2012; 119: 1295-1296.

124. Eydelman MB, Tarver ME, et al. *The Food and Drug Administration's Proactive Toxic Anterior Segment Syndrome Program.* Ophthalmology 2012; 119: 1297-1302.

125. Am Academy of Ophthalmology, *Basic Clinical Science Course Retina.* California; AAO, 2013, page 274- 276.

126. Am Academy of Ophthalmology, *Basic Clinical Science Course Retina*. California; AAO, 2013, page 95.

127. Early Treatment Diabetic Retinopathy Research Group. *Early Photocoagulation for Diabetic Retinopathy, ETDRS Report Number 9.* Ophthalmology 1991; 98: 766-785.

128. Am Academy of Ophthalmology, *Basic Clinical Science Course Retina*. California; AAO, 2013, page 104.

129. Gesternblith AT, Rabinowitz MP. *The Wills Eye Manual, 6th edition*. Pennsylvania; Lippincott, 2012, page 315.

130. Am Academy of Ophthalmology, *Basic Clinical Science Course Retina*. California; AAO, 2013, page 89.

131. Klein R, Klein BEK, et al. *The Wisconsin Epidemiologic Study of Diabetic Retinopathy*. Arch Ophthalmology 1984; 102: 520-532.

132. Am Academy of Ophthalmology, *Basic Clinical Science Course Retina*. California; AAO, 2013, page 90.

133. The Diabetes Control and Complications Trial Research Group (DCCT). *The Effect of Intensive Treatment of Diabetes on the Development and Progression of Long-Term Complications in Insulin-Dependent Diabetes Mellitus*. The New England Journal of Medicine 1993; 329: 977-986.

134. UK Prospective Diabetes Study Group. *Intensive Blood-Glucose Control with Sulphonylureas or Insulin Compared with Conventional Treatment and Risk of Complications in Patients with Type 2 Diabetes*. Lancet 1998; 352: 837-853.

135. The Diabetic Retinopathy Research Group. *Four Risk Factors for Severe Visual Loss in Diabetic Retinopathy*. Arch Ophthalmology 1979; 97: 654-655.

136. Gesternblith AT, Rabinowitz MP. *The Wills Eye Manual, 6th edition*. Pennsylvania; Lippincott, 2012, page 311 to 313.

137. Gesternblith AT, Rabinowitz MP. *The Wills Eye Manual, 6th edition*. Pennsylvania; Lippincott, 2012, page 374.

138. Gesternblith AT, Rabinowitz MP. *The Wills Eye Manual, 6th edition*. Pennsylvania; Lippincott, 2012, page 373.

139. Gesternblith AT, Rabinowitz MP. *The Wills Eye Manual, 6th edition*. Pennsylvania; Lippincott, 2012, page 370.

140. Gesternblith AT, Rabinowitz MP. *The Wills Eye Manual, 6th edition*. Pennsylvania; Lippincott, 2012, page 362.

141. Am Academy of Ophthalmology, *Basic Clinical Science Course Orbit, Eyelids, and Lacrimal System*. California; AAO, 2012, page 51-52.

142. Gesternblith AT, Rabinowitz MP. *The Wills Eye Manual, 6th edition*. Pennsylvania; Lippincott, 2012, page 154.

143. Gesternblith AT, Rabinowitz MP. *The Wills Eye Manual, 6th edition*. Pennsylvania; Lippincott, 2012, page 155.

144. Gesternblith AT, Rabinowitz MP. Th*e Wills Eye Manual, 6th edition*. Pennsylvania; Lippincott, 2012, page 155.

145. Srinivasan S, Mascarenhas J, et al. *The Steroids for Corneal Ulcers Trial (SCUT): Secondary 12 –Month Clinical Outcomes of a Randomized Controlled Trial*. Am J Ophthalmology 2014; 157: 327-333.

146. Am Academy of Ophthalmology, *Basic Clinical Science Course Retina*. California; AAO, 2013, page 276-278.

147. Am Academy of Ophthalmology, *Basic Clinical Science Course Retina*. California; AAO, 2013, page 55.

148. Comparison of Age Related Macular Degeneration Treatment Trials Research Group (CATT). *Ranizumab and Bevacizumab for Treatment of Neovascular Age Related Macular Degeneration.* Ophthalmology 2012; 119: 1388-1398.

149. Age-Related Eye Disease Study Research Group. *A Randomized, Placebo-Controlled Clinical Trial of High Dose Supplementation* with Vitamin C and E, Beta-Carotene, and Zinc for Age Related Macular Degeneration and Visual Loss, AREDS Report #8. Arch Ophthalmology 2001; 119: 1417-1436.

150. Ferris FL, Wilkinson CP, et al. *Clinical Classification of Age Related Macular Degeneration, the Beckman Initiative.* Ophthalmology 2013; 120: 844-851.

151. The Age Related Eye Disease Study 2 Research Group. *Lutein+ Zeaxanthin and Omega-3 Fatty Acids for Age Related Macular Degeneration.* JAMA May 15 2013; 309: 2005-2015.

152. Am Academy of Ophthalmology, *Basic Clinical Science Course Retina.* California; AAO, 2013, page 58.

153. Am Academy of Ophthalmology, *Basic Clinical Science Course Retina.* California; AAO, 2013, page 171-176.

154. Am Academy of Ophthalmology, *Basic Clinical Science Course Retina.* California; AAO, 2013, page 297-306.

155. Am Academy of Ophthalmology, *Basic Clinical Science Course Retina.* California; AAO, 2013, page 49.

156. Am Academy of Ophthalmology, *Basic Clinical Science Course Retina.* California; AAO, 2013, page 50.

157. Kolb H, Nelson R, et al. *Web vision: The organization of the Retinal Visual System, chapter on color vision by kalloniatis, figure 25.* mailto:webvision@hsc.utah.edu. 2014.

158. Sit AJ, Liu JHK, et al. *Asymmetry of Right Versus Left Intraocular Pressure over 24 Hours in Glaucoma Patients.* Ophthalmology 2006; 113: 425-430. Reprinted from this reference #158, figure 1 page 427, copyright 2006 with permission from Elsevier.

159. Lee AC, Mosaed S, et al. *Effect of Laser Trabeculoplasty on Nocturnal Intraocular Pressure in Medically treated Glaucoma Patients.* Ophthalmology 2007; 114: 666-670. Reprinted from this reference #159, figure 1 and 2 page 669, copyright 2007 with permission from Elsevier.

160. Bengtsson B, Heijl A. *A Visual Field Index for Calculation of Glaucoma Rate of Progression.* Am J Ophthalmology, 2008; 145: 343-353.

161. Heijl A, Patella VM, et al. *Effective Perimetry Fourth Edition.* California, Zeiss; 2012: page 68.

162. Heijl A, Patella VM, et al. Effective Perimetry Fourth Edition. California, Zeiss; 2012: page 62.

163. Nouri-Mahdavi K, Hoffman D, et al. *Prediction of Visual Field Progression in Glaucoma.* Invest Ophthalmology Vis Sci, 2004; 45: 4346-4351.

164. Gustavo de Moraes C, Furlanetto RL, et al. *A New Index to Monitor Central Visual Field Progression in Glaucoma.* Ophthalmology, published on-line 10 April 2014; manuscript number 2013-1407.

165. Heijl A, Bengtsson B, et al. *Prevalence and Severity of Undetected Manifest Glaucoma, Results from the Early Manifest Glaucoma Trial Screening.* Ophthalmology 2013; 120: 1541-1545. Reprinted from this reference #165, figure 1 page 1543, copyright 2013 with permission from Elsevier.

166. Heijl A, Buchholz P, et al. *Rates of Visual Field Progression in Clinical Glaucoma Care.* Acta Ophthalmologica, 2013; 91: 406-412.

167. Am Academy of Ophthalmology, Basic Clinical Science Course Ophthalmic Pathology and Intraocular Tumors. California; AAO, 2012, page 100.

168. Am Academy of Ophthalmology, Basic Clinical Science Course Ophthalmic Pathology and Intraocular Tumors. California; AAO, 2012, page 61-62.

169. Am Academy of Ophthalmology, Basic Clinical Science Course Ophthalmic Pathology and Intraocular Tumors. California; AAO, 2012, page 63-64.
170. Am Academy of Ophthalmology, Basic Clinical Science Course Ophthalmic Pathology and Intraocular Tumors. California; AAO, 2012, page 66.
171. Am Academy of Ophthalmology, Basic Clinical Science Course Ophthalmic Pathology and Intraocular Tumors. California; AAO, 2012, page 67.
172. Am Academy of Ophthalmology, Basic Clinical Science Course Ophthalmic Pathology and Intraocular Tumors. California; AAO, 2012, page 69.
173. Am Academy of Ophthalmology, Basic Clinical Science Course Ophthalmic Pathology and Intraocular Tumors. California; AAO, 2012, page 71.
174. Am Academy of Ophthalmology, Basic Clinical Science Course Ophthalmic Pathology and Intraocular Tumors. California; AAO, 2012, page 74.
175. Am Academy of Ophthalmology, Basic Clinical Science Course Ophthalmic Pathology and Intraocular Tumors. California; AAO, 2012, page 278-279.
176. Am Academy of Ophthalmology, Basic Clinical Science Course Ophthalmic Pathology and Intraocular Tumors. California; AAO, 2012, page 267.
177. Am Academy of Ophthalmology, Basic Clinical Science Course Ophthalmic Pathology and Intraocular Tumors. California; AAO, 2012, page 217.
178. Meyer CH, Rodriques EB, et al. *Grouped Congenital Hypertrophy of the Retinal Pigmented Epitheiliu Follows Development Patterns of Pigmentation Mosaicism.* Ophthalmology 2005; 112: 841-847.
179. Traboulsi EI, Krush AJ, et al. *Prevalence and Importance of Pigmented Ocular Fundus Lesions in Gardner's Syndrome.* New England Journal of Medicine. 1987; 316: 661-667.
180. Shields JA, Shields CL, et al. *Lack of Association among Typical Congenital Hypertrophy of the Retinal Pigment Epithelium, Adenomatous Polyposis, and Gardner Syndrome.* Ophthalmology 1992; 99: 1709-1713.
181. The Advanced Glaucoma Intervention Study Investigators. *Advanced Glaucoma Intervention Study.* Ophthalmology 1994; 101: 1445-1455.Reprinted from this reference #181, figure 1 page 1447, copyright 1994 with permission from Elsevier.
182. Musch DC, Gillespie BW, et al. *Visual Field Improvement in the Collaborative Initial Glaucoma Treatment Study.* American Journal of Ophthalmology. 2014; 158: 96-104. Reprinted from this reference #182, supplemental figure 2 page 104, copyright 2014 with permission from Elsevier.
183. Heijl A, Bengtsson B. *The Effect of Perimetric Experience in Patients with Glaucoma.* Acta Ophthalmol (Copenh). 1996; 11419-22.
184. Newman-Casey PA, Stern M, et al. *Risk Factors Associated with Developing Branch Retinal Vein Occlusion Among Enrollees in a United Sates Managed Care Plan.* Ophthalmology 2014; 121: 1939-1948.
185. Stern MS, Talwar N, et al. *A Longitudinal Analysis of Risk Factors Associated with Central Retinal Vein Occlusions.* Ophthalmology 2013; 120: 362-370.
186. Gesternblith AT, Rabinowitz MP. *The Wills Eye Manual, 6th edition.* Pennsylvania; Lippincott, 2012, page 143.
187. Watzke B, Burton TC, et al. *Ruby laser photocoagulation therapy of central serous retinopathy. A preliminary report.* Modern Problems in Ophthalmology 1974; 120: 242-246.
188. Gesternblith AT, Rabinowitz MP. *The Wills Eye Manual, 6th edition.* Pennsylvania; Lippincott, 2012, page 135.

189. Tsunoda K, Watanabe K, et al. *Highly Reflective Foveal Region in Optical Coherence Tomography in Eyes with Vitreomacular Traction or Epiretinal Membrane.* Ophthalmology 2012; 119: 581-587. Reprinted from this reference #189, figure # 1, copyright 2014 with permission from Elsevier.

190. Heijl A., Bengtsson B. *The Effect of Perimetric Experience in Patients with Glaucoma.* Arch Ophthalmology 1996; 114:19-22.

191. Shields CL, Furuta M, et al. *Metastasis of Uveal Melanoma Millimeter by Millimeter in 8033 Consecutive Eyes.* Arch Ophthalmology 2009; 127: 989-998.

192. Prescher G, Bornfeld N, et al. *Prognostic Implications of Monosomy 13 in Uveal Melanoma.* Lancet 1996; 347: 1222-1225.

193. Ip MS, Domalpally A, et al. *Long term Effects of Therapy with Ranibizumab on Diabetic Retinopathy Severity and Baseline Risk Factors for Worsening Retinopathy.* Ophthalmology 2015; 122: 367-374. Reprinted from this reference #193, figure # 4, copyright 2015 with permission from Elsevier.

194. Savatovsky E, Mwanza JC, et al. *Longitudinal Changes in Peripapillary Atrophy in the Ocular Hypertensive Treatment Study; A Case Control Assessment.* Ophthalmology 2015; 122: 79-86.

195. Stalmans P, Benz MS, et al. *Enzymatic Vitreolysis with Ocriplasmin for Vitreomacular Traction and Macular Holes.* N Eng J Med 2012; 367: 606-615.

196. Kim KE, Jeoung JW, et al. *Diagnostic Classification of Macular Ganglion Cell and Retinal Nerve Fiber Layer Analysis.* Ophthalmology 2015; 122: 502-510. Reprinted from this reference #196, figure # 3, copyright 2015 with permission from Elsevier.

197. Leske MC, Heijl A, et al. *Predictors of Long-term Progression in the Early Manifest Glaucoma trial.* Ophthalmology 2007; 114:1965-1972.

198. Bengtsson B., Leske MC, et al. *Disc Hemorrhages and Treatment in the Early Manifest Treatment Trial.* Ophthalmology 2008; 115: 2044-2048. Reprinted from this reference #198, figure # 1, copyright 2015 with permission from Elsevier.

199. Raiskup F, Theuring A, et al. *Corneal Collagen Crosslinking with Riboflavin and Ultraviolet-A Light in Progressive Keratoconnus: Ten Year Results.* J Cataract and Refractive Surgery 2015; 41: 41-46.

200. Brown DM, Michels M, et al. *Ranibizumab versus Verteporfin Photodynamic Therapy for Neovascular Age-Related Macular Degeneration: Two Year Results of the ANCHOR Study.* Ophthalmology 2009; 116:57-65.

201. Rosenfeld PJ, Brown DM, et al. *Ranibizumab for Neovascular Age-Related Macular Degeneration.* New England J Medicine 2006:355: 1419-1431.

202. Peden MC, Suner IJ, et al. *Long Term Outcomes in Eyes Receiving Fixed-Interval Dosing of Anti-Vascular Endothelial Growth Factor Agents for Wet Age-Related Macular Degeneration.* Ophthalmology 2015; 122: 803-808. Reprinted from this reference #202, figure # 4, copyright 2015 with permission from Elsevier.

203. Rofagha S, Bhisitkul RB, et al. *Seven Year Outcomes in Ranibizumab-Treated Patients in ANCHOR, MARINA, and HORIZON.* Ophthalmology 2013; 120: 2292-2299. Reprinted from this reference #203, figure # 2, copyright 2015 with permission from Elsevier.

204. Medeiros FA, Meira-Freitas D, et al. *Corneal Hysteresis as a Risk Factor for Glaucoma Progression: A Prospective Longitudinal Study.* Ophthalmology 2015; 120: 1533-1540. Reprinted from this reference #204, figure #3, copyright 2015 with permission from Elsevier.

205. Park JH, Jun RM, et al. *Significance of Corneal Biomechanical Properties in Patients with Progressive Normal-Tension Glaucoma.* British J of Ophthalmology 2015; 99: 746-751. Reprinted from this reference #205, figure #1, copyright 2015 with permission from Elsevier.

206. Ferris FL, Wilkinson CP, et al. *Clinical Classification of Age-Related Macular Degeneration.* Ophthalmology 2013; 120: 844-851. Reprinted from this reference #206, figure # 1, copyright 2013 with permission from Elsevier.

207. De Moraes CV, Gustavo H, et al. *Lower Corneal Hysteresis is Associated with More Rapid Glaucomatous Visual Field Progression.* J of Glaucoma 2012; 21: 209-280.

208. Sun L, Shen MS, et al. *Recovery of Corneal Hysteresis after Reduction of Intraocular Pressure in Chronic Primary Angle Closure Glaucoma.* Am J Ophthalmology 2009; 147: 1061-1066. Reprinted from this reference #208, figure #1 and figure #2, copyright 2009 with permission from Elsevier.

209. Hahn S, Azen S, et al. *Los Angeles Latino Eye Study Group. Central corneal thickness in Latinos.* Invest Ophthalmology Vis Science 2003; 44: 1508-1512.

BIBLIOGRAPHY

Books and Manuals

1. American Academy of Ophthalmology. *Basic Clinical Science Course* (BCSC). Volume 1 to 13. California;
   Section 2, Fundamentals and Principles of Ophthalmology, last revision 2010
   Section 4 Ophthalmic Pathology and Intraocular Tumors, last revision 2012
   Section 7 Orbit, Eyelids and Lacrimal System, last revision 2012
   Section 8, External Disease and Cornea, last revision 2014
   Section 10, Glaucoma, last revision 2013
   Section 12, Retina and Vitreous, last revision 2013
2. American Academy of Ophthalmology. *Preferred Practice Patterns* (PPP). California; American Academy of Ophthalmology
   Bacterial Keratitis, 2013
   Cataract Guidelines, 2011
   Idiopathic Macular Holes, 2013
   Primary Open Angle Glaucoma Suspect, 2010
   Primary Open Angle Glaucoma, 2010
   Primary Angle Closure, 2010
   Age-Related Macular Degeneration, 2008
3. Gerstenblith, Adam T. *The Wills Eye Manual.* 6th edition. Pennsylvania; Lippincott, 2012
4. Heijl, Anders. *Effective Perimetry.* 4th edition. California; Zeiss, 2012
5. Kline, Lanning B. *Neuro-Ophthalmology Review Manual.* 7th edition. New Jersey; SLACK, 2013